Geoinformatics
in **Applied**
Geomorphology

Geoinformatics in **Applied** Geomorphology

Edited by
Siddan Anbazhagan
S. K. Subramanian
Xiaojun Yang

CRC Press
Taylor & Francis Group
Boca Raton London New York

CRC Press is an imprint of the
Taylor & Francis Group, an **informa** business

CRC Press
Taylor & Francis Group
6000 Broken Sound Parkway NW, Suite 300
Boca Raton, FL 33487-2742

First issued in paperback 2017

© 2011 by Taylor & Francis Group, LLC
CRC Press is an imprint of Taylor & Francis Group, an Informa business

No claim to original U.S. Government works

ISBN 13: 978-1-4398-3048-2 (hbk)
ISBN 13: 978-1-138-07445-3 (pbk)

Visit the Taylor & Francis Web site at
http://www.taylorandfrancis.com

and the CRC Press Web site at
http://www.crcpress.com

Contents

v

Preface

Geoinformatics is the science and technology dealing with the acquisition, processing, analyzing, and visualization of spatial information. It includes remote sensing, photogrammetry, geographic information systems, global positioning systems, and cartography. The technologies in geoinformatics have been used in various disciplines such as geology, geography, urban planning, environmental science, and global change science. With recent innovations in data, technologies, and theories in the wider arena of remote sensing and geographic information systems, the use of geoinformatics in applied geomorphology has received more attention than ever. Nevertheless, there is no book dedicated exclusively to the use of geoinformatics in applied geomorphology, a field that examines the interaction between geomorphology and human activities.

Given the above context, a book discussing the roles of geoinformatics in applied geomorphology is timely. This book examines how modern concepts, technologies, and methods in geoinformatics can be used to solve a wide variety of applied geomorphologic problems, such as characterization of arid, coastal, fluvial, eolian, glacial, karst, and tectonic landforms; natural hazard zoning and mitigations; petroleum exploration; and groundwater exploration and management. In total, this book contains 19 chapters. Chapter 1 provides an overview of geoinformatics and some recent developments in this field. Chapter 2 introduces the airborne laser scanning technique applied to map fluvial landforms. Chapter 3 discusses some environmental issues in arid environments by remote sensing. Chapters 4 through 7 describe coastal zone management, coastal landform evolution, subsurface coastal geomorphology, and estuarine bathymetric change analysis. Chapters 8 through 10 deal with tectonic geomorphology. Chapters 11 through 14 discuss groundwater evaluation, artificial recharge for groundwater sustainability, deep fracture aquifer analysis, and aquifer vulnerability. Chapter 15 focuses on petroleum exploration. Chapters 16 and 17 deal with natural hazard assessment. Chapter 18 focuses on glacial landform mapping. Finally, Chapter 19 describes the development of sinkhole landforms in several different areas.

This book is the result of extensive research by interdisciplinary experts and will appeal to students, researchers, and professionals dealing with geomorphology, geological engineering, geography, remote sensing, and geographic information systems. The editors are grateful to all those who contributed chapters and revised their chapters one or more times, as well as to those who reviewed their chapters according to our requests and timelines. Reviewers who contributed their time, talents, and energies are as follows: Kwasi Addo Appeaning, Santanu Banerjee, T.K. Biswal, D. Chandrasekharam, Armaroli Clara, Damien Closson, Michael Damen, Daniela Ducci, G.S. Dwarakish, Mikhail Ezersky, B. Gopala Krishna, S. Muralikrishnan, Jeff Paine, Snehmani, S. Sankaran, Sridhar, K.P. Thrivikramji, P. Venkatachalam, and Weicheng Wu. This book would not have been possible without the help and assistance of several staff members at CRC Press, especially Irma Shagla and Stephanie Morkert. Thanks are also due to S. Arivazhagan and Ramesh for their invaluable help.

Siddan Anbazhagan
Salem, India

S.K. Subramanian
Hyderabad, India

Xiaojun Yang
Tallahassee, Florida

Editors

Dr. Siddan Anbazhagan is a director of the Centre for Geoinformatics and Planetary Studies and head of the Department of Geology at Periyar University, Salem, India. He received his PhD from Bharathidasan University (1995) and was awarded the Alexander von Humboldt fellowship for his postdoctoral research in Germany. Dr. Anbazhagan's research interests include remote sensing and GIS for applied geomorphology, hydrogeology, and disaster mitigation. His current area of interest is planetary remote sensing. His research has been funded by ISRO, DST, MHRD, and UGC. He has authored or coauthored more than 60 publications including an edited book entitled *Exploration Geology and Geoinformatics*. Dr. Anbazhagan serves as a reviewer for several remote sensing, environmental, and water resource journals. He has recently been made syndicate member and coordinator of research and development in Periyar University.

Dr. S.K. Subramanian is a senior scientist and head of the Hydrogeology Division at the National Remote Sensing Centre (NRSC), Indian Space Research Organization (ISRO), Hyderabad, India. He completed his higher studies at IIT Bombay and PhD from Indian School of Mines, Dhanbad. Dr. Subramanian has more than 33 years of professional experience in the field of remote sensing and geomorphology. He has coordinated a number of national mission projects, including Integrated Mission for Sustainable Development (IMSD), National (Natural) Resources Information System (NIRS), Rajiv Gandhi National Drinking Water Mission (RGNDWM), and National Agricultural Technology Project and an international project in Dubai. In addition, he was involved in several research projects, including Geomorphologic Evolution of West Coast, for hydrocarbon exploration, Mass Movement in Kosi Catchment, and Geomorphology of Nepal, Chambal Ayacut, Chandrapur district, Rajasthan State, and Northeastern states of India. Dr. Subramanian has authored or coauthored nearly 40 publications.

Dr. Xiaojun Yang is with the Department of Geography at Florida State University, Tallahassee, Florida. He received his BS in geology from the Chinese University of Geosciences (CUG) MS in paleontology from CUG's Beijing Graduate School, MS in applied geomorphology from ITC, and his PhD in geography from the University of Georgia. Dr. Yang's research interest includes the development of remote sensing and geographic information systems with applications in the environmental and urban domains. His research has been funded by EPA, NSF, and NASA. He has authored or coauthored more than 80 publications, including two journal theme issues and one book on coastal remote sensing. He was a guest editor for the Environmental Management; *ISPRS Journal of Photogrammetry and Remote Sensing*; *Photogrammetric Engineering and Remote Sensing*; the *International Journal of Remote Sensing*; and *Computers, Environment and Urban Systems*. Dr. Yang currently serves as chair of the Commission on Mapping from Satellite Imagery of the International Cartographic Association.

Contributors

K. Abilash
National Geophysical Research
    Institute
Hyderabad, India

A. Akilan
National Geophysical Research
    Institute
Hyderabad, India

Siddan Anbazhagan
Department of Geology
Centre for Geoinformatics and
    Planetary Studies
Periyar University
Salem, India

T.K. Biswal
Department of Earth Sciences
Indian Institute of Technology
Mumbai, India

Elmar Csaplovics
Institute of Photogrammetry and
    Remote Sensing
University of Technology Dresden
Dresden, Germany

Anup R. Gjuar
National Institute of
    Oceanography
Dana Paula, India

Balamurugan Guru
Jamsetji Tata Centre for Disaster
    Management
Mumbai, India

Vivekanand Honnungar
Department of Environmental
    Engineering
Texas A&M University-Kingsville
Kingsville, Texas

Fares M. Howari
College of Arts and Science
The University of Texas of the
    Permian Basin
Odessa, Texas

C. Jeganathan
School of Geography
University of Southampton
Southampton, United Kingdom

M. Kannan
School of Civil Engineering
SASTRA University
Thanjavur, India

Amal Kar
Central Arid Zone Research
    Institute
Jodhpur, India

V. Rajesh Kumar
School of Civil Engineering
SASTRA University
Thanjavur, India

N. Ravi Kumar
National Geophysical Research
    Institute
Hyderabad, India

Victor J. Loveson
Central Institute of Mining and
    Fuel Research
Dhanbad, India

E.C. Malaimani
National Geophysical Research
    Institute
Hyderabad, India

Debashis Mitra
Marine Science Division
Indian Institute of Remote Sensing
Dehradun, India

D.S. Mitra
Remote Sensing & Geomatics
    Division
KDM Institute of Petroleum
    Exploration
Oil & Natural Gas Corporation
    Limited
Dehradun, India

S. Neelamani
Environment and Urban
    Development Division
Coastal and Air Pollution
    Department
Kuwait Institute for Scientific
    Research
Safat, Kuwait

Pratima Pandey
Centre of Studies in Resources
    Engineering
Indian Institute of Technology
Mumbai, India

G. Philip
Geomorphology and
    Environmental Geology Group
Wadia Institute of Himalayan
    Geology
Dehra Dun, India

S.V.R.R. Rao
National Geophysical Research
    Institute
Hyderabad, India

G.S. Reddy
National Remote Sensing Centre
Hyderabad, India

Abdulali Sadiq
Department of Chemistry and
    Earth Sciences
Qatar University
Doha, Qatar

K.S. Sajinkumar
Geological Survey of India
Thiruvananthapuram, India

E. Saranathan
School of Civil Engineering
SASTRA University
Thanjavur, India

P.K. Srivastava
University of Petroleum and
    Energy Studies
Dehradun, India

S.K. Subramanian
National Remote Sensing Centre
Hyderabad, India

Marco Trommler
Institute of Photogrammetry and
    Remote Sensing
University of Technology Dresden
Dresden, Germany

Venkatesh Uddameri
Department of Environmental
    Engineering
Texas A&M University-Kingsville
Kingsville, Texas

S. Uddin
Environment and Urban
  Development Division
Environmental Sciences
  Department
Kuwait Institute for Scientific
  Research
Safat, Kuwait

G. Venkataraman
Centre of Studies in Resources
  Engineering
Indian Institute of Technology
Mumbai, India

Xiaojun Yang
Department of Geography
Florida State University
Tallahassee, Florida

Tao Zhang
Department of Fisheries and
  Wildlife
Michigan State University
East Lansing, Michigan

1

Geoinformatics: An Overview and Recent Trends

C. Jeganathan

CONTENTS

1.1 Introduction

Archeological evidences have unearthed the fact that the history of map making existed since ages. Humans have broadened their understanding over the years, about size, shape, and processes associated with earth, which in turn contributed in making sophisticated and accurate representation of the globe and its phenomena. Advancements in space technology, digital information, and communication technologies have stimulated the growth of earth-oriented information science/system, in short the development of geographical information system (GIS), which helps in representing and modeling earth's phenomena in an efficient way. Many new terminologies and terms, like geoinformatics, geomatics, geospatial systems, remote sensing (RS), GIS, etc., are often used when one deals with GIS. It has been generally felt that the term "GIS" restricts one to the idea of computer hardware and software, but the term "geoinformatics" was well received as it conveys and covers a broader meaning. There are many definitions coined for the term "geoinformatics." A simple way to understand this terminology would be to divide it as geo + informatics— the usage of information technology for geographic analysis. This chapter considers a definition on geoinformatics as "an integrated science and technology that deals with acquisition and manipulation of geographical data, transforming it into useful information using geoscientific, analytical, and visualization techniques for making better decisions." In this chapter, the term GIS is assumed to represent geoinformatics and vice versa. This chapter begins with a briefing on historical background about GIS; explains basic terminologies, concepts, and spatial database organization; gives a glimpse of the variety of spatial analytical functions and applications; and, finally, leads to a spectrum of issues, trends, and challenges in the geoinformatics domain.

1.2 Blossoming of Geoinformatics

Developments during the 1960s were caught up with many technical problems like converting analogue map into computer-compatible form, format for storage, display techniques, and more. The 1970s saw interests and participation of universities and the need for topology (spatial relation) was felt. The 1980s contributed for the major growth of GIS due to advancements in personal computers, cheaper hardware, and efficient software. This led to new initiatives, progress in spatial modeling, data structure issues, and RS linkages. This period also saw successful and reliable systems and government interests and investments. Further, a major leap was seen during the 1990s as more and more PCs, object-oriented architectures, networks, Internet, and mobile technology started taking day-to-day applications and hence benefited economic growth. Recently, the Internet has become a major medium of communication and data dissemination. Over the years, GIS has evolved into a geographical information science and at present it is a billion dollar market because it leads to geographical information services. Many software companies have also evolved with GIS and they have been playing a crucial role in making GIS a commercially viable solution, providing mechanisms for various domains of human activity. Readers are highly recommended to read a book named *The History of Geographical Information Systems: Perspectives from the Pioneers*, edited by Foresman (1998), to get a first-hand understanding from the words of the people who were originally involved in the development of modern GIS since its beginning.

1.3 Elements of a GIS

The GIS comprises of four elements. They are hardware, software, dataware, and humanware. Hardware refers to physical components like CPU, hard disk, monitors, digitizers, and printers. Software refers to programs, algorithms, and executable codes. Dataware refers to all possible input databases. Humanware refers to interaction of human to control and manipulate hardware, software, and dataware.

1.4 Geographic Phenomena, Types, and Its Representation

GIS is a computer-based tool, which helps in storing, retrieving, manipulating, analyzing, and producing maps about information related to geographic phenomena with the help of a human expert. Geographic phenomenon refers to a process associated with the earth. In order to represent geographic phenomenon in GIS, its position, its property, and time of occurrence must be known so that one can retrieve information about *what* has happened, *where* it has occurred, and *when* it has happened. In simple words, geographic phenomenon is nothing but what we see or observe about our earth, e.g., observation of daily temperature over a city, weather pattern, crop cycle, human settlement, mapping, etc. Many geographic phenomena are verbally easy to explain; some are easy to represent through drawings like buildings, but there are phenomena that are difficult to draw like temperature or elevation. So, in order to represent such diverse phenomena, some framework needs to be followed so that everybody represents the same thing in the same manner and, hence, it will be easy to understand by all, globally. In this regard, the geographic phenomena were divided into two major groups: *objects and fields* (DeBy et al., 2004). This chapter adopts this framework as it was logically easy to link with ground reality. Objects refer to the phenomena that are bounded by crisp boundaries, i.e., discrete existence. Fields refer to the phenomena that do not have sharp boundary, but are rather fuzzy in their presence and occur at all places, i.e., continuous existence. Field can be again categorized into two more categories: *continuous field and discrete field*, according to their fuzziness in representation. Examples of object-like phenomena are rivers, buildings, volcanoes, islands, etc. Examples of *discrete fields* are land use map, soil map, geology map, etc. Soil and geology occur everywhere and we cannot exactly see their starting and ending points on the ground, unless and until some sharp natural barriers occur. But for representing them, we have to consider some probable end point and introduce some artificial discreteness—hence, it is called discrete fields. Examples of *continuous fields* are temperature, elevation, humidity, etc. It is generally observed that man-made things are objects and natural things are fields, with exceptions like rivers, volcanoes, and islands that are natural things but are considered as objects too, as they do not occur everywhere. So we can say that "all man-made things are objects and all objects are not man made."

Point, line, and polygon are the basic building blocks for representing any phenomena in the computer. But representation of any phenomenon in computer depends upon the mapping scale, because at 1:1 million scale towns will become points, but at 1:10,000 scale they are polygons. Generally, land use, administrative boundary, and thematic maps are mapped through polygon. Roads, rivers, pipelines, and electricity lines are mapped as lines. Village locations, utility locations, and field observations are represented as points.

Apart from understanding their locational and conformal property, one must also understand their attribute properties, as every element, whether it is a point or line or polygon, needs to have some description about what it represents. The domain of attribute data is classified into four categories as *nominal, ordinal, rational,* and *interval.* Qualitative data are represented through nominal data, which cannot provide any quantitative meaning. Nominal refers to the data that are generally used for identification purposes. For example, the name of a person, the name of a road, telephone number, house number, etc. are nominal kinds of data. Ordinal data refer to ordered data in which we can infer order of importance, e.g., if we rank the people as per their exam score, then it is an ordinal data. Ratio data are the actual fact/data measured on the ground quantitatively, which has actual origin at zero; e.g., 0 mm is the same as 0 km; hence, distance or length is a "ratio" data. The term "ratio" does not convey any meaning about division; however, the term "rational" might have become "ratio." Interval data are data in which actual origin differs at different places and the value zero conveys different measures, e.g., 0° Kelvin is different from 0° Celsius. Hence, temperature is an interval data. Also, when we measure earthquake in Richter scale, the energy difference of two earthquakes of magnitude 5.1 and 5.2 is different than the energy released by 7.1 and 7.2. Although the difference in magnitude is 0.1 in both the cases, the actual quantitative meaning is completely different. Hence, earthquake measurement is an interval data. The user must be aware about these "types of data" while working on spatial operations, as all the operations/analyses are not possible with all types of data.

1.4.1 Spatial Data Structure

There are two broad types of data structures generally adopted to represent all geographic phenomena in computer, under GIS. The types are vector and raster. Vector in mathematical sense reveals a "quantity and direction." But in GIS, vector is used for referring to the basic mathematical elements or building blocks—point, line, and polygon. Any map that is prepared using these building blocks is called vector map. It is possible to represent roads, buildings, administrative units, land use, geology, and more using vector concepts, but there are many phenomena that cannot be represented in vector. For example, how can we record temperature or elevation or humidity or RS reflectance value? GIS has been molded to represent these types of phenomena using *tessellation* concept. Tessellation is nothing but the division of the space/area into uniform grid and finding the dominant phenomenon occurring within each grid. The space can be divided using uniform grid, also called regular grids like squares, rectangles, triangles, pentagons, or hexagons. Normally, a square grid is adopted in the tessellation process due to its simplicity with many hidden advantages; for instance, it is easy to find the location of any grid if we know the origin of the grid's location and size, and it is easy to store, retrieve, and analyze. If the size of the grid is smaller, then it occupies more storage space, and if we increase the size of the grid, then we may lose some information variability within that grid. Therefore, one has to come to a compromise in selecting the size of grid, which is called *resolution*.

In general, the tessellated space looks like a 2D matrix of grids/cells, which is called *raster*. Raster has greater advantage over vector, especially for spatial analysis and modeling, as we can deal with any part of the study area as every pixel is explicitly represented. But in vector, only the boundary of the phenomena is represented. Therefore, we can do only logical operations like AND, OR, NOT with vector layers and we cannot perform arithmetic operations like addition, subtraction, multiplication, or division. However, in raster, we can perform all kinds of operations like comparison, logical operations, arithmetic and trigonometric operations, and dynamic simulations. RS data are raster data and, hence, we can directly adopt RS into our GIS models, if our data are in raster structure. Vector data are good for printing accurate representation created in GIS (Figure 1.1).

If we adopt the regular tessellation for the phenomenon, which occurs over very large spatial area, then we will end up having the same cell

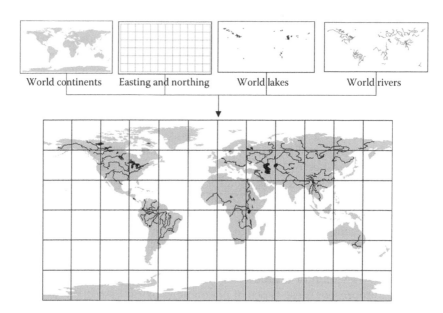

| World continents | Easting and northing | World lakes | World rivers |

FIGURE 1.1
Different layers and their integration in GIS.

value occurring repeatedly and hence occupy more storage space. In such redundant storage, another approach was adopted, which is called irregular tessellation. In irregular tessellation, the space is divided only if it has diverse features. By this approach, more storage space can be saved. One such irregular tessellation is Quadtree data storage structure. The examples of regular and irregular tessellation are shown in Figure 1.2.

There are many data storage formats that adopt various algorithms and some of them are even proprietary. The main objectives behind such algorithms are lossless compression, faster retrieval, and efficient manipulation. In order to represent elevation, a new kind of storage technique using an approach called triangulated irregular network (TIN) has been used. TIN is an irregular tessellation in which triangle is a basic building block in which three input reference height points are used. From the network of triangles, it would be easy to calculate elevation, slope, and any aspect for any point in the study area. Generally, GIS users adopt the vector-based approach for creating inputs, as it is more convenient. In vector mode, different software use different formats. Storing the data in vector format is a bit complex because there are many ways to save coordinates, attributes, data structure, spatial relationship (topology), and visualizing the stored information. Some of the most used vector formats are given in Table 1.1.

Raster format is not only used to save images obtained through scanning, digital photographs, or satellite images (RS data), but also to save

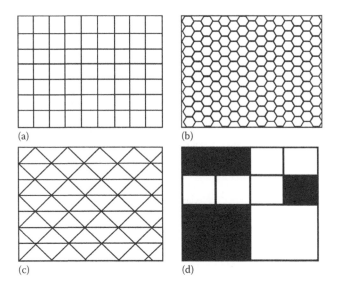

FIGURE 1.2
Regular and irregular tessellation: (a) square, (b) hexagonal, (c) triangular, and (d) quadtree.

geographic phenomena, which continuously vary in space—like topography and temperature. Table 1.2 gives the description about some of the widely used raster formats.

1.4.2 Spatial Layers or Geodatabase

The common requirement to access data, on the basis of the type of phenomena, has led to the creation of each type into a separate entity called layers, such as roads, rivers, or vegetation types, in which all the features of the same type are grouped within a so-called layer or map. The concept of layer is applicable in both vector and raster models. The layers can be combined with each other in various ways to create new layers that are a function of the individual input layers. For a given study area, the layers must have a positional information and it must belong to a specific geographic range, irrespective of whether it is a polygon bounded by lines in vector system or a grid cell in a raster system. Such database bundle is called geodatabase. The layers concepts in GIS are shown in Figure 1.1.

1.4.3 Planimetric Requirements

All the maps are used for some kind of measurement purposes in real-life applications. Hence, the measurement made on the maps has to be accurate. Since our earth is a 3D body and layers have to be in 2D mode,

TABLE 1.1

Widely Used Vector Formats

File Format	Name	Description
DGN	MicroStation design files	DGN is an intern format for MicroStation, a computer-aided design (CAD) software. DGN file contains detailed visualization information (like color for various layers, pattern, thickness, etc.) apart from the spatial, attribute information
DLG	Digital line graphs	DLG is used by the U.S. Geological Survey (USGS) for handling vector information from printed paper maps. It contains very precise coordinate information and sophisticated information about object classification, but no other attributes. DLG does not contain any visualization information (display)
DWG	Autodesk drawing files	DWG is an intern format for AutoCAD. Because of the lack of standards for linking attributes, problems may occur while converting this format between various systems
DXF	Autodesk drawing exchange format	DXF is a common transfer format for vector data. It contains visualization information and is supported by nearly all graphic programs. Nearly all programs can successfully import this format because of high standards
E00	ARC/INFO interchange file	E00 is a transfer format available both as ASCII and binary form. It is mainly used to exchange files between different versions of ARC/INFO, but can also be read by many other GIS programs
GML	Geography markup language	XML-standard for exchanging and saving geographical vector data. It is used in the Open GIS Consortium
MIF/MID	MapInfo interchange format	MIF/MID is MapInfo's standard format, but most other GIS programs can also read it. The format handles three types of information: geometry, attributes, and visualization
SDTS	Spatial data transfer system	SDTS is a transfer format developed in the United States and is designed for handling all types of geographical data. SDTS can be saved as ASCII or binary. In principle, all geographical objects can be saved as SDTS, including coordinates, complex attributes, and visualization information. These advantages nevertheless increase complexity. To simplify it, many standards have been developed as "coprojects" to SDTS. The first of these standards is Topological Vector Profile (TVP), used to save some types of vector data
SHP	ESRI shapefile	Shape is ArcView's internal format for vector data. Associated to the Shape file (*.shp), there is a file to handle attributes (*.dbf) and an index file (*.shx). Nearly all other GIS programs can import this format

TABLE 1.1 (continued)

Widely Used Vector Formats

File Format	Name	Description
SVG	Scalable vector graphics	XML-standard for presentation of vector on the Internet. It is approved in the World Wide Web Consortium
TIGER	Topologically integrated geographic encoding and referencing files	TIGER is an ASCII transfer format made by the U.S. Census Bureau to save road maps. It contains complete geographic coordinates and is line-based. The most important attributes include road names and address information. TIGER has its own visualization information
VPF	Vector product format	VPF is a binary format made by the U.S. Defense Mapping Agency. It is well documented and can easily be used internally or as a transfer format. It contains geometry and attribute information, but no visualization information. VPF files are also named VMAP product. The Digital Chart of the World (DCW) is published in this form
VXP	Idrisi32 ASCII vector export format	IDRISI 32's vector export format (ASCII)
WMF	Microsoft Windows metafile	WMF is a vector file format for Microsoft Windows Operation Systems

Source: CGISLU, GIS educational materials, Centre for Geographical Information System, Lund University, Lund, Sweden, 2003.

there is a need for a mechanism that can help in achieving this transition between 3D and 2D (Figure 1.3). This mechanism is called projection. The entire geospatial database generated under GIS must be in a planimetric coordinate system, i.e., it must be represented in a 2D reference frame so that we can find out the area and length correctly. There are many ways by which a 3D globe can be converted into 2D. Cylinder, cone, and plane are simple mathematical figures that can be utilized for this conversion. Conversion is done by wrapping the earth with the paper of these shapes and then making the imprint of global feature on the paper and then unwrapping the paper. Converting from 3D to 2D introduces some loss in either area or shape or direction measurement. Based on the property it preserves, the projection receives its name like equal area projection, conformal projection, etc.

Another major hurdle in the projection process is that the earth does not have smooth surfaces. It has gravitational undulations, which are visible from mean sea level plots. Because of the nonlinear complex undulations, it is very difficult to replicate the exact position and height of a location accurately on a map. Therefore, assumptions about the shape of our earth have to be made as either ellipsoid or spheroid so that mathematically it

TABLE 1.2

Widely Used Raster Formats

File Format	Name	Description
ADRG	Arc digitized raster graphics	ADRG is a format created by the U.S. military to save paper maps in raster format
BIL	Band interleaved by line	BIL is a computer compatible tape (CCT) format that stores all bands of remotely sensed data in one image file. Scanlines are sequenced by interleaving all image bands
BIP	Band interleaved by pixel	When using the BIP image format, each line of an image is stored sequentially, pixel1 all bands, pixel 2 all bands, etc.
BSQ	Band sequential	BSQ is a CCT format that stores each band of satellite data in one image file for all scanlines in the imagery array
DEM	Digital elevation model	DEM is a raster format created by the U.S.GS (U.S. Geological Survey) for saving elevation data
GTOPO30	Global 30 arc second elevation data set	GTOPO30 is a global, digital elevation model with a horizontal cell size of 30 s (approx. 1 km). GTOPO30 was created from different raster and vector sources
GeoTIFF	GeoTIFF	GeoTIFF is a form of tag image file format(TIFF) format for storing georeferenced raster data
GRIB	GRid in binary	GRIB is the World Meteorological Organisation's (WMO) standard for grid-based meteorological data
PCX	PC paintbrush exchange	PCX is a common raster format found in many scanners and graphic programs
SDTS	Spatial data transfer standard	SDTS is a format for transferring geographical information. An SDTS variant is specifically made for transferring raster data
TIFF	Tagged image file format	Like PCX, TIFF is a common raster format produced by drawing programs and scanners. TIFF format gives a relatively big data file, but compresses the data without loss of information

Source: CGISLU, GIS educational materials, Centre for Geographical Information System, Lund University, Lund, Sweden, 2003.

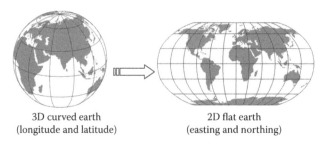

3D curved earth   2D flat earth
(longitude and latitude)   (easting and northing)

FIGURE 1.3
Map projection.

would lead to projection. Every country has adopted its local mean sea level surface reference for height measurement, which is called vertical datum and has adopted a mathematical surface (ellipsoid) that fits better for its portion of the globe, which is called horizontal datum. If we want to convert from one projection to another, we need to have seven parameters related to translation effect (dx, dy, dz), rotation effect (rx, ry, rz), and a scale factor.

1.4.4 Errors and Data Quality in GIS

In GIS, errors may intrude at various stages of operations starting from data generation to analysis. During vector-layer creation, one may encounter undershoots and overshoots, which refers to unclosed polygon lines and overdrawn lines, respectively. Undershoot is a major error for a polygon layer, which leads to under- or overestimation of polygon area. Errors may also arise if one misses a polygon attribute. Overshoot will lead to overestimation of length and will be a serious error for linear vector layers. Proper definition of tolerances, like snap tolerance for avoiding dangles at the time of digitization, fuzzy tolerance for removing dangles at the time of topology building after digitization, and tunnel tolerance for removing redundant vertices, would help in making quality input layers. Correctness of the feature codification needs to be checked, for obvious error, against laid down standards. When dealing with RS data, the error intrudes into georeferencing and classification. In multidate image registration, one has to make sure that all the images are accurately matched within a single-pixel accuracy. During classification, the training samples may contain misrepresentation and, hence, result in serious omission or commission errors. Proper field validation and cross-verification with existing toposheets/maps during training would help in reducing errors. By cross-tabulating field-check points or training sites versus classified images, one can make confusion matrix, and can, therefore, derive omission error, commission error, average accuracy, and Kappa measure. Some of the standard quality measures and mapping standards, as recommended by the National Natural Resources Management System (NNRMS, 2005) in India, are positional accuracy—*1 mm of scale*; coordinate movement tolerance—*0.125 mm of scale*; raster/grid resolution—*0.5 mm of scale*; weed tolerance—*0.125 mm of scale*; etc.

1.5 Spatial Analysis

Whether it is for natural resources or sustainable development, or natural disaster management, selecting the best site for waste disposal, optimum route alignment, or local problems, it will have a geographical component. GIS has power to create maps, integrate information, visualize scenarios, solve complicated problems, present powerful ideas, and develop effective solutions like never before. In brief, it can be said to be a supporting tool for decision-making processes. Only when all the maps are projected to an agreed uniform coordinate system, would it be possible to perform spatial analysis.

GIS is used to perform a variety of spatial analysis, using points, lines, polygons, and raster data sets. GIS operational procedure and analytical tasks that are particularly useful for spatial analysis include the following: single-layer operations, multilayer operations, measurement operations, neighborhood analysis, network analysis, 3D surface analysis, and predictive and simulation analyses. There are huge and diverse analytical groups in geoinformatics like point pattern analysis for sampled natural resources management; facility management in urban environment; decision-support system for planning support; population, health, and epidemiological modeling; time-series modeling in land–ocean–atmosphere; geomorphologic–geological analysis (landslide zonation, snowmelt runoff modeling, mineral exploration, hazard and risk assessment, etc.); and land cover change dynamics and predictive modeling. Out of all possible functionalities of GIS, two analytical perspectives mainly dominate the global GIS users: (a) geostatistics and (b) spatial decision support system.

1.5.1 Geostatistics

Most of the spatial analysis and modeling were done in the raster domain, where a value at each raster-pixel location plays a role most of the time— e.g., NDVI is derived by a "local" operation using two spectral bands; analysis using "weighted sum" does depend on individual raster values. However, the real power of GIS lies in its ability to carry out the analysis that takes into account the inherent spatial location along with its attributes, in a statistical manner, i.e., geostatistics. Geostatistics received more attention after its successful utilization in the domain of economic geology by Matheron (1963) and Journel (1974), and later improved its usability by many more researchers (Mark and Webster, 2006). Although the spatial coverage of an analysis can be extended through "focal" and "zonal" operations, "Kriging" is considered as one of the powerful geostatistical operation, which truly considers the locational information to calculate variability of pixel values through a semivariance measure, and it also provides the spatial variation of error in prediction. Kriging is a kind of interpolation technique to find out values at unknown locations using sparse sampled values. It falls under the category of two-point statistics (i.e., one single-pixel value is analyzed with reference to another single-pixel value) and many texture measures also fall under the two-point statistics. Fundamentals about Kriging can be found in Cressie (1990). Geostatistics has been utilized extensively in different scientific domains—e.g., in mining (Journel and Huijbriqts, 1978), for natural resources evaluation (Goovaerts, 1997), for reservoir modeling (Deutsch, 2002), to model spatial uncertainty (Chiles and Delfiner, 1999), in downscaling (Atkinson et al., 2008), and for environmental applications (Atkinson and Lloyd, 2010). Recently, due to a limitation of two-point statistics in revealing the complex earth surface pattern and landforms, many researchers attempted multipoint statistics (MPS) (Boucher, 2008). MPS looks into multiple values at multiple locations in a single operation and helps to simulate complex geological/geomorphological phenomena and patterns. It has been successfully attempted in modeling fluvial reservoir (Wong and Shibli, 2001), geological structure (Strebelle, 2002), and aquifer characterization (Mariethoz, 2009). It has also proven to be a very powerful tool for super-resolution mapping, where a coarse spatial resolution image is transformed into a fine spatial resolution image (Boucher, 2008; Mariethoz, 2009).

1.5.2 Spatial Decision Support System

In the past two decades, the concept of spatial decision-support system (SDSS) has taken an important role where the main thrust is placed on scenario building, data analysis, and decision-making techniques rather than mapping. SDSS is an interactive, flexible, and adaptable computer-based information system, especially developed for finding solutions for semistructured management problems (Sharifi and Marjan, 2002). Decision making in GIS has three perspectives (Malzewski, 1999; Sharifi and Marjan, 2002): multiattribute decision making (MADM) against multiobjective decision making (MODM), individual versus group decision problem, and decision under certainty versus decision under uncertainty. Although different researchers (Sprague and Carlson, 1982; Marakas, 1999; Haag et al., 2000; Power, 2002) proposed different components for SDSS, five-component architecture is generally accepted comprising the following: (i) database, (ii) model-base, (iii) knowledge-base, (iv) control-unit, and (v) user interface. *Database:* The database component is considered of warehouse in an SDSS. *Model-base:* Under the control of the model-base management system, it is able to handle models (both perspective and descriptive) that are designed to perform estimations and to solve specific optimization problems. *Knowledge-base:* The knowledge-base contains all the necessary information to handle models and data. *Control-unit:* The control-unit provides the necessary coordination between the various components. *User interface:* Finally, the user interface is considered as a separate component, since the dialogue with the controlling is personalized to adapt to a particular user/organization/project need. Analytical hierarchical process (AHP) is one of the widely used MADM techniques, and, recently, techniques like cellular automata and agent-based modeling tools are getting wider scope for analyzing spatiotemporal information from dynamical processes in the urban and natural environment in order to take quicker decisions in emergency situations and disaster relief modeling.

1.6 Recent Trends and Future Challenges in GIS

The number of GIS users has grown from mere thousands in the 1980s (Goodchild, 1998) to millions in 2005, and the annual revenue of GIS industry has grown from millions to billions of dollars in 2005 (Blake, 2006). Digital technology, considered once as an alien discipline of study, has now become a major topic of interest in India and all over the globe. Every graduate in India pursues computer course simultaneously along with their own discipline. Every geosciences student studies about GIS and uses it in his or her curriculum and projects.

With the rise of the World Wide Web (WWW), new Internet protocols such as the Hypertext Transfer Protocol (HTTP), new markup languages like HTML, DHTML, XML, GML, as well as easy-to-use interfaces (browsers), tools (Flash), and languages (.NET, Java, scripts), the Internet has become a powerful media for the future. Google Earth and Wikimapia are wonderful examples of such Internet-based geoinformation services. Furthermore, with the advancements in mobile communication technology, the world is moving toward having GIS in mobile phones, by which one can locate the nearest ATMs, nearest theatres, nearest hospitals, optimal route between source and destination, etc. Such applications are called location-based services (LBS), which will help in better e-governance and also during disaster relief operations (Bennett and Capella, 2006). GIS has grown from Desktop GIS to Enterprise GIS, and to Internet GIS, and today it is a Mobile GIS. Readers are recommended to read an article by Gupta (2006) titled "GIS in the internet era: What India will gain?" for a detailed outlook on the growth of Internet in India and the gift of its wedding with GIS.

Although the growth of the GIS is enormous, there are some vital challenges still to be answered or to be achieved by GIS developers. Satellite-based RS is providing a digital data at a resolution (in terms of spatial, spectral, and temporal) much higher than what is needed for many applications. This has pushed the GIS to look for more sophisticated and efficient algorithms, data models to store, retrieve, and analyze such huge volumes of data. This multidimensional growth in digital data has forced

GIS to incorporate additional information, apart from mere "where" and "what." GIS engineers have moved from making mere mapping tools to integrated application framework called "GeoDesign," which brings geographic analysis power into basic design (Dangermond, 2009). Some of the important challenges are discussed in the following sections.

1.6.1 2D GIS to 3D GIS

Today, GIS has pioneered the art of 2D representation, utilization, and dissemination. But there are very few GIS softwares that can deal with data in true 3D manner. In the current GIS, if one wants to represent the multifloor building and all rooms in it, then it will be very difficult and may not be possible in most of the so-called commercial GIS softwares. Also, if a geologist wants to represent an underground profile of a rockbed or any feature below the ground through sample points and at the same time he or she wants to represent surfacial thematic elements and perform 3D planimetric analysis, then it is almost near-impossible with the current softwares except for very few specialized softwares. In order to address such issues, one needs to have a true 3D GIS software.

A true 3D GIS will attach the Z-value along with every element/objects in the vector and at every pixel. In 3D GIS, each "pixel–picture element" will be called "voxel–volume element." The creation and maintenance of topology with third dimension is a tedious process, but will make the system self-intelligent at a real-world scale. Such systems will help to implement "agent-based modeling" in an easy manner.

1.6.2 2D, 3D GIS to 4D GIS

Over the past 30 years, huge volumes of geospatial data have been accumulated. If one wants to analyze the change scenario of an area in these years, then one needs to do a lot of digital jugglery in bringing all these temporal data into a single frame of spatial reference. Since the spatial resolution of RS data available 30 years back was coarser and it has been improving over time, the issue of "seamless database" is getting into GIS. Today, there are tools available to achieve this seamless database only through skilled manual interaction. Future GIS would be intelligent enough to deal with this without much interaction from users. Seamless occurs horizontally and vertically. "Horizontal" refers to the automatic linkages with areal extent and spectral match. "Vertical" refers to the scaling effect, i.e., ability to change to any scale without much hassle.

This leads to further requirement of intelligence at "graphical and conceptual generalization." The "conceptual generalization" seems to be easy at first look as it talks about change in the definition of thematic classes at various scales. This would be easy only if everybody agrees to the thematic change vis-à-vis scale. At present, every country/group has its own classification system, and hence conceptual generalization cannot be achieved very easily in an automatic manner, but semiautomatic system is already available. The "graphical generalization" deals with the reduction of geometric complexity in representation at the physical level, and it is much easier at raster level than at vector level. But, again, there are no uniform generalization rules at the global level, and very little at the national level. Hence, the future GIS will have to either force/direct each country toward a uniform understanding or to adapt each country's policy at software level to give a free hand to users to have their own choice.

1.6.3 Crisp Data to Fuzzy Data

A map created today is a very crisp data, i.e., it either shows the presence of an event or the absence of the event. There is no information about the associated uncertainty of the elements. But when the user advances from data level to knowledge level, his or her understanding in modeling a phenomenon is restricted by this "crispness" of the data. At this juncture, data need to have accuracy or uncertainty or fuzziness at individual element level or pixel level or at each processing level. Hence, efficient algorithms are needed to provide an uncertainty map along with each analysis/model. The current GIS packages lack tools for modeling uncertainty in a true manner.

1.6.4 Closed to Open Environment

Today, GIS is no longer a single user package. It is being procured and used at organizational level. Hence, GIS has grown from "desktop" to "enterprise wide." The Web has taken GIS a further step ahead in reaching a much wider audience located at farther places, and in making GIS a true global information system. In future, there may be supercomputers or intelligent workstations dedicatedly made to deal with national or global GIS layers. Users from anywhere can link and query. Processing will be handled at the central node, which may use "GRID"-based technology, in which each computer participating in the network will be utilized to increase the processing speed, based on its CPU availability. National Spatial Data Infrastructure (NSDI) and Global Spatial Data Infrastructure (GSDI) are initiated and implemented in some countries. Feasibility, adaptability, and interoperability of these infrastructures will further strengthen the development along these lines.

In India, major developments have taken place during the last decade with significant contribution coming from the Department of Space (DOS), emphasizing GIS applications for Natural Resources Management. Notable among them are Integrated Mission for Sustainable Development

(IMSD), Natural Resource Information System (NRIS), Biodiversity Characterisation at Landscape Level, Rajiv Gandhi Drinking Water Mission, and many others. Also, many challenging initiatives launched by DOS are in the pipeline. In addition, many government/nongovernmental organizations (Survey of India, Department of Science & Technology, Geological Survey of India, Forest Survey of India, National Informatics Center, Center for Spatial Data Management System, State Remote Sensing Centers, Indian Institute of Technology, universities, and colleges) have helped in expanding GIS usage in India. India's National Spatial Data Infrastructure (NSDI) is a major interinstitutional recent initiative that helps in establishing geospatial database infrastructure for providing easy and open access of geoinformation to the users, which may trigger further expansion and growth of GIS research and applications. A possible list of geoinformatics research domains are provided in Table 1.3.

1.7 Conclusion

It can be concluded that GIS has grown from a mere analytical engine into a science of its own, and has been influencing all thematic science disciplines to adapt itself. Internet applications and Internet GIS are the driving forces for the future. GIS along with GPS has brought, already, the geospatial applications to everybody's mobile phone and will make the GIS a part and parcel for the common man. The ultimate goal of GIS would be in helping the government to take appropriate spatial decisions and, hence, in leading the whole world toward sustainable existence—socially, environmentally, and economically. Recently, Google Earth was extensively utilized for mapping damage assessment due to a disastrous earthquake that struck Haiti on January 13, 2010, where hundreds of volunteers and organizations across the world (GEO-CAN: Global Earth Observation Catastrophe Assessment Network) shared the mapping process through Internet-assisted information system. Subsequently, the mapped layers were uploaded to a central GIS server and then results were available to the experts at field and at various decision-making levels through an Internet GIS-based virtual disaster viewer (http://www.virtualdisasterviewer.com/). The author participated as one of the volunteers in this exercise, which was carried out by ImageCat Inc., United Kingdom, as a noncommercial voluntary service to the United Nations and the World Bank. It would not be an exaggeration to say that the future national polls will be conducted through mobile phones, with GIS servers centrally managing and modeling the dynamic flow of spatial and nonspatial data.

TABLE 1.3

Challenging Research Areas in Geoinformatics

S. No.	Areas of Interest	Application	Possible Geoinformatics Research Areas
1.	Environmental sciences	Mapping and distribution analysis	Spectral reflectance characterization and its influence
			Information extraction techniques
			Data mining techniques
			Mapping level vs. resolution
			Subpixel characterization
			Scale variation vs. contextual relationships
			Multiresolution crisp/fuzzy analysis
			Generalization techniques
			Ontology (conceptual domain knowledge)
			Markov chain model
			Cellular automata approach
		Land dynamics	Agent-based modeling
			Time synchronization in high-level architecture
			Spectral library creation and matching
			Advanced classification algorithms
		Hyperspectral analysis	Models for zonation and prediction
			Simulation 2D, 3D, and 4D
		Disaster preparedness	Networked data accessibility (NSDI and natural resource repository [NRR]) of DST and DOS initiatives
2.	Urban planning and infrastructure development	Land information system	Automated feature extraction (object-oriented technique)
			Network concepts and analytical algorithms
		Planning support system	Optimal allocation/disposal algorithms
			3D visualization of urban environment
			Urban process reengineering in municipalities (geoinformation management)
			Demand/allocation/growth/migration influence on master plan development
3.	Socioeconomic development	Population migration	Simulation algorithms
		Economic distribution	Disease dispersion pattern characterization and analysis
		Health issues	Regional impact analysis (e.g., linking of rivers, etc.)
			Village resource information system
			Neighborhood mapping

TABLE 1.3 (continued)

Challenging Research Areas in Geoinformatics

S. No.	Areas of Interest	Application	Possible Geoinformatics Research Areas
4.	Business enterprise	Marketing strategies	Business processes and spatial integration
			Workflow management
			Potential resource identification and distribution strategies
5.	Technology	DBMS	Object-oriented data models
		GPS	Geodetic controls
		Mobile devices	Satellite-based automated real-time distributed processing
		Internet	XML, GML integration
		Location-based service	Geospatial data visualization algorithms for mobile devices
			Protocols, Ontology Web Language, Wireless Application Protocol (WAP)
6.	Standards	Software	Interoperability models
		Metadata	Certification algorithms
		Spatial data infrastructure	Accessibility issues in distributed environment
			Policies for geoinformation management

References

Atkinson, P. M. and Lloyd, C. D. 2010. *geoEnv VII: Geostatistics for Environmental Applications*, Quantitative Geology and Geostatistics Series, Springer, Berlin, Germany.

Atkinson, P. M., Pardo-Iguzquiza, E., and Chico-Olmo, M. 2008. Downscaling cokriging for super-resolution mapping of reflectance. *IEEE Transactions on Geoscience and Remote Sensing* 46(2): 573–580.

Bennett, V. and Capella, A. 2006. Location-based services: Wherever you are, wherever you go, get the information you want to know. http://www.ibm.com/developerworks/ibm/library/i-lbs/. Accessed on 27, January 2011.

Blake, V. 2006. The evolving GIS/geospatial industry. DaraTech Inc., Market Research and Technology Assessment, Cambridge MA.

Boucher, A. 2008. Super resolution mapping with multiple point geostatistics. In A. Soares et al., eds., *geoENV VI—Geostatistics for Environmental Applications*, pp. 297–305, Springer Science+Business Media, Berlin, Germany.

CGISLU. 2003. GIS educational materials. Centre for Geographical Information System, Lund University, Lund, Sweden.

Chiles, J. P. and Delfiner, P. 1999. *Geostatistics: Modeling Spatial Uncertainty*, John Wiley & Sons, Inc., New York.

Cressie, N. A. C. 1990. The origins of kriging. *Mathematical Geology* 22: 239–252.

Dangermond, J. 2009. A vision for geodesign. In *Plenary Session Presentation in the 2009 International ESRI Conference on GIS: Designing our Future*, San Diego, CA.

DeBy, R. A., Knippers, R. A., Weir, M. J. C. et al. 2004. *Principles of Geographic Information System: An Introductory Textbook*, ITC Educational Textbook Series, International Institute of Geo-Information Science and Earth Observation, Enschede, the Netherlands.

Deutsch, C. 2002. *Geostatistical Reservoir Modelling*, Oxford University Press, New York.

Foresman, T. W., ed. 1998. *The History of Geographical Information Systems: Perspectives from the Pioneers*, Prentice Hall, Upper Saddle River, NJ.

Goodchild, M. F. 1998. What next? Reflections from the middle of the growth curve. In T. W. Foresman, ed., *The History of Geographical Information Systems: Perspectives from the Pioneers*, Prentice Hall PTR, Upper Saddle River, NJ.

Goovaerts, P. 1997. *Geostatistics for Natural Resources Evaluation*, Oxford University Press, New York.

Haag, S., Cummings, M., McCubbrey, D. J., Pinsonneault, A., and Donovan, R. 2000. *Management Information Systems: For the Information Age*, pp. 136–140, McGraw-Hill Ryerson Limited, New York.

Journel, A. 1974. Geostatistics for conditional simulation of ore bodies. *Economic Geology* 69(5): 673–687.

Journel, A. and Huijbrigts, Ch. J. 1978. *Mining Geostatistics*, Academic Press Limited, New York.

Malzewski, J. 1999. *GIS and Multicriteria Decision Analysis*, John Wiley & Sons. New York.

Marakas, G. M. 1999. *Decision Support Systems in the Twenty-First Century*, Prentice Hall, Upper Saddle River, NJ.

Mariethoz, G. 2009. Geological stochastic imaging for aquifer characterisation. PhD dissertation, University of Neuchâtel, Neuchâtel, Switzerland.

Mark, R. M. and Webster, R. 2006. Geostatistical mapping of geomorphic variables in the presence of trend. *Earth Surface Processes and Landforms* 31: 862–874.

Matheron, G. 1963. Principles of geostatistcs. *Economic Geology* 58: 1246–1266.

NNRMS 2005. NNRMS standards: A national standard for EO images, thematic & cartographic maps, GIS databases and spatial outputs. ISRO:NNRMS:TR:112:2005, NNRMS Secretariat, Indian Space Research Organisation, Department of Space, Bangalore, India.

Gupta, R. 2006. GIS in the internet era: What India will gain? http://www.gisdevelopment.net/

Power, D. J. 2002. *Decision Support Systems: Concepts and Resources for Managers*, Quorum Books, Westport, CT.

Sharifi, A. and Herwijnen, M. van. 2002. Spatial decision support systems, Reading material, International Institute for Geoinformation Science and Earth Observation (ITC), Enschede, the Netherlands.

Sprague, R. H. and Carlson, E. D. 1982. *Building Effective Decision Support Systems,* Prentice Hall, Englewood Cliffs, NJ.

Strebelle, S. 2002. Conditional simulation of complex geological structures using multiple point statistics. *Mathematical Geology* 34(1): 1–22.

Wong, P. M. and Shibli, S. A. R. 2001. Modelling a fluvial reservoir with multi-point statistics and principal components. *Journal of Petroleum Science and Engineering.* 31(2–4): 157–163.

2

Airborne Laser Scanning and Very High-Resolution Satellite Data for Geomorphological Mapping in Parts of Elbe River Valley, Germany

Siddan Anbazhagan, Marco Trommler, and Elmar Csaplovics

CONTENTS

2.1 Introduction

Airborne laser scanning operates on the principle of light detection and ranging (LIDAR) using a pulse laser to measure the distance between the

sensor and the earth's surface (Flood and Gutelius, 1997). The time travel
of the laser pulse from the airborne platform to the ground and back is
measured by a precise time counter. Using the velocity of light and the
measured time, the distance from the laser scanner to the ground where
the laser pulse hits is determined. The laser scanner is supported by a
differential global positioning system (DGPS) and an inertial measurement unit (IMU), enabling the position of each ranged point to be identified (Wehr and Lohr, 1999). The importance of airborne laser scanning
(ALS) data has steadily increased in recent years and has been adopted
in various fields, such as in digital terrain modeling (DTM), in the generation of mapping corridors, 3D city modeling, etc. Optical remote
sensing data provide the various attributes of an environment, but such
data have limitations and represent spatial patterns only in 2D space,
whereas the LIDAR remote sensing provides 3D data through direct and
indirect retrievals (Kushwaha and Behra, 2002). The major advantage of
laser scanning is that it can operate even at night. In fact, the best results
are achieved during the night (Baltsavias, 1999). The scanning systems
furnish geometric results in terms of distance, position, altitude, and
coordinates (Ackermann, 1999). ALS altimetry is a highly efficient and
accurate method of obtaining data for the determination of visible surface
topography (McIntosh and Krupnik, 2002). This technology has made it
possible to calculate Terrain models with high vertical accuracy on the
order of 10–15 cm, and high spatial resolutions as small as less than 1 m
(Ackermann, 1996; Axelsson, 1998; Lohr, 1998). A DTM can be generated
automatically from the point-sample elevation data (Ackermann, 1999) of
a terrain or the superficial morphological features (Pereira and Wicherson,
1999). The spatial resolution of the data is dependent on several factors,
such as flying height, flying speed, and scanning frequency (Lemmens
et al., 1997). Several studies have shown that laser scanning is capable
of acquiring terrain information more rapidly than photogrammetry at
lower costs (Pereira and Wicherson, 1999).

Basic relations and formulas about laser ranging and airborne laser
scanning can be referred from Baltsavias (1999). Though laser data provide accurate points with high spatial frequency, break lines are not
explicitly present inside the data (Haala et al., 1997; Kraus and Pfeifer,
1998; Ackermann, 1999; Axelsson, 1999), and therefore integrating the ALS
data in combination with high-resolution multispectral satellite data provides helpful results in the delineation of various landform features. The
separation of surface covers, like trees and building heights, generates the
digital ground model (DGM). An interpolation step is required to calculate the digital raster models like DGM. The integration of high-resolution
multispectral satellite data with DSM and DGM data creates a synergy for
mapping and delineating geomorphic features in a 3D context and enables
the recognition and mapping of landscape objects clearly (Ackermann,

1999). The ALS technique has been successfully demonstrated in many applications including urban classification (Tao and Yasuoka, 2002) and landscape modeling (Hill and Veitch, 2002). Earlier, there was a discrepancy in studying geological structures and landforms under dense forest covers using remote sensing data. Now, with the advent of ALS data, the mapping of landforms and geological structures under forest cover can be achieved.

For barren terrain, very high-resolution satellite data like IKONOS provide valuable information on terrain morphology; however, it has limitations under forest cover. In such case, ALS data are complementary and can serve their purposes in getting 3D perspective model for an area. The present study area, the Elbe basin in parts of the National park area of Sächsische Schweiz, is mostly covered by forest. In this situation, except for the location of sandstone exposure, it is difficult to interpret the morphology of the other terrains. Under this circumstance, ALS data provide further details on terrain conditions under dense forest cover and enable the interpretation of the different types of landforms. The purpose of the study is to delineate the various geomorphic features in the Elbe river valley using ALS and very high-resolution satellite data, and their combined output. The results have shown that not only the required terrain features extracted easily, but also with better accuracy than required.

2.2 Study Area

The study area covers parts of the Elbe river valley and its surrounding mountain zone, situated in the southeast of Germany. It is located a few kilometers from Dresden, between the Erzgebirge and the Lausitz Mountain zone, and covers an approximate area of 97 km². This part of the Elbe sandstone mountain region is also called "Sächsische Schweiz." The majority of that fascinating landscape is today protected as the National Park, which is embedded in a landscape-protected area.

The national park is famous because of its geology (sandstone, basalt), its characteristic geomorphology (rocks, plateaus, gorges), and the interesting flora and vegetation close to the natural forest ecosystems. The sandstones are horizontal to the low dipping strata (flat lying sandstone formation) of the Cenomanian, of the Cretaceous age. The strata of the Cenomanian age crop out as "erosional outliers" in the Elbe valley between Meissen, Dresden, Pirna, and the boundary of the Czech Republic (Tröger, 2003). The typical vertical structure of the national park has been formed by intensive fluvial erosion at the end of the Tertiary and during the Quaternary periods that dissected the originally flat-topped sandstone massif. The Oberhäslich sandstone formation is typical of the area in between Dresden and Bad Schandau and in most of the erosional outliers (Tröger, 2003). A marginal trough lies south of the Lausitz fault between Meissen—Dresden—Bad Schandau. This marginal trough is connected with the N Bohemian basin south of Bad Schandau (Voigt, 1963).

2.3 Data Used

The ALS data were obtained from TopScan GmbH, Germany, as a subcontractor of Hansa Luftbild GmbH (TopScan, 1997). Data acquisition was made over 5 days in the year 1997 during spring. The laser point clouds were processed with reference to the national spatial reference systems RD/83 and HN76 (vertical reference). The classification of laser points and the interpolation of topographic models (digital terrain models and digital canopy models) were processed with the help of the SCOP software environment developed by the Institute of Photogrammetry and Remote Sensing (IPF) at the Vienna University of Technology. The density was 1 measurement point per $9\,m^2$. The accuracy of height was in between ±10.8 and ±15.1 cm (TopScan, 1997).

IKONOS satellite data, in digital format, acquired on August 1, 2000, from Hansa Luftbild GmbH, Germany, ordered by IÖR and IPF TUD, is used in the present study. The imagery covers $97\,km^2$ area in parts of the Sächsische Schweiz region. IKONOS satellite data comprise of a panchromatic band (PAN) with a 1 m spatial resolution and four multispectral bands with a 4 m spatial resolution. Out of four multispectral bands, three bands are in optical region and one in the infrared spectrum. In the present study, the multispectral bands were merged with the PAN data (1 m resolution). The output image has a 1 m spatial resolution with an eight bit format and processed into a true color composite to study the geomorphic features. In addition, the digital raster topographic map on 1:25,000 scale published by Landesvermessungsamt, Sachsen, is used in the study.

2.4 Methodology

2.4.1 Processing of Very High-Resolution Satellite Data

IKONOS multispectral data were geometrically merged with the rectified PAN data and resampled into a 1 m spatial resolution. Initially, the

IKONOS true color composite was used to understand overall land use, the land cover pattern, and the nature of the terrain condition. The high spatial resolution provided good information on land cover and landforms in the study area. Though very high-resolution satellite data provide information on landforms, most of the time a geomorphic study requires a regional perspective view. Most parts of the study area are covered by elevated hilly terrain or sandstone outcrop pillars, and the rest of the area is characterized by river valleys and undulating arable landscapes. In order to get impressions on the various geomorphic features, different enhancement and filtered techniques were adopted using the Erdas imagine 8.7 software. Out of various processed outputs, a 3×3 edge enhancement, a 3×3 high-pass filtered image, and histogram equalization provided contrast signatures on landforms, which were utilized for further interpretation.

2.4.2 Airborne Laser Scanning and DSM Generation

DSM was generated using airborne laser scanning measurement points, with the help of SCOP software using the linear prediction method. The prediction model is used in the final stage to attain DSM with cartographically appealing quality. The resulting digital model included terrain elevation for nonforest and forest regions. The hill-shaded DSM is useful to study the four most important morphometric parameters, such as slope, aspect, profile, and plain convexity. It is difficult to delineate the boundary between the different landforms for cloud-covered regions in the satellite data. Moreover, the forest cover spreads both on the plains as well as along the sloping grounds. Under these circumstances, the variation in altitude observed through ALS DSM provides valuable results.

2.4.3 Integration of Very High-Resolution Satellite Data with DSM

Very high-resolution IKONOS satellite data were superimposed over the ALS hill-shaded visualization of DSM using the "topographic analysis" module available in the Erdas imagine software. The output had the advantages of both airborne laser scanning and satellite data. The shaded relief output gave a 3D perspective view as well as the true color combinations of the terrain features. The combined data provided a more realistic picture of the geomorphic features than the individual data set. For example, the perspective view of cuesta with an asymmetrical sloping gradient is very clear in the integrated output. Similarly, the impression of first order drainages and the alternate arrangements of river terraces were contrasted in the output.

2.5 Results and Discussion

2.5.1 Land Use and Land Cover

The major land use and land cover in the study area are forest cover, farm land, pasture land, grass land, the rocky outcrop zone, the built-up area, and the riparian zone associated with the flood plain. The land use and land cover pattern mostly match the geomorphic setup. For example, the forest cover is always associated with relief topography, such as escarpments, sloping grounds, cuesta, hogbacks, mesa, and butte landforms. However, at some locations, the forest cover spreads down into other low relief landforms. Similarly, the boundary between escarpments and river terraces is difficult to trace in satellite data at many locations, because either the forest cover extends its limit up to river terrace or the grasslands continue up to the sloping ground. In such cases, the steep slope with dense single species forest cover gives uniform tonal and textural characteristics in the satellite data. Under this situation, the contour pattern observed in the topographic map is helpful to interpret the sloping condition of the ground.

2.5.2 Geomorphology

The typical vertical structure of the National Park area was evolved by intensive fluvial erosion at the end of the Tertiary and during the Quaternary periods, which dissected the originally flat-topped sandstone massif. Large morphological forms (macroforms) include rock plateaus, canyons, and rock walls, and rock cities or rock labyrinths are also present. Rock pillars, rock ledges, shelters, and chimney rocks also occur at some locations (Varilova, 2007). ALS data were successfully applied to geomorphological studies (Irish and Lillycrop, 1999; Pereira and Wicherson, 1999; Petzold et al., 1999; Davis et al., 2002; Hill and Veitch, 2002). Pereira and Wicherson (1999) assessed the feasibility of using laser data to generate a hydrodynamic model in the Netherlands to manage the fluvial zones. Their results show that laser data allow the recovery of relief information, with the desired quality needed for river management. The purpose of the present study is to understand to what extent airborne laser scanning data can be utilized for landform discrimination.

A part of the Elbe river valley is comprised of high-relief terrain conditions and suitable for applying ALS data. Very high-resolution satellite data are good enough to interpret various landforms in the area. However, the presence of dense forest covers at some locations subdues landform delineation. The third dimensional information, like elevation data obtained from ALS, is valuable for geomorphic study. The very high-resolution satellite data represent the objects with true color and pattern, whereas the ALS data provide height information for different landforms. A segmentation of this information, combined with optical remote sensing data, provide a nice perspective view for various geomorphological features in the study area.

Geomorphologically, the study has shown interesting landform features. The Elbe river flows across the "elbe sandstein geberg" and the various developed erosional outliers, like escarpments, plateaus, and valleys. The tributaries of the Elbe river, such as Polenz, Grünbach, Biela, Sebnitz, and Kirnitzsch, also flow along the major mountain areas and fabricated similar geomorphic patterns in this region. Landforms, such as escarpments and triangular facets, are clearly observed in the very high-resolution satellite data north of the Elbe river, from Wehlen in the west to Kurort in the east. With the help of processed satellite outputs and topographic maps, about 116 landform units were delineated and grouped into the following 10 geomorphic units (Figure 2.1):

- River terraces
- Talus/screes

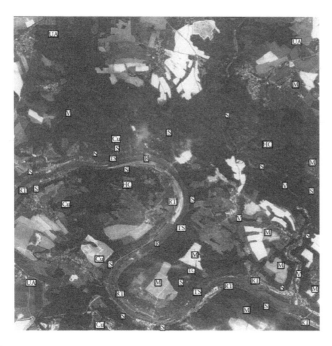

FIGURE 2.1
IKONOS satellite data showing various landforms in parts of the Elbe river valley (Cu, cuesta; HC, hogback complex; M, mesa; R, river; RT, river terrace; S, scarp; TS, talus scree; TC, talus cone; UA, undulating arable land; V, valley).

- Scarp faces and sloping grounds
- Mesa/mesa complexes
- Cuesta/escarpment complexes
- Hogbacks/hogback complexes
- Buttes
- Valleys/dissected valleys
- Terrace landforms
- Undulating arable landforms

2.5.2.1 River Terraces

River terraces are observed on both sides of the Elbe river bank at Wehlen, Rathen, and Bad Schandau, and it indicated that the river incisions might have taken place at different geological time periods (Figures 2.1, 2.2, and 2.4). The river terraces are nearly level surface, relatively narrow, bordering a stream or body of water, and terminating in a steep bank (Leet, 1982). They represent the remnants of stream channels or flood plains, when streams were flowing at a higher level. The subsequent downward cutting of streams left remnants of the old channels or flood plains standing as terraces above the present-day stream level. The river terraces can also indicate other types of regional change. For example, if the climate grows drier, a river will shrink in size. Along the Elbe river bank, the IKONOS satellite data show more than one river terrace at many locations indicating that the different stages of the river incision in the geological past. Moreover, the uneven distribution of terraces indicated that the river migration in one direction was due to some tectonic activity. In the IKONOS satellite data, river terraces are shown clearly on both sides of the river, arranged uneven in nature (Figure 2.1). In true color composite, it shows a light green color and is mostly covered with grasslands. River terraces are exceptionally visible in ALS–DSM, because it represents the elevation changes between the river and the terrace (Figures 2.2 and 2.3). This is the unique information one can obtain through ALS-based DSM. In Figure 2.3, at the west bank of the Elbe river, the Dresden and Bad Schandau railway line comes out clearly. From satellite and ALS data, at least two sets of river terraces were identified at Prossen, near Bad Schandau. Once again, the IKONOS and DSM combined output has shown the contrast signature between the river, the river terraces, the escarpments, and the talus/screes.

2.5.2.2 Talus/Screes

Talus/screes are the accumulation of broken rocks that lie on steep mountain sides or at the base of cliffs. Differentiating the talus/screes along scarp surfaces is significant in the study of slopes (Figure 2.2). Hill-shaded

FIGURE 2.2
Digital surface model (DSM, hill-shaded view) derived from airborne laser scanning point cloud data (Cu, cuesta; M, mesa; R, river; RT, river terrace; S, scarp; UA, undulating arable land; V, valley).

DSM derived from ALS point cloud data provides an enhanced perspective view on slope stability, like mass movements, landslides, etc. Talus/screes developed below the scarp face west of Grahlen and southeast of Lilienstein have clearly shown up in the IKONOS satellite data as a curvilinear break in the slope morphology (Figure 2.1). The distinction between scarp faces and talus/screes is also clearly indicated as sharp changes of shadow or a dull spectral signature in the satellite imagery. However, if the talus/scree zone is relatively old, it might be stablized by dense forest growth and would be difficult to interpret. The curvilinear break in the slope might be developed by the process of mass movements, such as landslides and rock falls in the sandstone formation. Again, the ALS hill-shaded DSM provides a perfect view of the talus/scree (Figure 2.3). In the combined output, the talus zone is very clear and could be delineated with the help of a break in the slope gradient. The topographic breaks along the slope morphology are normally curvilinear to linear pattern with abrupt changes in the pattern and tonal contrast.

2.5.2.3 Escarpments and Terrace Surfaces

"Escarpments or scarp faces" are the exposed rock surfaces with an almost vertical slope gradient. The highly sloping surface mostly with

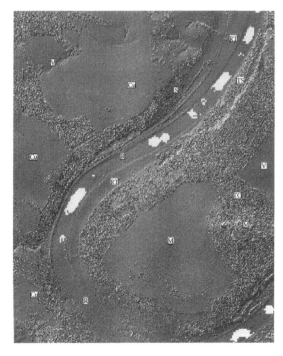

FIGURE 2.3
Combined IKONOS and ALS hill shaded view (Cu, cuesta; M, mesa; R, river; RT, river terrace; S, scarp; TS, talus scree; TC, talus cone; V, valley).

forest cover is named as "sloping ground." Csaplovics et al. (2003) have attempted to extract the scarp faces or rock edges in the study area using ALS data. In their studies, the semi-automatic approaches to laser data, supported by topographic information systems and digital image filtering analysis, show quite remarkable results.

At many locations, the linearity observed in the rock exposure helps to delineate the scarp faces developed in the sandstone region. In the imagery, the scarp faces are exposed as a series of rock cut surfaces. The 3×3 edge enhancement, the high-pass filter, and the histogram equalized images are shown as an enhanced signature on the escarpments. In the true color IKONOS imagery, escarpments are shown in a white contrast color. In most cases, the steep slope covered by dense forest and vegetation with single species gives uniform tonal and textural characteristics in the satellite data. In this case, a topographic map is helpful in identifying the steep sloping ground through dense contour patterns. However, the topomap may not provide the perspective view of escarpments. In that case, the height information collected from ALS data has given a perspective visualization by changes or breaks observed in the altitude, and sharp boundaries highlight the contact between steep slopes and gently sloping surfaces (Figure 2.3). Series of scarp faces developed by tributaries in the north eastern part of the study area are clearly observed with the help of combined IKONOS and ALS data. The series of topographic breaks observed in the gently sloping ground at Stadt Wehlen, north of the Elbe river, is expressed as "terrace landforms." It is observed both in the processed satellite imagery and DSM data. The topographic expression of such a terrace surface is clear in the combined terrain model, since it shows contrast spectral and elevation changes. The origin of such terrace surfaces is not exactly known.

2.5.2.4 Cuestas and Hogback Complexes

The south-western part of the study area is mostly covered by cuesta type of landforms. Cuestas are asymmetrical geomorphic features with steep escarpment faces in one direction and gently sloping surfaces in the other direction. The steep escarpments toward the Elbe river and the gently sloping terrain in the opposite direction show a nice panoramic view in the valley (Figures 2.2 and 2.3). Mostly dense forest covers associated with steep slopes and farmlands prevail on the gently sloping grounds and extend into undulating arable lands. Hogbacks are symmetrical landforms with a sharp crest flanked by steeply sloping surfaces. Hogbacks are arranged in series in the study area and are referred to as "hogbacks complexes." Such types of landforms are noticed near Oberrathen, in the west, and in between Hohnstein and Waltersdorf, in the east. The high-resolution filtered satellite data have shown enhanced boundary condition for cuestas and hogbacks (Figure 2.1). However, the continuation of similar land cover patterns from one landform to another landform gives unclear boundary conditions at certain locations. The discrimination of such landforms from the ALS-DSM output (Figure 2.3) is clear, since such landforms are topographically differentiated with the help of height, slope, and slope aspects. Similarly, the integrated output (DSM + IKONOS) has shown realistic pictures on land cover and slope configurations.

2.5.2.5 Mesa

The smooth landscape surrounded by steeply sloping ground is identified as a "mesa" in satellite and ALS data. At many locations, the flat landscape is surrounded by a sloping ground evolved by debris material brought down from the top, and which has formed a "talus cone." In the study area, two levels of erosional outliers or erosional surfaces are interpreted from the topographic map and ALS data south of Hohnstein and south of Waitzdorf. However, in the satellite data, the second level surface is not clear due to the forest cover. At most of the locations, mesas are associated with settlements and agricultural lands (Figure 2.4).

34

Geoinformatics in Applied Geomorphology

FIGURE 2.4
Terrace surface and river terraces observed in the hill-shaded visualization of DSM, the Elbe river valley (Cu, cuesta; R, river; RT, river terrace; S, scarp surface; TSu, terrace surface; UA, undulating arable land; V, valley).

2.5.2.6 Drainages and Valleys

ALS is more suitable for interpreting drainages and reveals the dynamic changes in the drainage pattern over a period of time. Watersheds delineated from the ALS elevation data appeared very sensitive to changes in terrain, particularly in areas of modified terrain (Hans et al., 2003). Hans et al. (2003) have attempted to trace the drainages from ALS data for highways drainage analysis. The tributaries of the Elbe river developed deeply engraved drainages and valleys in this region. The valleys are prominent in the very high-resolution satellite data through a dark tonal contrast, smooth texture, and contrast changes in the slope, tree cover, and moisture content. The forest cover also extends along drainages and is linearly arranged like branching trees. At few locations, the dense forest cover, without any rock exposure, obscures the demarcating boundary condition from the rest of the landforms. However, ALS data provide contrast view on the changes of the altitude and slope along the drainage valley.

FIGURE 2.5
Combination of IKONOS satellite data with hill shading visualization of DSM highlights various landforms (Cu, cuesta; M, mesa; R, river; RT, river terrace; S, scarp; TS, talus scree; TC, talus cone).

Once again, the drainages and valleys are more visible in ALS-DSM data when compared to very high-resolution satellite data. The impression of first order drainages is very clear in the DSM image. It exhibits broad "U"-shaped valleys along the river Polenz and the tributaries of the Elbe river. The drainages act as a boundary for different types of landforms (Figure 2.5).

2.6 Conclusion

The Elbe river valley in between the Erzgebirge and Lausitz mountain zone is known for its characteristic landscape and geomorphology. Very high-resolution satellite data (IKONOS) provide contrast signatures on land use and land cover practices in parts of the Elbe river basin. In case of geomorphology, ALS-DSM with height information is useful in delineating different geomorphic features in this part of the valley. Interestingly, the combined digital surface model generated from ALS and very high-resolution satellite data provide an excellent perspective view on geomorphology. The ALS technique and its terrain models provide additional quality information and could be utilized for many types of natural resource mapping.

Acknowledgment

The first author acknowledges the Alexander von Humboldt foundation for sponsoring his research visit to Technical University, Dresden, Germany.

References

Ackermann, F. 1996. Airborne laser scanning for elevation models. *GIM* 10(10): 24–25.

Ackermann, F. 1999. Airborne laser scanning—Present status and future expectations. *ISPRS Journal of Photogrammetry and Remote Sensing* 54: 64–67.

Axelsson, P. 1998. Integrated sensors for improved 3D interpretations. *International Archives of Photogrammetry and Remote Sensing* 32: 27–34, part 4.

Axelsson, P. 1999. Processing of laser scanner data—Algorithms and applications. *ISPRS Journal of Photogrammetry and Remote Sensing* 54(1): 138–147.

Baltsavias, E.P. 1999. Airborne laser scanning: Basic relations and formulas. *ISPRS Journal of Photogrammetry and Remote Sensing* 54: 199–214.

Csaplovics, E., Naumann, K., and Wagenknecht, S. Beiträge zur Extraktion von Felskanten aus Airborne Laser Scanner Daten am Beispiel der Elbsandsteininformationen im National park Sächsische Schweiz. *Photogrammetrie-Fernerkundung-Geoinformation* 2/2003: S.107–116, 9 Abb.

Davis, P.A., Mietz, S.N., Kohl, K.A., Rosiek, M.R., Gonzales., Manone, M.F., Hazel, J.E., and Kaplinski, M.A. 2002. Evaluation of lidar and photogrammetry for monitoring volume changes in riparian resources within the Grand Canyon, Arizona. In *Pecora 15/Land satellite Information IV/ISPRS commission I/FIEOS 2002 Conference Proceedings*, Denver, CO.

Flood, M. and Gutelius, B. 1997. Commercial implications of topographic terrain mapping using scanning airborne laser radar. *Photogrammetry Engineering and Remote Sensing* 63: 327–329, 363–366.

Haala, N., Brenner, C., and Anders, K.H. 1997. Generation of 3D city models from digital surface models and 2DGIS. *International Archives of Photogrammetry and Remote sensing* 32(3-4W2): 68–75.

Hans, Z., Tenges, R., Hallmark, S., Souleyrette, R., and Sitansu, P. 2003. Use of LiDAR based elevation data for highway drainage analysis: A qualitative assessment. In *Proceedings of the 2003 Mid-Continent Transport Research Symposium*, Ames, IA, August 2003, pp. 1–19.

Hill, R.A. and Veitch, N. 2002. Landscape visualization: Rendering a virtual reality simulation from airborne laser altimetry and multi-spectral scanning data. *International Journal of Remote sensing* 23(17): 3307–3309.

Irish, J.L. and Lillycrop, W.J. 1999. Scanning laser mapping of the coastal zone: The SHOALS system. *ISPRS Journal of Photogrammetry and Remote Sensing* 54: 123–129.

Kraus, K. and Pfeifer, N. 1998. Determination of terrain models in wooded areas with airborne laser scanner data. *ISPRS Journal of Photogrammetry and Remote Sensing* 53(4): 193–203.

Kushwaha, S.P.S. and Behra, M.D. 2002. Lidar remote sensing and environment. Employment news, New Delhi, pp. 1–3.

Leet, L.D. 1982. *Physical Geology*, 6th edn. Englewood Cliffs, NJ: Prentice-Hall.

Lemmens, M., Deijkers, H., and Looman, P. 1997. Building detection by fusing airborne laser altimeter DEMs and 2D digital maps. *International Archives of Photogrammetry and Remote Sensing* 32(3-4W2): 42–49.

Lohr, U. 1998. Laserscan DEM for various applications. *International Archives of Photogrammetry and Remote Sensing* 32: 353–356, part 4.

McIntosh, K. and Krupnik, A. 2002. Integration of laser-derived DSMs and matched image edges for generating an accurate surface model. *ISPRS Journal of Photogrammetry and Remote Sensing* 56: 167–176.

Pereira, L.M.G. and Wicherson, R.J. 1999. Suitability of laser data for deriving geographical information: A case study in the context of management of fluvial zones. *ISPRS Journal of Photogrammetry and Remote Sensing* 54: 105–114.

Petzold, B., Reiss, P., and Stössel, W. 1999. Laser scanning—Surveying and mapping agencies are using a new technique for the derivation of digital terrain models. *ISPRS Journal of Photogrammetry and Remote Sensing* 54: 95–104.

Tao, G. and Yasuoka, Y. 2002. Combining high resolution satellite imagery and airborne laser scanning data for generating bareland DEM in urban areas, International archives of the photogrammetry, Remote sensing and spatial information science 30.

TopScan. 1997. *Projektbericht Sächsische Schweiz*, TopScan GmbH, Rheine, Germany.

Tröger, K.A. 2003. The Cretaceous of the Elbe valley in Saxony (Germany)—A review. Notebooks on Geology, Online manuscript, April 26, 2003. http://paleopolis.rediris.es/cg/uk-bookmarks.html

Unbenannt, M. 1999. Generation and analysis of high-resolution digital elevation models for morphometric relief classification, represented at a Cuesta Scarp Slope on the Colorado Plateau, USA. http://www.geographie.uni-halle.de/phys/UNBENANN/genera.htm

Varilova, Z. 2007. Occurrence of Fe-mineralization in sandstones of the Bohemian Switzerland National Park (Czech Republic) in; Haertel, H., Cilek, V., Herben T., Jackson, A., Williams, R. (eds.) Sandstone landslides, Academia, Prague, pp. 25–33.

Voigt, E. 1963. Über Randtröge vor Schollenrändern und ihre Bedeutung im Gebiet der Mitteleuropäischen Senke und angrenzender Gebeite. *Zeitschrift der deutschen Geologischen Gesellschaft, Stuttgart* 114(Teil 2): 378–418.

Wehr, A. and Lohr, U. 1999. Airborne laser scanning—An introduction and overview. *ISPRS Journal of Photogrammetry and Remote Sensing* 54: 68–82.

3

Geoinformatics in Spatial and Temporal Analyses of Wind Erosion in Thar Desert

Amal Kar

CONTENTS

3.1 Introduction

The Thar Desert (TD) in the hot arid tropics lies in the eastern margin of the world's largest contiguous arid land between the Atlantic coast of North Africa and the Aravalli Ranges of western India. The Aravalli Ranges mark the eastern margin of the present-day aridity in India (i.e., moisture availability index at −66.6%) and TD. The area to the east of this boundary has many fields of gullied stable sand dunes and other low aeolian bedforms, which bear testimony to the desert's former extensions during drier climates (Goudie et al., 1973), especially before and after the last glacial maxima (Kar et al., 2001, 2004; Singhvi and Kar, 2004).

To the west, the boundary is along the fertile plains of the Indus River, while in the north the Thar fades into the fertile Indo-Gangetic plain where numerous low and stabilized source-bordering-dunes occur along a maze of paleo-channels connected to the Himalayan streams, viz., the Sutlej and the dry Ghaggar (or the legendary Saraswati River) (Oldham, 1893). The southern limit is along the Great Rann of Kachchh, which is a vast salt playa arising from the withdrawal of the Arabian Sea. The sandy desert formerly extended far to the southeast of this boundary also (Juyal et al., 2003).

The vestiges of aeolian sediments beyond the present TD constitute the Megathar, where lower wind velocity, lesser dryness, and better vegetation cover make the landscape less vulnerable to wind erosion (with the exception of the very dry and arid west). Yet, factors like high human pressures on the land resources as well as projected episodic expansion of aridity may lead in future to enhanced wind erosion from the stabilized-aeolian-landscapes.

The TD can be divided into the much drier western Thar, where aeolian processes are more active even without human pressures and the wetter eastern Thar, where the aeolian processes are less active except in segments that are under high human pressure (Figure 3.1). Although

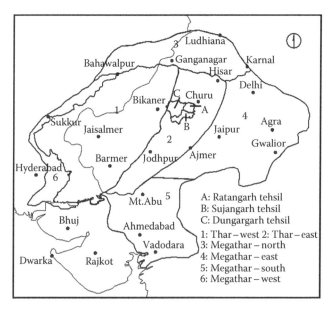

FIGURE 3.1
South Churu tehsils in the Thar–Megathar region.

dominantly sandy, the Thar is thickly populated, where agriculture is the major occupation of the inhabitants.

3.1.1 Issues

Wind erosion is a major process of desertification in TD, where prime agricultural lands and infrastructural facilities get regularly threatened by either aeolian sand deposition or sediment deflation during the summer months. Several studies have revealed that high human and livestock pressures on a poor resource base are the major causes of accelerated wind erosion in the region (Ghose et al., 1977; Kar, 1986, 1996; Singh et al., 1994; Kar and Faroda, 1999; Kar et al., 2007).

For a proper assessment and monitoring of the magnitude of the threats, studies are periodically undertaken to map the extent of wind erosion and degree of sand reactivation, which are principally based on visual interpretation of remote sensing data products (aerial photographs till the 1970s and false color composites of satellite images since then), and supplemented by adequate field checks. The precision of such studies depends on the knowledge and skill of the interpreter and hence no two interpreters may agree with the boundaries of variously reactivated sand patches. The way out from this situation is to develop a system of digital mapping of the sand reactivation (SR) pattern, based on remote sensing signature of the pixels. Once such a system is developed it might be possible to quantitatively relate the SR pattern with pressure variables over large areas. Despite some commendable attempts to link state of degradation with pressure variables under GIS (Grunblatt et al., 1992), the results are not yet very satisfactory.

3.2 Objectives

The objectives of this study are (1) to develop a scientific and quantitative method for identification and mapping of different SR units, (2) to establish a means of using the available data in a GIS environment for quick and large-area mapping, and (3) to develop a system for integrating the different variables to focus on degraded areas.

In order to minimize the operator bias and to maintain certain degree of precision while mapping large areas and integrating information from different variables, this study was carried out in the sandy terrain of three southern tehsils (i.e., administrative blocks, consisting of several contiguous villages), viz., Ratangarh, Sujangarh, and Dungargarh in Churu district, northeastern TD, Rajasthan state (Figure 3.1). Here, the landscape is characterized by 10–30 m high semistabilized sand dunes, separated by interdune plains and occasional salt playas. Wind erosion is a dominant process of land degradation here.

3.2.1 Climate and Population

The average annual rainfall in the three tehsils is as follows: 354.0 mm (Ratangarh), 372.0 mm (Sujangarh), and 261.0 mm (Dungargarh). Much of the rainfall is received during the summer monsoon between June and September. The population is dominantly rural, and occupation overwhelmingly agriculture (~90%). The average population density increased from 36.0 km^{-2} in 1971 to 87.0 km^{-2} in 2001. Population density in the easternmost tehsil of Ratangarh was 106.0 km^{-2} during 2001, while in Sujangarh tehsil to the west it was 101.0 km^{-2}, and in Dungargarh tehsil further west it was 64.0 km^{-2}. The corresponding figures for 1971 were 45, 44, and 24 km^{-2}, respectively. Paralleling the rise in human population, there was a corresponding rise in livestock population by 15%–20%, as well as an expansion of croplands (i.e., from 88.6% in 1971 to 92.5% in 2001), while the extent of grazing lands dropped from 7.9% to 4.4% during the same period.

In order to measure the parameters of desertification and to establish causal processes, the conceptual model of driving force—pressure—state—impact (for a discussion on the concept, utility, and limitations, see e.g., Enne and Zucca, 2000; Kar and Takeuchi, 2003; Svarstad et al., 2008) is used in this study. For this study, 339 contiguous villages (polygons) were selected for analysis. The data set had the variables like spatial pattern of human and livestock population densities (driving forces), quanta of cultivation and grazing pressures (pressure variables), digital analysis of orthorectified, georeferenced, and atmosphere-corrected Landsat data for a measure of sand reactivation (state variable), and analysis of NDVI condition and dust load (impact variables).

3.3 Methodology

3.3.1 Data Sources

Two sets of cloud-free Landsat scenes of dry-cool winter season were used for this study; one set (dated January 9, 1973) was of Landsat-1 MSS (path-row numbers 159-040 and 159-041) of 70 m spatial resolution, resampled at 57 m resolution, and the second set (dated January 22, 2001) was of Landsat-7-ETM + (path-row 148-40 and 148-41) at 30 m spatial resolution. The study area experienced moderate drought during 1972 and 2000, i.e., the years preceding the imaging, and therefore had almost identical meteorological controls on land resources.

Radiometric and geometric corrections were carried out on the images during the preprocessing stage. Although conventionally, atmospheric correction of the scenes is carried out using dark object subtraction of digital numbers, or subtraction from invariant features (Chavez, 1988), the method does not yield absolute radiance (Ra) values. In order to use the near-accurate Ra from a given object at a given time of the year for this study, an atmospheric correction was carried out on all the images by first converting the digital numbers into absolute Ra values per wavelength band, and then converting the Ra values into at-sensor Re, or albedo, using the following formulae (see LANDSAT-7 Guide):

$$Ra = (DN * gain) + bias \tag{3.1}$$

where DN is digital number.

$$Re = \frac{(\Pi * Ra * d^2)}{(e\,sun * \cos(solar\ zenith\ angle))} \tag{3.2}$$

where
  d is normalized Sun–Earth distance
  e sun is mean solar exoatmospheric irradiance
  Re is in W m^{-2} sr^{-1} μm^{-1}

All the images were corrected geometrically with respect to Survey of India (SOI) topographic maps, and co-registered using first affine transformation and then UTM (Zone 43 N) projection.

To derive the digital signatures of SR, the radiometrically and geometrically corrected images were analyzed for principal component analysis (PCA), soil brightness index (SBI) classification from the tasseled cap transformation coefficients, and colorimetric brightness index (CBI) classification. PCA was calculated using the procedures in ERDAS (1997).

SBI values were calculated using the following formulae (Kauth and Thomas, 1976, for MSS; Huang et al., 2001, for ETM+):

$$SBI(MSS) = 0.406MSS4 + 0.600MSS5 + 0.645MSS6 + 0.243MSS7 \tag{3.3}$$

$$SBI(ETM) = 0.356ETM1 + 0.397ETM2 + 0.390ETM3$$
$$+ 0.697ETM4 + 0.229ETM5 + 0.160ETM7 \tag{3.4}$$

To convert the SBI values into SR categories, the DN ranges of some known sand reactivated areas were measured from standard FCCs, and their corresponding SBI values were estimated and used for calibration.

CBI was calculated using Mathieu et al.'s (1998) formulae:

$$CBI(MSS) = \sqrt{\frac{MSS4^2 + MSS5^2 + MSS6^2}{3}} \qquad (3.5)$$

$$CBI(ETM) = \sqrt{\frac{ETM1^2 + ETM2^2 + ETM3^2}{3}} \qquad (3.6)$$

The raw CBI values were converted into sand reactivation categories using the protocol used for SBI.

Additionally, supervised classification was performed using maximum likelihood algorithm (Lillesand and Kiefer, 1987) for which site-based information on the land cover characteristics, degree of reactivation, etc., for a number of localities were gathered through field campaigns. Results from PCA, SBI, and CBI classifications were evaluated with respect to the output from supervised classification.

3.3.2 Measurement of Driving Force and Pressure Variables

For linking SR pattern with the socioeconomic variables, information on human and animal population, land uses, and fallow land were gathered from all the 339 villages in the three tehsils (Census of India 1971 and 2001). The chief occupations in the three tehsils were cropping and animal husbandry. The 339 village boundaries in Census maps were digitized and their boundaries rectified with known control points in SOI maps, followed by preparation of attribute tables on decadal land use and livestock data for the villages. The livestock data were then converted into adult cattle units (ACU), using conversion factors (Table 3.1).

TABLE 3.1

Livestock Type and ACU Conversion Factors

Type	ACU Conversion Factor	Type	ACU Conversion Factor
Cattle, >3 years	1.00	Sheep	0.15
Cattle, 1–3 years	0.75	Goat	0.15
Calf, 1 year or lower	0.25	Lamb	0.06
Buffalo, >3 years	1.30	Kid	0.06
Buffalo, 1–3 years	0.75	Camel	1.00
Calf, 1 year or lower	0.50	Donkey/horse	0.75

Intensity of cropland use was measured using data on single cropped land (mostly the summer-monsoon-based Kharif crops), double cropped land (i.e., both Kharif and Rabi or winter), and fallow land. In almost all cases, rabi cultivation relied on groundwater irrigation. The intensity indices for single cropped-fallow land (SCI) and double cropped land (DCI) were first estimated. The index for total cropland use intensity (TCI) is the product of SCI and DCI.

$$SCI = \frac{(C_T - F)}{C_T} \tag{3.7}$$

$$DCI = \frac{(C_T + C_D)}{C_T} \tag{3.8}$$

where
C_T is total cropland, including fallow land
F is fallow land
C_D is total double cropped land

The carrying capacity (CC) of the land was estimated from average fodder production from the different land uses and demand for the ACUs (Table 3.2). Mostly the pastures of the region are degraded, yielding average plant biomass of ~0.2 t ha^{-1}, while few good pastures yield ~0.9 t ha^{-1}.

Based on data in Tables 3.1 and 3.2, grazing pressure index (GPI) of permanent pasture (PP) was estimated, as shown in the following:

$$GPI \text{ for unmanaged } PP = \frac{(ACU - (PP/CC))}{(PP/CC)} \tag{3.9}$$

TABLE 3.2

Averages of Fodder Production and Carrying Capacity in Different Land Uses (Based on Dry Fodder Need of ACU = 2.5 t year^{-1})

Land Use Type	Average Fodder Production (t ha^{-1} year^{-1})	Carrying Capacity (ha ACU^{-1} year^{-1})
Unmanaged permanent pasture	0.22	11.4
Fallow land	0.22	11.4
Cultivable waste	0.19	13.2
Culturable waste	0.21	12.0
Grazing land (incl. all the above)	0.21	12.0
Cropland stubble	0.15	16.7

Similarly, GPI estimates for cultivable waste, culturable waste, other grazing lands, and cropland-stubble were made for each village polygon, and their values used for calculating average GPI of the village.

All the input variables of driving force and calculated pressure variables were tagged to the respective village polygons as attribute tables in the maps. Such maps were then rasterized in the same resolution as the Landsat imageries to build an interrelationship between the map-derived pressure variables (e.g., cultivation pressure and grazing pressure) and the image-derived state variable (e.g., sand reactivation). The variables were then co-processed using ordered weighted average technique and fuzzy set membership function in a coupled remote sensing-GIS mode for multicriteria modeling of the driving force, pressure, and state variables. In order to minimize interpreter bias, the results were first placed in a 1–255 DN space, and then regrouped as per our perception of the ground situation and information into categories of desertification.

An analysis of the AVHRR-derived NDVI for the month of February was then performed for the Thar–Megathar region (sourced from GIMMS; 8 km resolution) covering 1982–2002, along with a complementary analysis of the dust load over TD using the aerosol index (AI) measured by the total ozone mapping spectrometer (TOMS) satellite sensor for the same time span, i.e., 1982–2002.

3.4 Results and Discussion

3.4.1 Sand Reactivation

Digital maps of surface brightness of PCI, SBI, and CBI broadly yielded similar results. The results of PCA, especially FCC of PC bands 1, 2, and 3, reasonably discriminated aeolian sand from fluvial sand, possibly because the former rests in a matrix of fine silty-clay sediment where individual sand grains often have a thin coating of fine silt or clay of lower Re compared to the cleaner, uncoated sand. On the contrary, the fluvial sand gets washed out from the mixed deposits and appears brighter. PCA, however, did not help in recognition or classification of categories of reactivated sand.

Results of classifications based on SBI and CBI were nearly the same (<5% difference between the two), yet both suffered from slight mixing of signatures with non-aeolian reactivation units, especially near the maximum of the range. The Ra values of riverine sand, dry gypsiferous surface, other dry salt-affected plains, and dry surfaces of salt lakes, begin near the maximum of the reactivated-aeolian-sand, and continue with still higher values. During the field campaign in parts of the TD, a similar pattern was also detected in the measured—visible as well as near and middle infrared—wavelengths of the variously reactivated aeolian sand surfaces, gypsiferous surfaces, salt-affected plains, natural vegetation at various states of degradation, agricultural fields with different crop varieties, etc.

The possibility of density slicing of the SBI and CBI images at critical limits of separation between aeolian-bright and other bright areas was explored based on the knowledge of tonal characteristics of a typical sand-reactivated area and a typical gypsiferous surface on standard FCC. A very thin pixel continuum was found between the highly reactivated sand category and the bright salt-affected surface, but the mode of the two distributions could be separated appreciably. The brightness of the dry floor of salt lakes was eliminated by digitizing the lake boundaries in a vector environment and rasterizing the resulting map to fit into the image, and then masking the brightness. The alkaline surfaces are differentiated better from the saline and non-affected surfaces in the middle infrared band (10.4–12.5 μm).

In spite of the near-similitude of results from SBI and CBI, the latter was preferred, as SBI calculation depended not only on calibration constants of the sensor for different wavelength bands and was difficult to acquire but also on the age of the sensor and the satellite, warranting repeated ground monitoring. CBI calculation, on the other hand, does not require the satellite-dependent coefficients. Its construction is based on a simple premise that aeolian sand surfaces change in color with the degree of reactivation. Supervised classification needs accurate ground information for the training sites, and hence is ruled out for areas where ground information is scarce or difficult to obtain.

So the SR pattern was mapped using CBI classification for January 1973 and January 2001 (Figure 3.2) followed by a comparative study of the changes in the three tehsils. As the area experienced moderate to severe drought in the years 1972 and 2000, with rainfall of 184 and 273 mm respectively, against an annual average of 366 mm, both scenes were similarly predisposed and vulnerable to wind erosion. Image differencing for the reactivated sand categories did not, however, reveal much change in the area statistics (Table 3.3), but significant spatial variability was noticed in SR categories for 1973 and 2001 (e.g., in Sujangarh tehsil; Figure 3.3). Co-evaluation of the Ra images of the area with the CBI images showed that much of the eastern half, recording Ra values of <11 mW cm^{-2} Sr^{-1} μm^{-1}, had lower SR in 2001 than in 1973; perhaps a result of more than a decade-long practice of groundwater-irrigation and resulting stabilization of the landscape to a certain degree, but notwithstanding the increase in cultivation pressure since 1973.

The westernmost parts of Dungargarh and Sujangarh tehsils, on the other hand, showed large areas under moderate to severe SR. The higher Ra (>11 mW cm^{-2} Sr^{-1} μm^{-1}) was especially concentrated in the northwest,

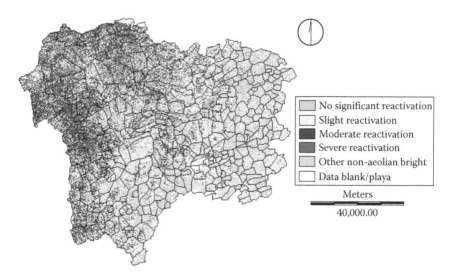

FIGURE 3.2
South Churu: aeolian sand reactivation, January 2001 (based on digital interpretation of Landsat ETM+bands 1, 2, 3).

TABLE 3.3

SR Categories from CBI Analysis

CBI Class	Category	January 1973		January 2001	
		(km²)	(%)	(km²)	(%)
1	Non-aeolian bright surface (Tal)	1.57	0.03	0.14	0.01
2	No significant reactivation	1390.73	24.30	1316.92	22.98
3	Slight reactivation	3191.34	55.76	3196.98	55.80
4	Moderate reactivation	922.23	16.11	1057.86	18.46
5	Severe reactivation	217.26	3.80	157.35	2.75
	Total	5723.13	100.00	5729.11	100.00

where groundwater irrigation started late, which ushered in the land leveling and the deep ploughing of sandy soils with tractors. Overall, 36% area did not show any change in SR categories during the period, while 52% showed positive changes in reactivation and 12% had negative changes (chiefly in the eastern part).

The relative stability of the landscape in the eastern part in comparison with the western part, despite groundwater irrigation and ploughing with tractors in both the parts, may be attributed to the lag in landscape responses to human actions. The sandy terrain in the eastern part saw use of new agricultural technology at least three decades earlier, while

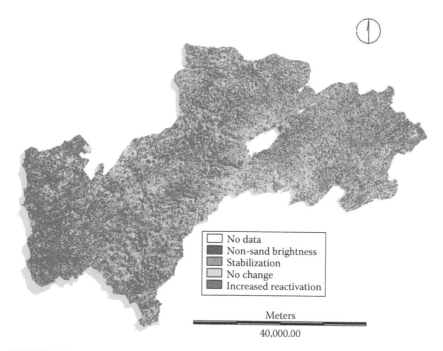

FIGURE 3.3
Sujangarh tehsil: changes in sand reactivation (1973–2001) (based on digital interpretation of satellite data).

the western part had such changes only from the mid-1990s, and that too only with the discovery of a good source of groundwater in parts of Sardarshahr, Dungargarh, and Sujangarh tehsils.

Vast sandy landscape of the interdune plains in the eastern part appears to have slowly adjusted to the changes brought in by irrigation despite tractors ploughing. Intensive cropping and application of water and fertilizer gradually helped to minimize sand reactivation from the interdune plains, although pressure on dune slopes continued. Using this argument, and provided irrigation and tractorization continue to expand together at the same pace, and other pressures do not accelerate, it is likely that the sandy landscape in the western part will take at least another three decades to adjust to such a scenario of irrigated agriculture. Since the sandy terrain in the west has higher wind erosivity than in the east (Kar, 1993), the operations related to expansion of irrigated cropping there would most likely increase the vulnerability to wind erosion for a longer period of time before landscape stability sets in.

It is also logical to hypothesize that in the near future, the aforesaid response time or time lag for adjustment may get severely modulated under the following conditions: a sharp drop in groundwater resource and higher incidence of moderate to severe drought predicted under the new climate change scenario. Depleted groundwater availability and dryness may trigger a shift back to the dominant land use from irrigated agriculture to rain-fed cropping, when the lagged response of vegetation in coming up in the rain-fed crop fields might induce widespread sand reactivation, threatening the food and water security of the human and livestock populations. The response time for the society to get adjusted to such changes will be longer.

3.5 Quantification of Cultivation and Grazing Pressures

3.5.1 Cultivation Pressure

Cultivation pressure was measured through the TCI. It was revealed that between 1970–1971 and 2000–2001 both the intensity and the magnitude of cropland use had changed significantly (Table 3.4). During 1970–1971, despite a favorable rainfall regime, TCI was moderate in 60.9% area and high in 24.1% area. During 2000–2001, in spite of the mild drought, the region had 60.5% area under high TCI and only 37.8% area under moderate TCI (Figure 3.4). Although other areas in the west experienced positive shifts in TCI, the northwest registered a fivefold increase. By contrast, 35.7% area, mainly in the eastern part of Ratangarh and Sujangarh tehsils, showed no change in TCI.

3.5.2 Grazing Pressure

Increased animal pressure on sandy terrain is usually mentioned as a major factor of land degradation. This view is partly supported by the

TABLE 3.4

Intensity and Change Magnitude of Cropland Use

TCI Classes	Area (%)		Magnitude of Change			
	1970–1971	2000–2001	Positive	Area (%)	Negative	Area (%)
1. No cropland use	3.55	0.58	+1	42.34	−1	10.71
2. Very low use	3.33	0.18	+2	4.50	−2	0.51
3. Low use	8.06	0.96	+3	3.18	−3	0.49
4. Moderate use	60.87	37.79	+4	1.81	−4	0.55
5. High use	24.19	60.49	+5	0.23		
			No change: 35.70			

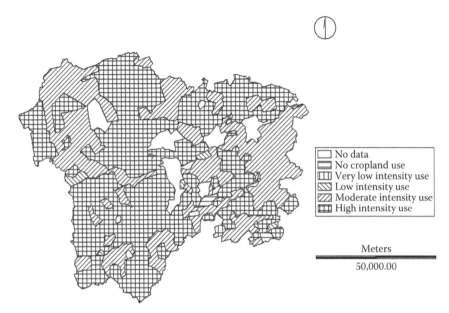

FIGURE 3.4
South Churu: intensity of cropland use, 2000–2001.

villagers also. ACU density per ha in a village provided a misleading view of the animal pressure on land. Hence, density per ha of culturable waste was taken as a partial signature of the pressure. The culturable wastes in the three tehsils varied between 1% and 17%, and are declining gradually. Between 1971 and 2001, the decline in some villages is by ~99%.

In Rajasthan, livestock census is semidecadal (e.g., like 1972, 1977, …, 1997, etc.), and does not coincide with the decadal Census of India. Archival, village-wise data of cattle census, and ACU density for past years are not available as it is destroyed soon after calculating tehsil-wise summaries. Consequently, this study used only 1997 data. Mapping of the 1997 ACU density per ha of culturable wastes in the villages (polygons) revealed that an ideal density of 1–5 ACU ha^{-1} existed mostly in the northwestern part of Dungargarh tehsil, and in the south-central part of Sujangarh tehsil. The eastern part of Ratangarh tehsil had a density of 6–10 ACU ha^{-1}, while the western part of it had a range of 11–15 ACU ha^{-1}. By contrast, the eastern part of Dungargarh tehsil, especially the strip through Jetasar, Riri, Indpalsar, and Keu, the adjoining Reda area of Sujangarh tehsil, and the NE–SW trending belt through Malaksar, Bhasina, and Bhom Telap in southern part of Sujangarh had extreme high density of >30 ACU ha^{-1}. Density in the area to its northwest ranged between 21 and 25 ACU ha^{-1}. Such high densities of >20 ACU ha^{-1} were also registered in the eastern part of Sujangarh tehsil, especially in Khaliya-Ghotra-Sobhasar-Parbatisar tract. The rest of the area had either 11–15 or 16–20 ACU ha^{-1}. On superposition of this pattern over the pattern of degradation, the areas with >20 ACU ha^{-1} generally matched with the areas under moderate to severe wind erosion. However, the matching was not very perfect, suggesting the role of other factors.

Decline of grazing lands in the three tehsils (Ratangarh tehsil: from 23.7% in 1970–1971 to 7.4% in 2000–2001; Dungargarh tehsil: 35.7% to 2.1%; and Sujangarh tehsil: from 24.8% to 3.7%) possibly had little effect on the fodder availability, as attested by high ACU densities in 1997 census.

It was observed that dry fodder from croplands and other bioresources of culturable wastes also helped sustain the livestock. Assuming an average of ~12 ha^{-1} ACU^{-1} year as the optimum carrying capacity in the region (Table 3.2), the calculated GP for 1997 was moderate in 15.1% area and high in 14.6%, especially in the west (Figure 3.5). This pattern was mimicked during 2000–2001.

In order to decipher the influence of the two pressure variables (TCI and GPI) on the SR pattern observed during January 2001, cross-tabulation of the pressure variable maps with the SR maps was carried out. This showed good agreement of the moderate and high cropland use intensities with

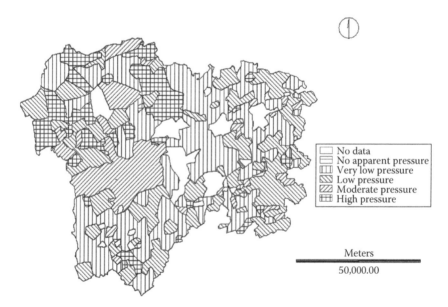

No data
No apparent pressure
Very low pressure
Low pressure
Moderate pressure
High pressure

Meters
50,000.00

FIGURE 3.5
South Churu: grazing pressure, 1997.

TABLE 3.5

Index of Agreement between Sand Reactivation Classes and Pressure Variables

CBI Class	TCI Class	Index of Agreement	CBI Class	GP Class	Index of Agreement
1	1	−0.0008	1	1	0.0000
2	2	0.0026	2	2	0.3474
3	3	0.0060	3	3	0.2353
4	4	0.3088	4	4	0.1579
5	5	0.5395	5	5	0.2832

moderate and high SR categories. The agreement was poor in the case of grazing pressure (Table 3.5), suggesting that intensity of cropland use was a better determinant of degradation than animal grazing.

3.5.3 Multicriteria Evaluation of Desertification

Using ordered weighted average technique and fuzzy set membership function for multicriteria evaluation of desertification, we co-evaluated the human and livestock densities, cultivation, and animal pressures, as well as the satellite-derived wind erosion pixels for January 2001 to unravel the magnitude of desertification in a 1–255 DN space. Based on ground observations at selected sites, the DN values were reclassified into five subjective groups of severity (Figure 3.6). The results suggested that 3% area, located mostly in the northwest in Dungargarh tehsil, had severe risk of desertification, 20% moderate (largely in the western part of Sujangarh and Dungargarh tehsils), and 62% slight.

Since at-satellite Ra or albedo is considered to be a measure of barrenness, and by implication land degradation (Otterman, 1974; Jackson et al., 1975), the possibility of correlation between the pattern of red-band Ra for January 2001 and the desertification index from multicriteria evaluation for the same period came for scrutiny. The at-satellite Ra from the land surfaces in red band was $11\,mW\,cm^{-2}\,Sr^{-1}\,\mu m^{-1}$ or higher in large parts of Dungargarh tehsil and in the western part of Sujangarh tehsil where the desertification index was also higher, especially around the villages where groundwater irrigation was introduced in recent years (Figure 3.7). This could be attributed to increased activities of land leveling and clearing of vegetation cover (both trees and bushes) in the fields to ensure efficient irrigation, which, instead of stabilizing the landscape, encouraged SR. Much of the eastern half, registering radiances of $<11\,mW\,cm^{-2}\,Sr^{-1}\,\mu m^{-1}$ had slight to no desertification hazard, despite the high cultivation pressure due to extensive groundwater irrigation.

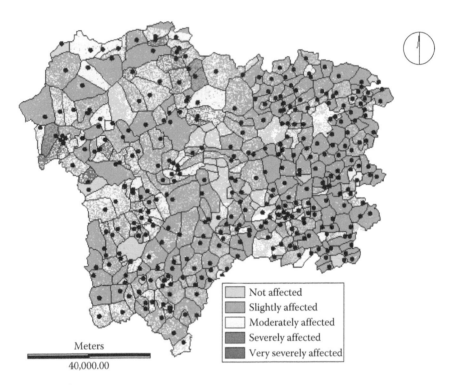

FIGURE 3.6
South Churu: multicriteria evaluation of desertification in sandy terrain, 2001.

3.6 From Local to Regional: Flagging the Key Variables

This study indicated that by choosing the right type of data, some of the causative variables of desertification are amenable to analysis based on remote sensing and GIS. Driving-force variables like human population density, livestock density, etc., can be easily accessed from census data. Equally feasible is accessing the secondary data on land use to decipher the cultivation and grazing pressures (pressure variables), provided key data like fallow land, biomass yield, etc. are available.

Processed digital signatures of SR, both in local and regional scales, are suitable state variables of wind erosion. Perhaps the only constraint is a knowledge of the nature of terrain, based on which the processed values are grouped into aeolian/non-aeolian brightness and magnitude of erosion. Impact variables for wind erosion are several, like degradation of land quality due to sand advance and reduced crop yields, or increased dust load in the atmosphere, which need better measurement tools for an

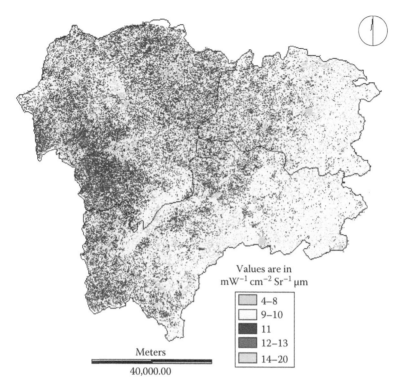

Values are in
$$mW^{-1}\,cm^{-2}\,Sr^{-1}\,\mu m$$

	4–8
	9–10
	11
	12–13
	14–20

Meters

40,000.00

FIGURE 3.7
South Churu: red band radiance categories (January 2001).

improved understanding. Also, there are some emerging pressure variables like groundwater irrigation and tractor use that need monitoring and integration with SR and atmospheric dust load. Perception of the spatial patterns of dust emission and atmospheric concentration is better at regional, rather than at local scale, hence the data strings on tractor use, SR, dust emission, and atmospheric dust load, calculated at different spatial scales, need to be upscaled/downscaled for integration. Some preliminary findings on groundwater use and atmospheric dust concentration now follow.

3.6.1 Will Groundwater Irrigation Lead to Wind Erosion?

In an earlier section, the signature of landscape stability in the eastern half of the three tehsils was projected as a product of more than three decades of groundwater-irrigation. It was cautioned that greater use of groundwater might deplete the aquifer, force the area to rain-fed cropping, but without any benefit of the traditional agroforestry that was destroyed during the onset of irrigated cultivation, and cause a severalfold rise in the potential of wind erosion and rise of atmospheric dust load. This paradigm calls for proper monitoring of the areas irrigated with groundwater in the sandy desert.

In order to examine the likely impact of groundwater irrigation in the northeastern Thar, where no other major sources of irrigation exists, a time-series analysis was carried out using the GIMMS satellite-drift-corrected NDVI data of AVHRR (8 km resolution), sourced from the Global Land Cover Facility, University of Maryland (Pinzon et al., 2005; Tucker et al., 2005). In large parts of the TD, NDVI increases in direct response either to rainfall (mostly during summer monsoon, June–September) or to irrigation (largely during winter, December–March, when there is hardly any rainfall). Data for the second fortnight of February (hereafter called 02b), when the annual plants from the previous summer monsoon become dry, were used. Since common pastures are always in moderate to severely degraded state, and a few forest areas on the Aravalli Ranges register low NDVI values in February due to leaf fall (typically <380; known forest boundaries could be masked out), the rich NDVI signatures (mostly 0.400 and above) represent irrigated croplands only. This analysis showed that the irrigated area expanded between 1982 and 2002, but there were distinct declines in NDVI values in some areas, which could be related to abandoning of irrigated winter crop due to depletion of the groundwater (Figure 3.8). Such changes in some localities are so conspicuous that these could also be noticed easily on standard FCCs (Figure 3.9a and b).

To assess the general pattern of change in irrigated area in the Thar–Megathar region between 1982 and 2002, a standardized classification image of the NDVI changes in 02b from 1982 to 2002 (or the standard deviation map) was designed. The word Megathar is coined for the past extent of the TD—deciphered from surface and subsurface aeolian sand thickness beyond the present arid boundary. The standard deviation (sd) mapping revealed an increase in NDVI value by > +2 sd in large parts of eastern Thar (mostly groundwater-irrigated), eastern part of Megathar (irrigated by both groundwater and canal systems), southern part of Megathar (mostly under groundwater irrigation), as well as in northern and western parts of Megathar (mostly canal irrigation). In western Thar, increase in NDVI values was noticed in the range of +1 to +2 sd, especially in the north where canal irrigation is dominant, and in the south where groundwater irrigation is spreading. A decline by −2 sd and beyond was conspicuous in the northern part of eastern Thar, but in majority of the areas the decline was by < −1 sd, which included both croplands and pastures (Figure 3.8). Further studies are continuing.

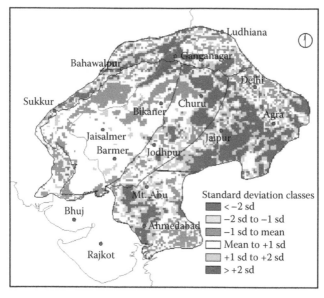

FIGURE 3.8
Thar–Megathar region: standardized classification image for AVHRR-NDVI from GIMMS (second fortnight of February; 1982–2002).

3.6.2 Tractor-Ploughing and Atmospheric Dust Load: Is There a Pattern?

Apart from groundwater-irrigation, the other major threat in the region is deep ploughing of the sandy soils using tractors, which needs to be factored in the modeling. In the early 1970s, the three tehsils had only 51 tractors, which rose to ~2500 by 2002! Use of wooden and iron ploughs, numbering ~0.1 million in the early 1970s, is now almost discarded. In western Rajasthan, the number of tractors increased from ~15,000 in 1980 to ~0.2 million in 2002 (>1000% increase).

Deep ploughing of the sandy tract, including the upper and middle slopes of the 15–30 m high sand dunes, at the beginning of summer monsoon and/or before the winter sowing, now turns up sand over a larger tract in the TD. This increases the vulnerability of the sandy terrain to wind erosion, and could be a major cause of increased atmospheric dust load in recent years.

In order to decipher the changes in the pattern of atmospheric dust load in relation to meteorological parameters in TD, a range of data types for the period 1981–2001 were used. As wind erosion and atmospheric dust load in the desert are very high during March to July (Kar, 1993), the average AI for June (1980–2001) from TOMS sensor was collected (*Source*: NASA; for methodologies and some results, see Washington et al., 2003;

(a)

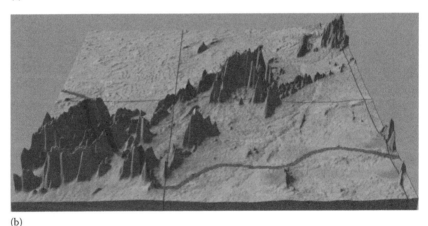

(b)

FIGURE 3.9
(a) Standard FCC of Landsat MSS images (March 2, 1975) for Udaipurvati and adjoining areas in Jhunjhunun district, showing extensive groundwater-irrigated winter crops in the intermontane plains of the Aravalli Ranges, and fewer such areas in the sandy plains to the west. (b) Standard FCC of MODIS-Terra images (February 14, 2007) of the same area as in Figure 3.9a, showing very few areas under groundwater-irrigated winter crops in the intermontane valleys, and notable increase in such practice in the sandy plains in the west.

Torres et al., 2002; Dey et al., 2004; El-Askary et al., 2005; Habib et al., 2006). Data on the U and V components of wind for the period came from NOAA's NCEP/NCAR site, and were converted into resultant wind speed (km h^{-1}). Annual rainfall (in mm) data were also sourced from NCEP/NCAR. All the basic data were averaged for TD. Plotting revealed almost synchronous changes in the curves for previous year's total rainfall, current year's June wind speed, and current year's June AI till the late 1980s. Since then, rainfall showed a rising trend during the 1990s, but AI showed

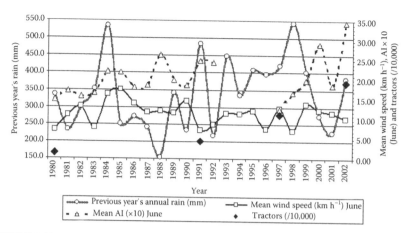

FIGURE 3.10

Time series of the spatial mean of TOMS-derived aerosol index (AI) for June over TD, compared with the time series of spatial mean of the total annual rainfall (previous year), June wind speed, and the number of tractors used.

a falling trend, before the instrument failed. However, a new instrument since 1997 recorded steep increase in AI despite a drop in wind strength and variable rainfall (Figure 3.10). This possibly reflected a greater control on AI, in recent times, of the critical land use changes in the sandy tract that was brought about by large-scale use of tractors for ploughing of the sandy terrain, although other factors might also be involved.

The above two case studies tend to justify the argument that if groundwater-irrigated area proliferates without the aquifers getting adequately recharged and land conservation measures not strengthened, there is every possibility that groundwater irrigation system will gradually collapse, and will pose a major threat of wind erosion, leading to much higher atmospheric dust load. If that is so, we may put groundwater irrigation on the watch list of indicators for wind erosion.

3.7 Conclusion

Following are the major conclusions from this study.

It is possible to classify the sand reactivated areas from medium-resolution satellite images, but that a successful classification also requires interpreter's knowledge of the terrain, and the ability to identify the critical limits of separation between a bright aeolian surface and other bright surfaces. The critical boundaries between the different reactivation/non-such classes cannot be expected to be the same in different scenes of the same period/different periods. Deflation areas are as yet difficult to estimate.

It is also possible to use the secondary data on human and animal densities, as well as land uses to estimate the cultivation and grazing pressures, and then to establish relationship among the variables in GIS. Similar quantitative procedures can be developed for relating the actual occurrence of other kinds of degradation (in contrast to the potentials of degradation of any kind) with the human use systems.

In the TD, technology is fast becoming a major threat of degradation. In the context of sand reactivation, groundwater irrigation and use of tractors for ploughing have started to play some crucial roles. Therefore, procedures for estimating the impacts of groundwater irrigation on sand reactivation in a longer time frame, and of tractor-ploughing on dust emission need attention.

References

Chavez, P.S. 1988. An improved dark object subtraction technique for atmospheric scattering correction of multispectral data. *Remote Sensing of Environment* 24: 459–479.

Dey, A., Tripathi, S.N., and Singh, R.P. 2004. Influence of dust storms on the aerosol properties over the Indo-Gangetic basin. *Journal of Geophysical Research* 109: doi:10.1029/2004JD004924.

El-Askary, H., Gautam, R., Singh, R.P., and Kafatos, M. 2005. Dust storms detection over the Indo-Gangetic basin using multi-sensor data. *Advances in Space Research* 37: 728–733.

Enne, G. and Zucca, C. 2000. Desertification Indicators for the European Mediterranean Region: State of the Art and Possible Methodological Approaches. ANPA—National Environmental Protection Integrated Strategies, Promotion, Communication Department, Rome, and NRD—Nucleo di Ricerca sulla Desertficazione, Universita degli Studi di Sassari, Sassari, 260 pp.

ERDAS Field Guide. 1997. ERDAS Inc., Atlanta, GA, 656 pp.

Ghose, B., Singh, S., and Kar, A. 1977. Desertification around the Thar: A geomorphological interpretation. *Annals of Arid Zone* 16: 290–301.

Goudie, A.S., Allchin, B., and Hegde, K.T.M. 1973. The former extensions of the Great Indian Sand Desert. *The Geographical Journal* 139: 243–257.

Grunblatt, J., Ottichilo, W.K., and Sinange, R.K. 1992. A GIS approach to desertification assessment and mapping. *Journal of Arid Environments* 23: 81–102.

Habib, G., Venkataratnam, C., Chiapello, I. et al. 2006. Seasonal and interannual variability in absorbing aerosols over India derived from TOMS: Relationship to regional meteorology and emissions. *Atmospheric Environment* 40: 1909–1921.

Huang, C., Wylie, B., Yang, L., Homer, C., and Zylstra, G. 2001. Derivation of a Tasseled Cap Transformation based on Landsat-7 at-satellite reflectance. U.S. Geological Survey, 10 pp. http://landcover. usgs.gov/tasseled.pdf

Jackson, R.D., Idso, S.B., and Otterman, J. 1975. Surface albedo and desertification. *Science* 189: 1012–1015.

Juyal, N., Singhvi, A.K., and Rajaguru, S.N. 2003. Chronostratigraphic evidence of episodes of desertification since the Last Glacial Epoch in the southern margin of Thar desert, India. In: *Desertification on the Third Millennium*, (ed.) Alsharhan, A.S., Wood, W.W., Goudie, A.S., Fowler, A., and Abdellatif, E.M. pp. 123–128, Swets & Zeitlinger Publishers, Lisse, the Netherlands.

Kar, A. 1986. Physical environment, human influences and desertification in Pushkar-Budha Pushkar lake region, India. *The Environmentalist* 6: 227–232.

Kar, A. 1993. Aeolian processes and bedforms in the Thar Desert. *Journal of Arid Environments* 25: 83–96.

Kar, A. 1996. Desertification: The scenario for arid western India. *Journal of Geography and Environment* 1: 16–24.

Kar, A. and Faroda, A.S. 1999. Desertification in drylands of India: Processes and control. In: *Fifty Years of Dryland Agricultural Research in India*, eds. Singh, H.P., Ramakrishna, Y.S., Sharma, K., and Venkateswarlu, B., pp. 93–104. Central Research Institute for Dryland Agriculture, Hyderabad.

Kar, A. and Takeuchi, K. 2003. Towards an early warning system for desertification. In: *Early Warning Systems. UNCCD Ad Hoc Panel*, Committee on Science and Technology. UN Convention to Combat Desertification, Bonn, Germany, pp. 37–72.

Kar, A., Singhvi, A.K., Rajaguru, S.N. et al. 2001. Reconstruction of the late Quaternary environment of the lower Luni plains, Thar Desert, India. *Journal of Quaternary Science* 16: 61–68.

Kar, A., Singhvi, A.K., Juyal, N., and Rajaguru, S.N. 2004. Late Quaternary aeolian sedimentation history of the Thar Desert. In: *Geomorphology and Environment*, eds. Sharma, H.S., Singh, S., and De, S., pp. 105–122. ACB Publications, Kolkata.

Kar, A., Moharana, P.C., and Singh, S.K. 2007. Desertification in arid western India. In: *Dryland Ecosystem: Indian Perspective*, eds. Vittal, K.P.R., Srivastava, R.L., Joshi, N.L., Kar, A., Tewari, V.P., and Kathju, S., pp. 1–22. Central Arid Zone Research Institute, Jodhpur, and Arid Forest Research Institute, Jodhpur.

Kauth, R.J. and Thomas, G.S. 1976. The Tasseled Cap—A graphic description of the spectral temporal development of agricultural crops as seen by Landsat. In: *Proceedings of the Symposium on Machine Processing of Remotely Sensed Data*, Purdue University, West Lafayette, IN, pp. 4B41–4B51.

Lillesand, T.M. and Kiefer, R.W. 1987. *Remote Sensing and Image Interpretation*. John Wiley & Sons, New York, 721 pp.

Mathieu, R., Pouget, M., Cervelle, B., and Escadafal, R. 1998. Relationships between satellite-based radiometric indices simulated using laboratory reflectance data and typic soil color of an arid environment. *Remote Sensing of Environment* 66: 17–28.

Oldham, C.F. 1893. The Saraswati and the lost river of the Indian Desert. *Journal of the Royal Asiatic Society* 34: 49–76.

Otterman, J. 1974. Baring high-albedo soils by overgrazing: A hypothesized desertification mechanism. *Science* 186: 531–533.

Pinzon, J., Brown, M.E., and Tucker, C.J. 2005. EMD correction of orbital drift artifacts in satellite data stream. In: *Hilbert-Huang Transformation Introduction and Applications*, ed. Huang, N. and Shen, S.P.S., pp. 167–186. World Scientific, Singapore.

Singh, S., Kar, A., Joshi, D.C., Kumar, S., and Sharma, K.D. 1994. Desertification problem in western Rajasthan. *Annals of Arid Zone* 33: 191–202.

Singhvi, A.K. and Kar, A. 2004. The aeolian sedimentation record of the Thar Desert. *Proceedings of the Indian Academy of Sciences (Earth & Planetary Science)* 113: 371–401.

Svarstad, H., Petersen, L.S., Rothman, D., Siepel, H., and Watzold, F. 2008. Discursive biases of the environmental research framework DPSIR. *Land Use Policy* 25: 116–125.

Torres, O., Bhartia, P.K., Herman, J.R., Sinyuk, A., Ginoux, P., and Holben, B. 2002. A long term record of aerosol optical depth from TOMS observations and comparisons to AERONET measurements. *Journal of the Atmospheric Sciences* 59: 398–413.

Tucker, C.J., Pinzon, J.E., Brown, M.E. et al. 2005. An extended AVHRR 8-km NDVI data set compatible with MODIS and SPOT Vegetation NDVI data. *International Journal of Remote Sensing* 26: 4485–4498.

Washington, R., Todd, M., Middleton, N.J., and Goudie, A.S. 2003. Dust-storm source areas determined by the total ozone monitoring spectrometer and surface observations. *Annals of the Association of American Geographers* 93: 297–313.

4

Remote Sensing and GIS for Coastal
Zone Management: Indian Experience

Debashis Mitra

CONTENTS

4.1 Introduction

The coastal zone is the interface where the land meets the ocean, encompassing shoreline environments as well as adjacent coastal water. Its components can include river deltas, coastal plains, wetlands, beaches and dunes, reefs, mangrove forests, lagoons, and other coastal features. The limit of the coastal zone is often arbitrarily defined, differing widely among nations, and are often based on jurisdictional limits or demarcated by reasons of administrative ease. It has often been argued that the coastal zone should include the land area from the watershed to the sea, which theoretically would make sense, as this is the zone where biophysical interactions are strongest.

For practical planning purposes, the coastal zone is a *special area*, endowed with special characteristics, whose boundaries are often determined by the specific problems to be tackled. Its characteristics are

- It is a dynamic area with frequently changing biological, chemical, and geologic attributes.
- It includes highly productive and biologically diverse ecosystems that offer crucial nursery habitats for many reasons.
- Coastal zone features such as coral reefs, mangrove forests, and beach and dune systems serve as critical natural defenses against storms, flooding, and erosion.
- Coastal ecosystems may act to moderate the impacts of pollution originating from land (e.g., wetlands absorbing excess nutrients, sediments, human waste).
- The coast attracts vast human settlements due to its proximity to the ocean's living and nonliving resources as well as for marine transportation and recreation.

4.1.1 Problems and Issues in Coastal Developments

Coastal zones throughout the world are very precious and have delicate ecological environments, both for man and for nature. Since they often have fertile soils, and are favored by man through their location near the sea (ports, fisheries), the pressure on the yet undisturbed coastal zones is great. In addition, the coastal zones already inhabited and cultivated often encounter difficulties caused by the complex nature of the environment and conflicts of interest between the different inhabitants and users. There are many coastal activities laying their own claims to the coastal zone. The main activities are transport, aquaculture, fishery, agriculture, forestry, human settlement, mining, recreation, and tourism. To sustain coastal resources, the development of activities and their effects on the coastal zone, planning, and management needs to be guided and monitored.

Some major problems encountered in coastal developments are

- Deterioration of coastal resources by destruction, overexploitation and uneconomical use
- Development activities along the coast that create many adverse effects on coastal resources

- Upland development activities having a negative impact upon the downstream coastal areas
- Sea-level rise and landfall resulting in inundation of coastal lowlands

4.1.2 Value of Coastal Resources

Coastal resource systems are valuable natural endowments that need to be managed for the present and future generations. The coastal zone offers physical and biological opportunities for human use, and Integrated Coastal Zone Management tries to find the optimum balance between these uses based on a given set of objectives. Concern is growing, in particular, about the destruction of natural coastal ecosystems by the demands placed upon them by population and economic growth. These natural ecosystems have considerable value for sustainable extractive and nonextractive use that is often undervalued in comparison with other often nonsustainable uses.

In nature, the coastal system maintains an ecological balance that accounts for shoreline stability, beach replenishment, and nutrient generation and recycling, all of which are of great ecological and socioeconomic importance. These natural systems are under increasing threat from unmanaged human activities such as pollution, habitat destruction, and overexploitation of resources. In coastal rural areas, fishing of nearshore water and farming of coastal lowlands are the major economic activities supplying fish and agricultural products for subsistence of the nearby residents and those in the urban centers. Activities that add further value to coastal resources include recreation and tourism, which have become major sources of domestic and foreign earnings in many coastal nations.

4.1.3 Growth in Coastal Population

Population growth in the coastal zone is a major concern. The world population is expected to grow at an exponential rate from 5.8 billion in 1995 to 8.5 billion by the year 2025. It is projected to reach 11 billion in a century's time, with 95% of the growth occurring in developing countries. More than 50% of the world population is already concentrated within 60 km of the coast, while there is considerable migration of population to the coast from the inland areas. In developing countries, by the turn of this century, two-thirds of the population (3.7 billion) are expected to live along the coast.

This growth will exacerbate already severe coastal-use conflicts in terms of land and water space and resource utilization. The negative impacts of increased human settlements and industrial development are also more acutely felt in the coastal zone since it is at the receiving end of land- and water-based pollution. Compounding the problem, the coastal zone is often subject to the overlapping governance of local, provincial, and central governments resulting in interagency conflicts and unclear policy concerning resource development and management and environmental protection. In many countries, large parts of the coastal zone are privately owned. Stabilizing population through family planning programs is an integral part of integrated coastal zone management (ICZM); therefore, it is of crucial importance to maintain the quality and productivity of the coastal zone, as indeed of the rest of the planet.

Increased coastal resource use conflicts will inevitably intensify social and economic development problems. Problems of multiple jurisdiction and competition between users of resources without the benefit of a conflict resolution mechanism, inadequate regulations for protecting resources, and the lack of nationally or locally adapted coastal policies for informed decision making will translate into a loss of capability for future sustainable development. As the resource base is depleted, conflicts may reach alarming dimensions to the point of threatening human life and public order.

4.2 Integrated Coastal Zone Management

Coined during the UNCED in 1992, ICZM has become a concept, which coastal nations around the world are adopting to wisely plan and manage the use of coastal resources. "Integrated" is used to describe the bringing together of participants, initiatives, and government sectors (Kay and Alder, 1999). While integrated means an adaptive decision-making process, a coastal management program cannot be thought of as "integrated" unless these goals are addressed:

- Sustainable development of coastal areas
- Reducing the vulnerability of coastal areas to natural hazards
- Maintaining essential ecological processes, life support systems, and biodiversity (National Research Council, 1995)

Once these goals have been addressed, the next step of an ICZM program should focus on techniques for administration and implementation. The techniques should focus on ways of bringing disparate planning and management tactics together to form a holistic and flexible coastal management system (Kay and Alder, 1999). The challenge of an integrated coastal management program is to thus combine the development of an adaptive, integrated, environmental, and socioeconomic management system that focuses on coastal areas (Kay and Alder, 1999).

To achieve integrated coastal management, several dimensions of integration, as shown in Table 4.1, will need to be addressed. These integrations are mainly integration among sectors (intersectoral integration); integration between the land and water sides of the coastal zone (spatial integration); integration among scientific aspects of coasts, levels of government, and among agencies within each level of government (interagency integration); and integration among disciplines and policy-making and implementation (National Research Council, 1995). An effective coastal plan for sustainable coastal development needs a thorough knowledge of short- and long-term processes and their response to natural disasters and anthropogenic hazards. A well-conceived coastal management strategy ensures sustainable development of the coastal area, the protection of marine resources, the reduction of the vulnerability of coastal habitats from natural hazards, and the conservation of marine biodiversity.

Table 4.1 demonstrates achieving a successful ICZM means more than just having an innovative partnership among involved stakeholders. It is rather more complex because it means accounting for all the various parameters such as those listed above. Moreover, given the set of complex parameters needed for ICZM, the role of remote sensing will increasingly become an indispensable tool for coastal planning and management (O'Reagan, 1996).

TABLE 4.1

Parameters That Need to Be Considered for ICZM

- *Topography and terrain*: Periodic beach surveys, aerial and ortho-photographs, bathymetric charts, soil maps, watershed/catchment information
- *Morphological data*: Side-scan sonargraph sediment samples, geological bore log data
- *Major infrastructure*: Inventory of shore protection structures, roads, and marinas
- *Forestry and conservation*: Forest reserves, forest types, natural vegetative species, conservation areas, marine reserves
- *Coastal fisheries*: Licensing zones, pelagic and demersal fish distribution, commercial aquaculture
- *Oceanography*: Variety of physical, chemical, and biological oceanongraphic data
- *Environment*: Point pollution sources, water quality data, industrial site locations, sensitivity analyses
- *Socioeconomic data*: Housing location, valuation data, demographic structure, census information
- *Planning*: Past and present land use information, administrative boundaries, coastal hazard zones, development pressure, land use capability, environmental constraints

---

4.2.1 Critical Issues of Coastal Zone Management in India

India has a coastline of about 7500 km, of which about 5400 km belong to peninsular India and the remaining to the Andaman, Nicobar, and Lakshadweep Islands. With less than 0.25% of the world coastline, India houses 63 million people, approximately 11% of the global population living in low elevation coastal areas (MOEF, ICZM report, 2009). India's coastal zone is endowed with a wide range of mangroves, coral reefs, sea grasses, salt marshes, sand dunes, estuaries, lagoons, and unique marine and terrestrial wildlife. The abundant coastal and offshore marine ecosystems include 6740 km² of mangroves, including part of the Sundarbans in West Bengal and the Bhitarkanika in Orissa, which are among the largest mangroves in the world. Despite the ecological richness and the contribution to national economy, the coastal and marine areas are under stress. Rapid urban-industrialization, maritime transport, marine fishing, tourism, coastal and sea bed mining, offshore oil and natural gas production, aquaculture, and the recent setting up of special economic zones have led to a very significant increase in demand for infrastructure, resulting in exploitation of natural resources. About 34% of mangroves of India have been destroyed in the last 40 years; 66% of the coral areas are threatened; marine fish stocks are declining; and aquarium fish and sea cucumbers are fast disappearing. Such depletion and degradation, unless arrested, will impact the livelihood, health, and well-being of the coastal population, affecting, in turn, prospects for sustained economic growth.

The management regime for coastal and marine areas of this country suffers from the lack of an integrated and coordinated decision-making system. This is reflected in a multiplicity of institutional, legal, and economic planning frameworks, all narrow and sector driven. Consequently, sectoral activities and interventions in coastal and marine areas work in isolation from each other, at times with conflicting objectives and outputs. The overall policy and plan responses are further crippled by lack of knowledge on coastal resources, processes, impact analyses, and management options.

The following issues are critical in the context of coastal zone management in India (Nayak, 1996):

1. Coastal habitat conservation related
   a. Availability of benchmark or reference data (base line data)
   b. Preservation, conservation, and monitoring of vital and critical habitats, e.g., coral reefs, mangroves, etc.
   c. Appropriate site selection for industries, landfall points, aquaculture, recreational activities, etc.

    d. Assessment of conditions in regulation zones, areas under construction set-back lines, megacities

    e. Reclamation of wetland for agricultural and industrial purposes

2. Coastal processes related

    a. Planning and implementation of coastal protection work (erosion, flood protection, saltwater intrusion, etc.)

    b. Interactions between developmental activities and the modification of coastal processes

    c. Impact of dam construction on shoreline equilibrium

    d. Suspended sediment dynamics

    e. Changes in bottom topography

3. Coastal hazards

    a. Cyclones, storm surges

    b. Coastal erosion

    c. Sea-level rise and possible effects

    d. Nonpoint and point pollution

    e. Phytoplankton blooms

4. Availability of resource and its utilization (sand mining, fisheries)

4.2.2 Coastal Geomorphology and Coastal Management

Coastal geomorphology maps help to understand the geomorphic process of erosion, transportation, and deposition. They also depict the trend of shoreline changes and coastal evolution in any particular region. The study is important as tides, waves and currents provide energy that is constantly working to change the landforms.

The morphology and sedimentary facies of the coastal environment are the product of interactions between relative sea levels, coastal processes, and sediment supply (Pedro et al., 2000). Tidal coastal plains develop under conditions of rising sea level (transgressive coasts), or stable or falling sea level (prograding coasts), in response to different combinations of sea-level changes history (Dalrymple et al., 1992). These settings are responsible for the landward migration of the shoreline (transgression) and the seaward migration of the shoreline (regression) and may be identified by geomorphologic mapping. Therefore, the recognition of shoreline deposits may be crucial to understanding sea-level changes and the stratigraphic sequence (Reading and Collinson, 1996).

A study of coastal geomorphology has paramount importance on the decision making of any coastal area. It has impact on vulnerability assessment, environmental impact statements, coastal management practices, coastal planning, coastal policy, legislation and regulation, and communication.

Satellite remote sensing data, due to its repetitive, multispectral, and synoptic nature, provide a unique view to recognizing various features on land and sea. Multispectral imagery produces comprehensive images of the land and sea surface displaying medium- to large-scale phenomena and relationships that often cannot be observed from a low altitude (aircraft) or from a surface perspective. Variation in color, brightness, texture, and contrast of the imagery represents various land/water features and assists in the visual interpretation of the data.

There is no ideal classification system for coastal geomorphology as there are different perspectives in the classification process, and the process itself tends to be subjective. Each classification is made to suit the needs of the user, which does not meet the need of another user who wants to apply the same classification system in some other study area. The nature of coastal geomorphology is dependent on the climatic conditions of the area and the properties of the land. So, all the classification systems are regional in nature, and one system will not fit another region perfectly. Mapping of coastal geomorphology has been carried out in the Bay of Bengal coastal area using the Landsat TM data of 1998 (Figure 4.1).

A detailed analysis of the spectral signatures of terrain associated with important elements like geometric, tonal and textural characteristics of the coastal landforms helped to identify different geomorphologic units. Fieldwork was carried out to cross check the features observed in the image interpretation. During geomorphologic mapping vegetation, sediments and landforms of different coastal geomorphologic units were analyzed.

4.3 Coastal Habitats

Coastal habitats, especially wetlands, coral reefs, mangroves, salt marshes, and sea grasses, are rapidly being cleared for urban, industrial, and recreational growth as well as for aquaculture ponds. The actual estimates of coastal habitat loss are not available. In most areas, topographical maps prepared during the 1960s and 1970s are being used. In tropical countries, the loss of mangroves is well over 50% of the pre-agricultural area.

Optical remote sensing data of various resolutions have been used for mapping all *coral reef* areas of India. There are mostly fringing reefs in the Gulf of Kachchh, the Gulf of Mannar Indian Remote Sensing (IRS), and

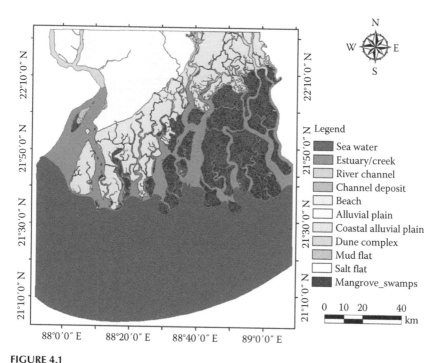

FIGURE 4.1

The geomorphology map of the Bay of Bengal coastal area prepared from the Landsat TM data of 1998.

the Andaman & Nicobar Islands, with a few platform, patch, and atoll reefs and coral pinnacles. The Lakshadweep Islands are mostly atolls with few coral heads, a platform reef, and sand cays (Figure 4.2).

Mangroves are very important as they help in the production of detritus, organic matter, recycling of nutrients, and thus enrich the coastal waters and support the benthic population of the sea. They support the most fundamental needs of the coastal people—food, fuel, shelter, and monetary earnings. At many places, mangroves are degraded and destroyed due to conversion of these areas for agriculture, aquaculture on the East Coast, and industrial purposes on the West Coast. Satellite remote sensing data can be of huge help in studying the inventory of mangroves, monitoring their degradation in the time domain, carrying out community zonation, and sometimes mapping up species level using high-resolution aerial photography.

Mangrove vegetation shows distinct zonation and is characterized by the presence of particular species, with a specific physicochemical environment and related dominant genus, being dependent on the extent and frequency of inundation under tidal waves, salinity, and

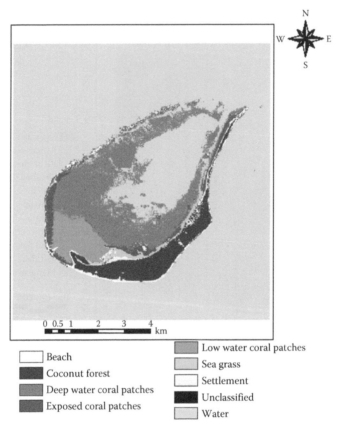

FIGURE 4.2
Coral reef distribution at Minicoy Island. (*Source*: IRS P6/LISS IV, 2005.)

soil characteristics. Information regarding different mangrove community zonations is a vital remote sensing-based input for biodiversity assessment and for preparing management plans for conservation
(Figure 4.3).

Recently, there has been much interest in using synthetic aperture radar
(SAR) remote sensing to retrieve the biophysical characteristics from
forest targets. The characteristics of the plant (density, distribution, orientation, shape of the foliage, dielectric constant, height, and branches),
the ground (dry, moist, and flooded), and the sensor (polarization, incidence angle, and wavelength) are important in determining the radiation
backscattered toward the radar antenna (Dobson et al., 1992). It has been
observed that amplitude information of ERS SAR data is more sensitive to
different mangrove communities than coherence information of the tandem pair. Consequently, the mapping potential of individual community

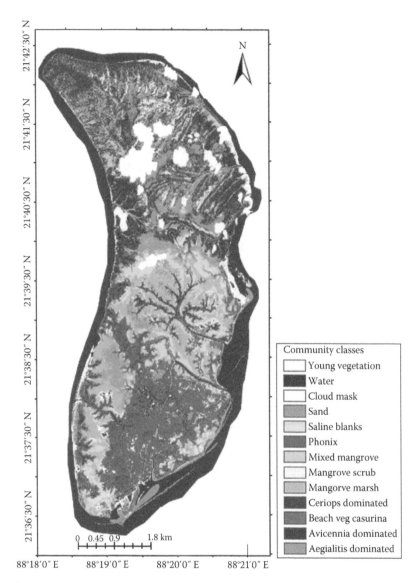

FIGURE 4.3
Mangrove community zonation map of Lothian Island, Bay of Bengal (IRS P6/LISS IV, 2006).

classes is higher in amplitude information of ERS SAR than that in coherence information. Also, the mapping potential of individual community classes is the maximum in the HH-polarization amplitude image followed by the VV and HV polarized amplitude images of ENVISAT ASAR data (Chatterjee et al., 2004).

Brackish water aquaculture has tremendous potential for economic development of any region because of ever-increasing demands of prawn. In India, aquaculture development started essentially to provide employment in rural coastal areas as well as to increase export to developed countries. Aquaculture development and planning require comprehensive data on land use and water resources. Remote sensing data provide information on these aspects due to their repetitive and multispectral character and synoptic view. Intensive commercial aquaculture, practised both in developed and developing countries and growing in popularity among many Asian and Latin American countries as an export industry, could have harmful consequences as it replaces coastal mangrove habitats, reducing breeding grounds of wild stocks (FAO, 1991). Large-scale encroachment has been noticed in many mangrove areas. In order to plan aquaculture at the national level, a multi-objective land allocation method needs to be followed to avoid land conflicts in the subsequent years among different stakeholders. Important parameters, such as accessibility to the site, topography, water source, tidal amplitude, soil type, availability of seed, marketing, etc., are sufficient to identify suitable areas for aquaculture (Figure 4.4). However, the number of parameters to be considered will also depend upon the type of farming system, viz., freshwater or brackish water aquaculture and extensive or modified extensive or semi-intensive or intensive aquaculture. Aquaculture site selection involves more of GIS than only remote sensing.

4.4 Coastal Processes

Across the world, many coastal areas are being eroded, which threatens the life and property of the local population. One of the major requirements of planning coastal protection work is to understand the coastal processes of erosion, deposition, sediment transport, flooding, and sea-level changes that continuously modify the shoreline. Human-induced activities like coastal construction work, coastal mining, dredging activities, etc., continuously change the natural coastal processes and aggravate coastal hazards (Pranesh, 2000). Multidate satellite data have been used to study shoreline change and coastal landforms that in turn help to understand coastal processes.

The shoreline is one of the rapidly changing landforms. The accurate demarcation and monitoring of the shoreline (long-term, seasonal, and short-term changes) are necessary for an understanding of coastal processes. The rate of shoreline change varies depending up on the intensity of causative forces, the warming of oceanic waters, the melting of

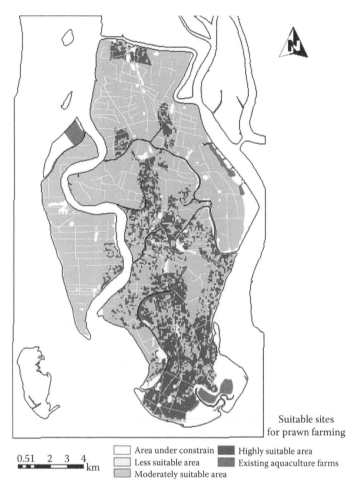

Suitable sites
for prawn farming

0.51 2 3 4
▬▬▬▬ km

☐ Area under constrain ■ Highly suitable area
☐ Less suitable area   ■ Existing aquaculture farms
☐ Moderately suitable area

FIGURE 4.4
Aquaculture site selection in Namkhana Block, South 24 Pargana, West Bengal.

continental ice, etc. Monitoring the shoreline in an area will tell you the erosion/accretion process and will help in the development of a management plan. Shoreline change can be studied from the automated image processing of multitemporal satellite data or the manual digitizing of the temporal land/water boundary in a GIS platform (Figure 4.5) where the IRS 1C LISS III images of 1996 and 2000 have been used. One thing that has to be kept in mind while attempting a change detection study in a coastal area is that both the images should be of the same tidal level. Otherwise, you have to perform some tidal corrections to minimize the tidal differences. The delineation of a shoreline is a difficult task due to the presence of intertidal mudflats and marshy/swampy areas along the coast

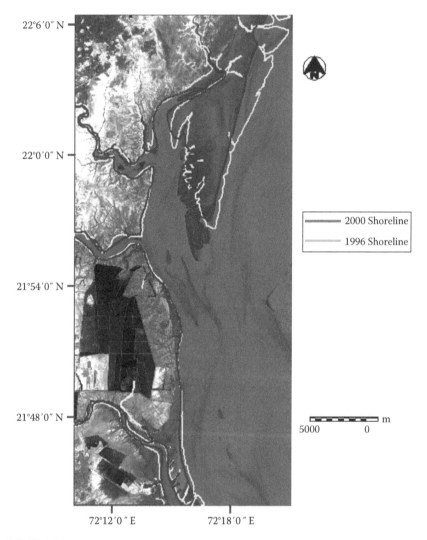

FIGURE 4.5

Shorelines extracted from multitemporal images. Shorelines of 1996 and 2000 were super-imposed over the IRS IC LISS III satellite data acquired in 2000.

most of which are usually misclassified as part of water. Pixels quite often represent the mixture of different spatial classes. Intermixing the pixels of water-saturated land that represent shallow water bodies in the satel-lite imagery affects the discrimination of an accurate shoreline boundary (Muralikrishna et al., 2005).

Most areas on the East Coast show depositional activities. The Chilka lake (Orissa), the Pulicat Lake (Andhra Pradesh), and the Hooghly estuary on the East Coast of India are being silted up. The Krishna and the Gautami–Godavari delta are prograding. Progradation of the coast is also noticed near Vedaranyam in Tamil Nadu. Most of the spits on the East Coast of India are growing. However, spits near the Mahanadi and the Vasistha–Godavari are being eroded.

The West coast is experiencing both erosion and accretion in several places. Near Honavar, Karnataka, it was observed that the southern spit is growing while the northern spit is being eroded (1971–1989). On account of this, the mouth of the river Sharavati has been shifted further northward (Chauhan and Nayak, 1996). Shifting of river mouths, the formation of shoals, and the growth of spits have been noticed along the Maharashtra and Goa coasts. The Narmada estuary in the Gulf of Khambhat (Cambay) is silted up while erosion was noticed in the Mahi estuary in the Gulf of Khambhat.

Industrial development in the coastal zone is a burning issue nowadays for management purposes. The coastal zone is always on the forefront of civilization and has been by far the most exploited geomorphical unit of the earth. Its easy access and resourcefulness have always attracted human activities, but its complexity in understanding has led to misuse and abuse. In recent times, the coastal zone of the world is under increasing pressure due to the high rate of human population growth, the development of various industries, fishing, mining, the discharge of municipal sewage and industrial waste effluents. This industrial development on the coast has resulted in the degradation of coastal ecosystems and in diminishing living resources. Satellite remote sensing, because of its synoptic view and repetitive coverage, helps to monitor a particular area before industrialization and the changes after industrialization. Groins, seawalls, breakwaters, and other protective structures have secondary effects resulting in downstream erosion. Erosion has been observed north of Visakhapatnam, Paradip, Ennore, north of Madras, near Nagapattiam and Kanyakumari, ports on the East Coast of India, while deposition has been observed south of these ports (Figure 4.6). The knowledge about suspended sediment movement helps in understanding near-shore water flow. Sediment plume, which may act as a barrier for sediment movement, causes erosion and deposition at various places. The recently available IRS P4 Ocean Color Monitor (OCM) data are extremely useful for studying sediment dispersal and sediment transport due to their 2-day repetivity.

Salt water ingress is an alarming problem in many parts of coastal India. The continued and unconcerned use of the groundwater in the coastal belts has led to alarming situations in many parts of our country. Continued human interference into the coastal hydrologic system has led to the pollution of the coastal groundwater aquifers by saltwater. Groundwater pollution incidents due to saltwater intrusions have increased manyfold in the last couple of decades. Generally, one comes to know about groundwater pollution due to saltwater mixing only after the incident has occurred.

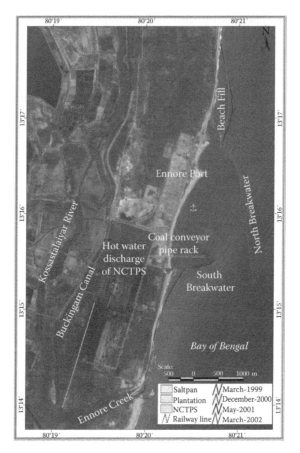

FIGURE 4.6
Shoreline change due to breakwater at Ennore Port (ICMAM, Chennai).

Experience shows that the remediation of the groundwater system, which has undergone saltwater intrusion, is rather difficult and uneconomical in most cases. Change in groundwater levels with respect to mean sea elevation along the coast largely influences the extent of seawater intrusion into freshwater aquifers.

Salt-encrusted mudflat and soil salinity can be monitored through remote sensing data. Soil salinity has a direct role over the vegetation in that region. Nowadays, several vegetation indices give information about the soil and the related properties. Using the Landsat TM and IRS sensor, coastal soils have been classified as highly saline, saline, and nonsaline depending upon the condition of vegetation, color, association, location, salt-encrustations, etc.

Hydrogeological conditions and human activities close to the coast mainly affect groundwater pollution due to seawater mixing. There has been no methodology for evaluating the spatial distribution of the seawater intrusion potential, which essentially takes into account hydrogeological factors, and allows the seawater intrusion of the coastal hydrogeological setting to be systematically evaluated in any selected coastal area where hydrogeological information is available. Therefore, it is necessary to adopt a mapping system that is simple enough to apply using the data generally available, and yet is capable of making best use of that data in a technically valid and useful way. Salt water intrusion can be mapped and modeled using several commercially available models like MUDFLOW, DRASTIC, etc.

4.5 Coastal Hazards

The coastal zone is subject to various cyclic and random processes, both natural and anthropogenic, that continuously modify the region. Protection of human life, property, and natural ecosystems from various hazards is a major concern. The major hazards are cyclones and associated tidal floods, coastal erosion, pollution, and sea-level rise and its impact.

Tropical cyclones constitute one of the most destructive natural disasters that affect India, especially its East Coast. Its impact is greatest over coastal regions, which bear the brunt of strong winds, heavy rainfall, and flooding. Remote sensing data have been utilized for the tracking, monitoring, and forecasting of cyclones, the assessment of damage, and for taking preventive measures. INSAT data have been regularly utilized to monitor the track of cyclones and to forecast its crossing point on land. IRS 1C/1D data have been utilized to assess the damage and inundation caused by a cyclone (Nayak and Bahuguna, 2001). SAR data seem to be the best to study cyclone-affected places because of its cloud-penetrating capabilities (Figure 4.7). Recently, lots of studies have been carried out to study tsunami using satellite data. Satellite remote sensing data are also very helpful for locating inundations due to a tsunami. Penetration of a water body because of a tsunami toward the land can be easily traced from suitable remote sensing data. Also, the inundation line derived from ground survey can be superimposed over the taluk or village map using GIS to get an idea about the affected agricultural areas, human population, and infrastructure.

Pollution of coastal water may be caused due to various reasons, e.g., industrial affluent, domestic sewage, agricultural waste, oil spill, thermal and radioactive waste, etc. Turbidity/suspended sediments and color/ chlorophyll are indicators of water quality. Chlorophyll indicates the trophic status, the nutrient load, and the possibility of pollutants in coastal waters. Suspended sediments affect navigation, fisheries, aquatic life, and

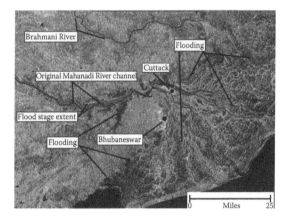

FIGURE 4.7
Inundated areas due to flooding as seen from RADARSAT satellite data.

the recreation potential of sea resorts. They also carry absorbed chemicals and create an environmental problem. Suspended sediments and pollutant outflow are easily observed on satellite images (Figure 4.8). They help in studying the dynamic relationship between sediment input, transport, and deposition. High sediment load is hazardous to many developmental activities. The knowledge about suspended sediment movement helps in predicting the transportation path of waste effluents. In this regard, IRS P4 OCM images will be extremely useful as it is a dedicated satellite for

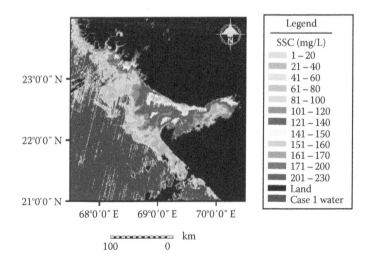

FIGURE 4.8
Suspended sediments as seen from IRS P4 OCM satellite data acquired on November 21, 2001, for the Mumbai coastal zone.

studying suspended sediments and chlorophyll. It has a 2-day temporal resolution that is important to understand highly dynamic features, like sediment plume, coastal pollution, etc.

4.6 Potential Fishing Zone

Knowledge of particular conditions and processes affecting fish population, however, may often be deduced using remote sensing technique. Conceptually, the parameter providing information on these environmental factors coupled with other appropriate biological data may allow a forecast of fish distribution generally in relative terms. But this in no way implies that space technology offers an alternative to fishery survey. Both sea truth information and satellite data, when judiciously applied, can be said to be mutually complementary. Sea surface temperature (SST) was retrieved with an accuracy of ±0.80°C over cloud-free areas on an operational basis from NOAA-AVHRR since 1991, at the National Remote Sensing Agency (NRSA) under Marine Remote Sensing Information Services (MARSIS) Programme of the Department of Ocean Development (DOD), coordinated by the Department of Space (DOS) involving many national organizations. Potential fishing zone (PFZ) maps are generated based on oceanographic features, such as thermal boundaries, fronts, eddies, rings, gyres, meanders, and upwelling regions visible on a 3–4 day composite map of SST. Currently, efforts are ongoing to provide integrated fishery forecasting with the scenario of the availability of new sensors optimized for deriving various other ocean features in the coming years. The relationship between chlorophyll and SST has been established. The data from the OCM sensor, launched in May 1999 on board the IRS-P4 (Oceansat-1), have been used to develop a model to integrate chlorophyll information along with SST as a first step toward providing fishery forecast to predict the likely availability of fishes more accurately (Figure 4.9).

4.7 Primary Applications of Remote Sensing for Coastal Zone Study

Remote sensing is used to address a wide variety of management and scientific issues in the coastal zone. Due to its repetitive, multispectral, and synoptic nature, remote sensing data have proved to be extremely

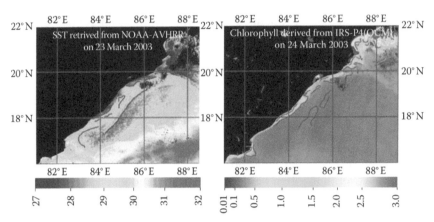

FIGURE 4.9
Synergy of SST and chlorophyll derived from NOAA and OCM, respectively.

useful in providing multispectral information on various components of the coastal environment, viz., coastal wetland conditions, mangrove and coral reef degradation, coastal landforms and shoreline changes, tidal boundaries, brackish water areas, suspended sediment dynamics, coastal currents, air pollution, etc.

Different wavebands of light penetrate water to varying degrees; red light attenuates rapidly in water and does not penetrate deeper than 5 m or so, whereas blue light penetrates much further (15 m), and in clear water, the seabed will reflect enough light to be detected by a satellite sensor even when the depth of water approaches 30 m. The green light penetrates as far as 15 m in clear waters. NIR (0.7–0.8 µm) penetrates to a maximum depth of 0.5 m and IR (0.8–1.1 µm) is fully absorbed (Mumby and Edwards, 2000).

The importance of remotely sensed data for inventorying, mapping, and monitoring of coastal zone was realized early. Due to its repetitive, multispectral, and synoptic nature, remote sensing data have proved to be extremely useful in providing information on various components of the coastal environment, viz., coastal wetland conditions, mangroves and coral reefs degradation, coastal landforms and shoreline changes, tidal boundaries, suspended sediments dynamics, coastal currents, etc., as shown in Table 4.2. IRS-1C/1D has LISS III and PAN that have proved to be extremely useful in the discrimination of dominant mangrove community zones, in the mapping of details regarding ports and harbor areas, as well as in assessing damage due to cyclones in the coastal areas, the delineation of the coastal regulation zone, shoreline changes, etc. (Nayak, 1996; Mitra et al., 2000).

TABLE 4.2

Status on Utilization of Remote Sensing Data for Coastal Studies in India

Resources/Parametres/ Processes	Remote Sensing Compliances	Status
Mangrove, coral reefs, salt pans, aquaculture, wetlands, other coastal inland resources	Mapping and monitoring on a different scale	Operational using high resolutions multispectral sensors data from IRS series
Fisheries	Forecasting and monitoring	Semi-operational with NOAA and IRS-P4
Mineral and energy	Exploration and monitoring	R & D stage with existing RS data
Coastal geomorphology and shoreline changes	Mapping and monitoring on different scales	Operational high resolution data from IRS series
SST, winds, waves, water vapor content, etc.	Fishery forecasting, monsoon, ocean and atmospheric studies	Operational with IRS-P4 and other foreign satellites
Upwelling, eddies, gyres, etc.	Fishery and ocean dynamics studies	Operational with IRS-P4 and others
Coastal regulation zone	Mapping and monitoring in 1:50,000 and 1:25,000 scale	Operational using IRS 1C and IRS 1D
Suspended sediment concentration	Mapping and monitoring	Semi-operational with IRS-P4
Oil slicks	Mapping and monitoring	Semi-operational with IRS series and other foreign satellites
Chlorophyll concentration	Mapping and monitoring	Semi-operational with IRS-P4
Currents and surface circulation patterns	Mapping and monitoring	Semi-operational with IRS series and other foreign satellites

4.7.1 Major Projects Carried Out along the Indian Coast Using Remote Sensing Data

With the availability of Landsat sensors data in India, the Government of India initiated a major program for the mapping of coastal resources and for sustainable utilization. With the launch of IRS satellites, coastal zone mapping and monitoring at the national level became an imperative for planning and administrative purposes. The Space Application Centre (SAC) in association with regional remote sensing service centers and state remote sensing application centers have carried out several projects as follows:

1. Coastal zone mapping for the entire country on a 1:250,000 and 1:50,000 scale
2. Wetland mapping on a 1:250,000 scale for the entire country and on a 1:50,000 scale in selected areas
3. Coral reefs and mangroves area mapping on a 1:50,000 scale
4. Shoreline changes for the entire Indian coast on a 1:250,000 scale and on a 1:50,000 scale in selected areas
5. Coastal land form studies on a 1:250,000 scale for the entire Indian coast and on a 1:50,000 scale in selected areas
6. Lagoonal/lake studies on a 1:50,000 scale
7. Mapping of the coastal regulation zone (CRZ) on a 1:25,000 scale for the entire Indian coast using high-resolution data from IRS and SPOT
8. Flood map zoning on a 1:50,000/25,000 scale
9. Integrated coastal management studies
10. Critical habitat analysis mapping
11. Identification of suitable sites for brackish water aquaculture
12. Estuarine and river morphological studies

4.8 Conclusion

The coastal zone in India assumes importance because of the high productivity of its ecosystems, the development of industries, the concentration of the population, the exploration of living and nonliving resources, the discharge of industrial waste effluents and municipal sewage, and the spurt in recreational activities. It is important to understand the effect of these activities on the coastal zone and its consequent ecological degradation. Remote sensing data, especially IRS data, with moderate (23–36 m) and high (6 m) spatial resolutions, have been used to generate a database on the various components of the coastal environment of the entire country.

References

Chatterjee, R.S., Sarkar, T., Roy, P.S., Rudant, J.P. et al. 2004. Potential of ERS and Envisat SAR data for environmental monitoring of Kolkata (Calcutta) city and coastal region of West Bengal, India. In: *Presented in ENVISAT Symposium (Organized by European Space Agency)*, Salzburg, Austria, September 6, 2004.

Chauhan, P. and Nayak, S. 1996. Shoreline-change mapping from space: A case study on the Indian coast. In: *The Proceedings of the International Workshop on International Mapping from Space*, Institute of Remote Sensing, Anna University, Madras and International Society of Photogrammetry and Remote Sensing, pp. 130–141.

Dalrymple, R.W., Zaitlin, B.A., and Boyd, R. 1992. Estuary facies models: Conceptual basis and stratigraphic implications. *Journal of Sedimentary Petrology* 62:1130–1146.

Dobson, M.C., Ulaby, F.T., Le Toan, T., Kasische, E.S., and Christensen, N. 1992. Dependence of radar backscatter on coniferous forest biomass. *IEEE Transactions on Geosciences and Remote Sensing* 30:412–415.

FAO. 1991. *Environment and Sustainability of Fisheries*. Committee on Fisheries, FAO, Rome, p. 7.

Kay, R. and Alder, K. 1999. *Coastal Planning and Management*. Routledge, London.

Ministry of Environment and Forest. 2009. *World Bank Assisted Integrated Coastal Zone Management Project*, Ministry of Environment and Forest, pp. 1–22.

Mitra, D., Sudarshana, R., and Mishra, A.K. 2000. Rapidly changing coastal environment: Hidden risks in Gulf of Cambay, Gujarat, India. In: *Subtle Issues in Coastal Management*, eds. R. Sudarshana, D. Mitra, A.K. Mishra et al., pp. 149–157. IIRS (NRSA) Publication, Dehradun, India.

Mumby, P.J. and Edwards, A.J. 2000. *Remote Sensing Handbook for Tropical Coastal Zone Management*. UNESCO Publications, Paris.

Muralikrishna, G., Mitra, D., Mishra, A.K., Oyuntuya, Sh., and Nageswra Rao, K. 2005. Evaluation of semi-automated image processing techniques for the identification and delineation of coastal edge using IRS—LISS III Image: A case study on Sagar island, East Coast of India. *International Journal of Geoinformatics* 1:1–11.

National Research Council. 1995. *Science, Policy and the Coast: Improving Decision Making*. National Academy Press, Washington, DC.

Nayak, S. 1996. Monitoring of coastal environment of India using satellite data. In: *Science, Technology & Development, Special Issue on The Environment and Development in India*, Vol. 14, eds. D. Ghosh and N. Kundu, pp. 100–120. Frank Cass & Co. Ltd., Essex, U.K.

Nayak, S. and Bahuguna, A. 2001. Application of remote sensing data to monitor mangroves and other coastal vegetation of India. *Indian Journal of Marine Sciences* 30:195–213.

O'Reagan, P.R. 1996. The use of contemporary information technologies for coastal research and management—A review. *Journal of Coastal Research* 12:192–204.

Pedro Walfir Martins E. Souza Filho, Maâmar El-Robrini, and Setembro de. 2000. Geomorphology of the Braganca Coastal Zone, Northeastern Para State. *Revista Brasileira de Geociências* 30:518–522.

Pranesh, M.R. 2000. Dredging activities along the coastal zone and its impact on the coastal morphology. In: *Symposium on Management Problems in Coastal Areas*, Ocean Engineering Centre, IIT, Madras, pp. 217–281.

Reading, H.G. and Collinson, J.D. 1996. Clastic coasts. In: *Sedimentary Environments: Processes, Facies and Stratigraphy*, 3rd edn., ed. H.G. Reading, pp. 154–231. Blackwell Science, Oxford.

5

Kuwait Coastline Evolution during 1989–2007

S. Neelamani, S. Uddin, and Siddan Anbazhagan

CONTENTS

5.1 Introduction

More than 50% of the world's population lives within 60 km of the coastline and this will rise to almost 75% by 2020 (Anon 1992). In Asian countries, the increase of coastal population is significant within 100 km from the coastline (Duedall and Maul 2005). In general, most of the coasts around the world are dynamic and are undergoing significant changes due to natural and man-made influences. A proper understanding of the coastal morphological changes is essential for integrated management and sustainable development of the coastal zone. Kuwait is a coastal country and has 496 km of total coastline including its islands. The country is situated on the northeastern corner of the Arabian Peninsula, covering an area of about 17,800 km², extending between latitudes 28° 30′ N and 30° 05′ N and longitudes 46° 3′ E and 48° 35′ E. Kuwait is bordered by the Arabian Gulf on the eastern side.

The area of Kuwait territorial waters is estimated at about 7611 km². They can be divided into two parts: the shallow northern area, which is less than 5 m deep in most places with a muddy bed, and the relatively deep southern area, which has a bed of sand and silica deposits. The northeastern part of Kuwait is closer to the drainage basin of the Tigris and Euphrates rivers (e.g., Boubyan Island), less inhabited, and the coast mostly evolving by natural processes (Figure 5.1). The southeastern part of Kuwait and the southern part of Kuwait Bay is thickly populated and the coast is evolving mainly due to man-made processes like the construction of marinas, ports and harbors, seawater intakes, jetties, seawalls, moles, slipways for boats, private and public chalets, etc. A sustainable and holistic development of all coastal engineering activities in Kuwait is essential for meeting the future requirements of the increasing population. The coastal zone of Kuwait has some very vital activities, such as

- Power and desalination plants built on the coastal area for ease of drawing seawater for cooling and desalination
- Navigation, which is one of the lifeline activities in Kuwait, since import and export of most of the cargoes are made via sea transports
- Some of the prominent housing schemes for the present and future populations are located on the coast (e.g., Pearl City in Al-Khiran)

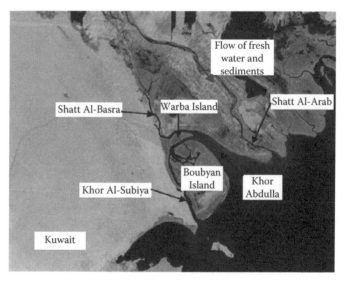

FIGURE 5.1
Satellite data show the north and northeast Kuwait coastal zones.

- Tourism and recreation activities that have been gaining momentum in recent times
- Environmental-friendly methods of disposal of treated sewage from residential areas and effluent from industry into deeper Arabian Gulf waters for better dispersal
- New mega ports planned at Boubyan Island to meet the future demands of this region
- Protection of Kuwaiti islands that are strategically important for the country, and protection from erosion and other natural processes

It is essential to understand the temporal changes in the coastal morphology of the entire Kuwait coast, either for selecting a site for a new coastal project or for the proper planning and management of the existing coastal projects. For example, the development of housing infrastructure in a coastal location historically known for erosion is not desirable. Similarly, it is not preferable to select an accreting site for construction of a port, where the annual maintenance cost is high for dredging of sediment accumulation. Coastal morphological changes, including area of accretion, erosion, or stable areas, over a period of 10–15 years were studied. Similar coastal morphological changes were investigated in various parts of the world using bathymetric maps, topographic information, remote sensing, and GIS techniques for the time span of 18–60 years (Galgano and Douglas 2000; Zujar et al. 2002; Seker et al. 2003; Appeaning Addo 2009).

Some of the coastal stretches in Kuwait are very dynamic with reference to sediment transport, like the coastline around the Boubyan islands due to the sediment supply from Shatt Al-Basra and the draining of Iraqi marshes (Al-Ghadban et al. 1998). Identification of erosion and accretion zones in the coastal sector is essential for proposing effective coastal zone management. Remote sensing techniques provide an important means to study the spatiotemporal changes in coastal morphology. In addition, field investigations were carried out to verify remote sensing–based information, on the ground. The results of coastline evolution for the south eastern part of Kuwait and around the Sabiya power plant area are highlighted in this chapter.

5.2 Remote Sensing and Data Sources

Remote sensing technique has been successfully used to monitor coastline changes using multitemporal satellite images (Zujar et al. 2002; Seker et al.

2003; Wu 2007; Zhou and Xie 2009; Ezer and Liu 2010). In this study, the Landsat TM satellite of 1989 and the Cartosat 1 or IRS P5 (Indian Remote Sensing Satellite) of 2006–2007 were used to analyze the spatiotemporal changes of coastal morphology. The band 4 of Landsat TM data is in the near infrared spectral range and is suitable for demarcating the outlines of water bodies from satellite images. In addition to band 4, bands 5 and 7 were also used. Band 4 data are normally used for coastlines with vegetation, while bands 5 and 7 are useful for coastlines with sandy beaches or beach rocks. The procedure involved the determination of the border value for coastlines, where the histogram of the segment is parallel to the coastline. The waterline extraction technique is used for studying changes in the coastline and the tidal zone (Ryu et al. 2002; Yamano et al. 2006). It was understood that most of the accretion zone developed either due to natural processes or man-made coastal construction activities. For example, the northwestern part of the Arabian Gulf used to receive millions of cubic meters of sediments from the Euphrates and Tigris rivers, whereas the southern part of Kuwait has many hundreds of boat ramps, which act as groin fields to trap sediments and sediment deposition.

The waterline is defined as the boundary between the water body and an exposed landmass in the satellite image. It is assumed to be a line of equal elevation. Satellite images were stacked to generate elevation information, where water lines are observed under different tidal conditions. The tidal lines were referenced and stored in PCI focus as overlays. The extracted water line is subjected to tidal correction using Tide Calc software developed locally. The naval hydrography chart provides the contour of bathymetry and the distance between contours.

The selection of suitable satellite data is of critical importance for particular application. The basic characteristics like spatial, spectral, radiometric, and temporal resolutions are important for specific application. Landsat TM and Cartosat 1 satellite data were used in this study. Landsat TM was used to delineate the coastal morphological changes for a duration of 15 years between 1989 and 2003. The spatial resolution of TM data is 30 m, with a swath width of 185 km.

Cartosat 1 data were used in this study for coastline mapping during 2006–2007. The Cartosat satellite has two panchromatic (PAN) cameras for in-flight stereo viewing and a wide swath mode of data acquisition for 30 km with a spatial resolution of 2.5 m. The specification of satellite data is given in Table 5.1. These data basically cater to the need of cartographers and for terrain modeling applications and is used to establish the standardized base map for demarcation of present day coastlines. This high-resolution dataset provides a higher level of confidence on interpretation. Cartosat 1 data were obtained from Global Scan Technologies, Dubai, for the entire Kuwait coastal zone. The details of data acquisition including path, date, and time of Cartosat 1 are given in Table 5.2.

TABLE 5.1

Specifications of Cartosat 1 Satellite Data

Specifications	PAN-Fore	PAN-Aft
Tilt along track (°)	+26	−5
Spatial resolution (m)	2.5	2.5
Swath-width (km)	30	27
Radiometric resolution, quantization (bit)	10	10
Spectral coverage (nm)	500–850	500–850
Focal length (mm)	1945	1945
CCD arrays (no. of arrays * no. of elements)	1 * 12,000	1 * 12,000
CCD size	$7\,\mu m \times 7\,\mu m$	$7\,\mu m \times 7\,\mu m$
Integration time (ms)	0.336	0.336

TABLE 5.2

Details of Cartosat 1 Data Acquisition

Path	Date	Year	Time
Path 1	October 16	2006	10:00:00 AM
Path 2	September 19	2006	10:07:11 AM
Path 3	September 24	2006	09:58:44 AM
Path 4	January 1	2007	09:58:14 AM
Path 5	January 17	2007	09:57:09 AM

Paths 4 and 5 cover the southeastern part of Kuwait from Ras Al-Julia to the southern border of Kuwait. The tidal levels during these different time periods were estimated using the "TIDECALE" program and tidal corrections for the coastline were incorporated before comparison with the coastline morphology of the year 1989.

5.3 Methodology

There are several ways to map coastlines. The waterline method is one way and can be extracted from satellite data. In this particular study, the integrated approach is adopted by a combined man and machine processing of the datasets that include cartographic base mapping using digitization and visual interpretation, density slicing, edge enhancement, and change detection to a single band data and image classification of multiband data. The interpreted outputs are summarized as vector overlays.

Remote sensing is an excellent tool for mapping spatial changes, particularly coastal morphological changes. Theoretically, Landsat TM coverage is completed with a 16 day temporal resolution and makes it feasible to have snap shots almost twice every month. However, there are some limitations in terms of the cost of multiple datasets, image corrections, and the resolution and availability of colorful data. In fact, satellite data are rarely available on a real-time basis. There are many workers who have attempted to map change detections in coastal zones (Hopley and Catt 1989; Loubersac et al. 1989; Ibrahim and Yosuh 1992; Zainal et al. 1993).

5.3.1 Correction of Datasets

Geometric correction is basically restoring the geometry of the image so that it represents accurate ground information. It includes correcting geometric distortion due to the sensor—the Earth geometry variation. It requires ground control points (GCP) from the map along with their coordinates. The GCPs in the map were registered with pixels of the same points from the image. In this study, the geometric corrections and image registrations were carried out by the spatial interpolation of pixels through the nearest neighbor procedure. The removal of atmospheric effect due to absorption and scattering is necessary for any remote sensing study. These corrections are time consuming and often problematic since the atmospheric conditions at specific date are sometimes unknown, especially in the case of historical analysis. However, calibration is essential for the spatiotemporal correlation of images, for the use of band ratios for image interpretation, and for modeling the interaction between an object and electromagnetic radiation. In the present study, a radiometric normalization of the images was done using PCI Geomatica software.

5.3.2 Cartographic Base Mapping

The primary concern is the delineation of landmasses and water bodies, which usually requires very high-resolution data. High-resolution satellite data were accommodated with few ground control points (GCP) to generate considerably accurate spatial data. In the present study, the cartographic base map was generated from both the Landsat TM and Cartosat 1 satellite data. The outputs were stored as a vector layer in the PCI Geomatica Focus software.

5.3.3 Density Slicing

In this study, the density slicing method was effectively utilized for waterline demarcation (Ryu et al. 2002). It is one of the more popular and effective methods of enhancing the coastlines for delineation. The accuracy of the method depends on the selection of an appropriate threshold value. The threshold value is selected from the histogram analysis without any bias. It is the critical DN value between the seawater and the exposed tidal flat. Though the technique is simple, it is difficult to determine the threshold value in highly turbid waters. The mere use of the density slicing method on an short wave infrared (SWIR) band can be slightly deceptive in a tidal flat with ebb tides. The waterlines were extracted from the density sliced output of the Landsat TM and Cartosat 1 images for the coastal zone. In this process, bands 4, 5, and 7 of Landsat TM and bands 2 and 3 of Cartosat 1 were utilized.

5.3.4 Band Ratios

The band ratios were performed to achieve image enhancement for interpretation. The ratioed outputs were generated using the following band combinations:

$$\text{Ratio}_{\text{Landsat TM}} = \frac{\text{Band } 4 - \text{Band } 3}{\text{Band } 4 + \text{Band } 3} \tag{5.1}$$

$$\text{Ratio}_{\text{Cartosat 1}} = \frac{\text{Band } 3 - \text{Band } 2}{\text{Band } 3 + \text{Band } 2} \tag{5.2}$$

The ratioed outputs were useful in considerably reducing the effect of suspended particulates near the coastline and thus made it easier to demarcate the coastal features. The rationale for using these ratio images is derived from an experimental work carried out by Lodhi et al. (1997). They have reported an increased reflectance in wavelength ranges corresponding to 580–690 and 760–900 nm as the amount of suspended particulate increases. They have also emphasized the effectiveness of visible bands in discriminate silt and clay in the coastal zone. The wavelength range of 710–880 nm is useful for obtaining information on total suspended sediments.

These observations were validated and it was found that the findings hold well in the study area. The TM band 4 data show high reflectance over both the tidal flat and turbid waters, while band 3 shows low reflectance over the exposed tidal flat. Thus, the difference between bands 4 and 3 enhances the exposed tidal flat. The morphological feature was vectorized and stored in the PCI focus software. The overlays for the years 1989 and 2006–2007 were integrated using the overlay analysis to characterize the coastline as an accreting, an eroding, or a stable zone. The vector output is split into a 3′ × 3′ zone for detailed representation and assessment of each segment along the coastline.

5.3.5 Field Verification

Most of the remote sensing exercises require field verification at some stage. Field survey can be taken up for defining training sites, for real-time calibration, and accuracy assessment. In the present study, at selected sites, field investigations were carried out and validated and the satellite data interpreted. Ground truth helps in understanding the ongoing process in a particular location, the reasons for coastal morphological changes in that location, and for accuracy assessment. Interaction with the local community is also useful to confirm morphological changes in the coastal zone.

5.4 Results and Discussion

The spatial integration of coastal morphology derived from the Landsat TM and Cartosat 1 dataset is not so easy, because both the data are at different levels of resolution. However, the images were co-registered quite precisely with an RMS error of 0.341. The accuracy of the digital coastline vector is within ±10 and <1 m, respectively, for coastlines in the years 1989 and 2007. The vector overlays representing coastlines for the years 1989 and 2006–2007 are considered for further analysis.

The digital coastline datasets derived from satellite data can be used as a basic data source for coastal zone management and planning. The adopted methodology in the present study provides an economical and efficacious means of monitoring coastline changes. The database provides complete information on the morphological changes of the Kuwait coastline and has categorized it as an eroding, an accreting, or a stable zone. The coastal morphological changes for over the period of 18 years have indicated that there is a significant accretion process taking place in some parts of the southern Kuwait coast, in a few spots inside the Kuwait Bay, and in parts of Khor Sabiya. The entire Kuwait coastal zone is divided into 3′ latitude × 3′ longitude and a total of 61 pictorial outputs were prepared to show the eroding, accreting, and stable coastlines. The accretion process favors the formation of sand spits along the southern Kuwait coast. The length of such spit formations is in the order of 1500 m (Figure 5.2).

The sand spit formation has blocked the mouth of Khor Iskandar for the last 18 years and this has been verified during field investigation (Figure 5.3). Beach house owners in these areas are not able to take their boats out into the sea due to the formation of spit. However, they use the advantages of high tide conditions for such operations.

The information collected during field visits in the southern Kuwait coastal zone is represented in Figure 5.4. Complete ground truth

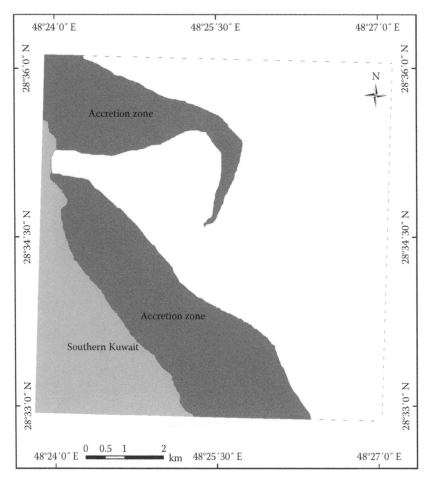

FIGURE 5.2
Spit formation due to accretion processes at the southern Kuwait coast (between 28° 33′ N and 28° 36′ N and 48° 24′ E and 48° 27′ E).

information for locations 1–22 is available in Neelamani et al. (2007). Location 23 is dominated by the erosion process in the order of 150 m. Locations 24, 25, and 26 are dominated by the accretion process. The accretion process is due to the fact that there are many boat ramps built by beach house owners, which has resulted like a groin field for coastal protection against sea wave–induced erosion. A similar trend is also noticed at location 25. The erosional process is significant at Khor Al-Mufatah, the southern Kuwait coastal zone (Figure 5.5). This was also ascertained by field verification and was confirmed through discussions with the local community.

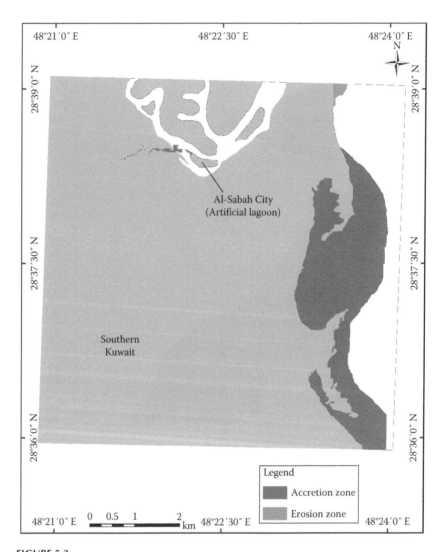

FIGURE 5.3
Sand spit formation blocked the mouth of Khor Iskandar, in the southern Kuwait coastal zone (between 28° 36′ N and 28° 39′ N and 48° 21′ E and 48° 24′ E).

The stable and accreting coastal zone was demarcated around the Az-Zour power plant intake location (Figure 5.6). The coastal zone was stabilized using ripraps. However, one can notice the sand spit formation at the Ras Az-Zour area. The coast running in the east–west direction shows accretion due to the construction of a number of slipways, which works as a groin field.

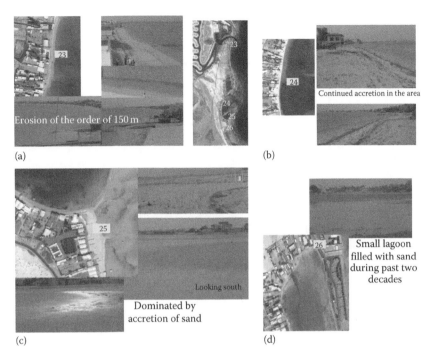

FIGURE 5.4
Satellite image and field photographs show erosion and accretion processes at the southern
Kuwait coastal zone: (a) erosion process, (b) continued accretion, (c) dominated by accretion
of sand, and (d) small lagoon filled with sand deposits.

The northern part of the Kuwait coastal zone is influenced by the periodic discharge of the dredged materials from the seawater intake structure of the Sabiya Power Plant (Figure 5.7). This site is located about 50 km
from the estuary of the Tigris–Euphrates river called "Shatt Al-Arab." The
tide-induced anticlockwise global current is mainly responsible for push-
ing the silt and clay around the Sabiya area.

The coastal area around the seawater intake of the Sabiya Power Plant
is shown in Figure 5.8. It is noticed that there is deposition of sediment
on both sides of the intake. The projected intake structure acts as a
groin.

A recent ground truth of this site has indicated that the coast has
advanced to an extent of 1400 m from 1989 to 2007 due to the continuous
deposition of suspended sediments. The deposition process started due
to the construction of the seawater intake structure in the year 1989. The
guide wall has been acting as a groin, which accelerated the deposition of
sediments both at the upstream and downstream side of the area. There
are plans to increase the quantity of seawater intake for the proposed

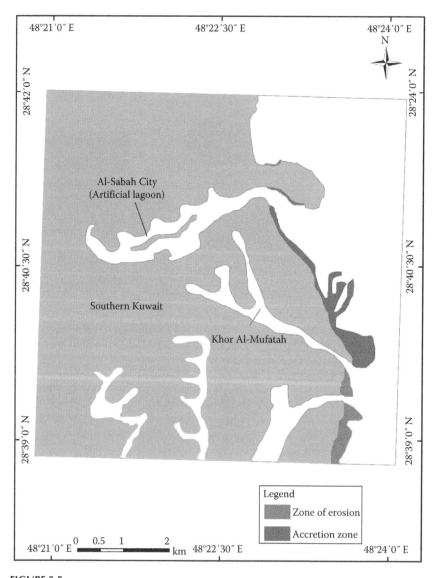

FIGURE 5.5

Erosional process and coastline changes at the Khor Al-Mufatah region, of the southern Kuwait coastal zone (between 28° 39′ N and 28° 42′ N and 48° 21′ E and 48° 24′ E).

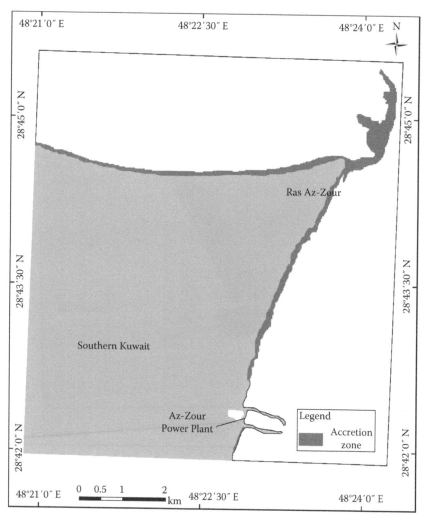

FIGURE 5.6
Stable and accreting coastal zone around the intake of the Az-Zour power plant location (between 28° 42′ N and 28° 45′ N and 48° 21′ E and 48° 24′ E). Sand spit formation can be noticed in the Ras Az-Zour area.

additional 2050 MW power plant. Based on this study and field investigation, it is strongly suggested that the existing intake not be used, since it will only release more sedimentation into the intake channel. This will increase dredging, and hence, more sediment accumulation. Overall, the comparison of the remote sensing images of 1989 and 2006–2007, and the ground truth of this area, showed significant morphological changes due to man-made effects during the last two decades. The knowledge gained

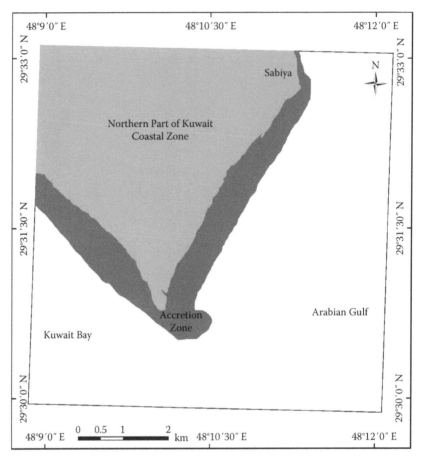

FIGURE 5.7
The northern part of the Kuwait coastal zone influenced by the periodic discharge of the
dredged materials from the seawater intake structure of the Sabiya Power Plant (between
29° 30′ N and 29° 33′ N and 48° 09′ E and 48° 12′ E).

through this study is useful for the integrated coastal zone management
of the Kuwait coastal zone.

5.5 Conclusions

The coastal areas are dynamic and undergo changes due to natural and
man-made processes. It is essential to understand the trend of the coastal
morphological changes with respect to time for right decision making for
the sustainable development of coastal projects. The rate of natural changes

Kuwait Coastline Evolution during 1989–2007

Kuwait Coastline Evolution during 1989–2007 101

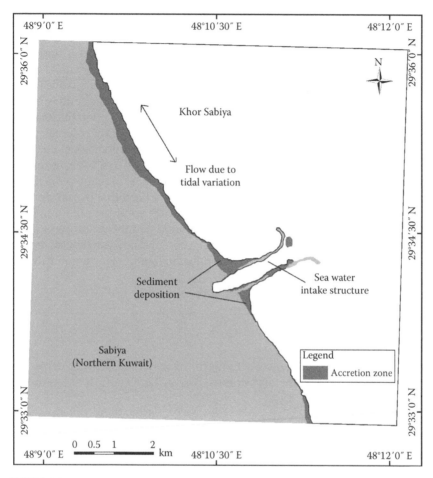

FIGURE 5.8
The coastal area around the seawater intake of the Sabiya Power Plant. Sediment deposition noticed on both sides of the intake structure.

depends upon the intensity of forces from natural elements, such as wave heights and periods, current velocity, wind speed, etc. In this study, satellite data like Landsat TM of 1989 with a spatial resolution of 30 m and Cartosat images of 2006–2007 with a spatial resolution of 2.5 m were used. Detailed field investigations were carried out to check the remote sensing-based interpretation for the Kuwait coastal area. In Kuwait, most of the man-made coastal development projects are concentrated in between the southern coastal boundary and the Shuwaikh port zone. The coastal zone in between the Shuwaikh port and the northern coastal boundary had fewer man-made activities and was dominated by natural tidal flats. It is also found that in many places, starting from the southern Kuwait border to Ras Al-Ardh, groups of boat ramps constructed by the public act as groin fields and stabilize the coast. The entire northern part of Kuwait Bay and the coast from the north of the Sabiya Power Station is undergoing changes of natural deposits of silt and mud, supplied by Shatt Al-Basra and Shatt Al-Arab. The periodic dredging activity for maintenance of the intake of the Sabiya Power Station and the deposition in the southern side has caused significant coastal morphological changes. Though the Kuwaiti coastal zone is relatively small when compared to many other coastal countries around the world, it undergoes substantial morphological changes due to both man-made and nature-induced forces. Extensive spit formations have been observed during the last two decades in many parts of the southern Kuwait coastal areas. It is also clearly noticed that the sand accretion in most parts of the southern Kuwait coasts are due to the influence of slipways. Ground-truth information has validated these findings for the southern Kuwait coast. The results of the present study can be utilized for future holistic and sustainable coastal developmental planning in Kuwait.

Acknowledgments

The authors acknowledge V. Ramesh, research scholar, Centre for Geoinformatics and Planetary Studies, Periyar University, for supporting the GIS work.

References

Al-Ghadban, A.N., Saeed, T., Al-Dousari, A.M., Al-Shemmari, H., and Al-Mutairi, M. 1998. Preliminary assessment of the impact of draining of Iraqi marshes on Kuwait's northern marine environment. Part I. Physical manipulation. In: *The Third Middle East Conference on Marine Pollution and Effluent Management*, November 23–25, 1998, State of Kuwait, pp. 18–38.

Anon. 1992. Agenda 21. In: *United Nations Conference on Environment and Development*, Rio de Janeiro, Brazil, June 3–14, pp. 215–249.

Appeaning Addo, K. 2009. Detection of coastal erosion hotspots in Accra, Ghana. *Journal of Sustainable Development in Africa* 11 (4): 253–265.

Duedall, I.W. and Maul, G.A. 2005. Demography of coastal populations. In: *Encyclopaedia of Coastal Science*. Springer, Dordrecht, the Netherlands, pp. 368–374.

Ezer, T. and Liu, H. 2010. On the dynamics and morphology of extensive tidal mudflats: Integrating remote sensing data with an inundation model of Cook Inlet, Alaska. *Ocean Dynamics*. Online publication date: July 23, 2010.

Friel, C. and Haddad, K. 1992. GIS manages marine resources. *GIS World* 5(9): 33–36.

Galgano, F.A. and Douglas, B.C. 2000. Coastline position prediction: Methods and errors. *Environmental Geosciences* 7 (1): 23–31.

Hopley, D. and Catt, P.C. 1989. Use of near infrared aerial photography for monitoring ecological changes to coral reef flats on the Great Barrier Reef. In: *Proceedings of the 6th International Coral Reef Symposium*, Townsville, Australia, vol. 3, pp. 503–508.

Ibrahim, M. and Yosuh, M. 1992. Monitoring the development impacts on coastal resources of Pulau Redang marine park by remote sensing. In: *Conference Proceedings of 3rd ASEAN Science and Technology Week*, eds. L.M. Chou and C.R. Wilkinson. *Marine Science Living Coastal Resources* 6: 407–413.

Lodhi, M.A., Rundquist, D.C., Han, L., and Juzila, M.S. 1997. The potential of remote sensing of loess soils suspended in surface water. *Journal of the American Water Resources Association* 33 (1): 111–127.

Loubersac, L., Dahl, A.L., Collotte, P., LeMaire, O., D'Ozouville, L., and Grotte, A. 1989. Impact assessment of cyclone Sally on the almost atoll of Aitutaki (Cook Island) by remote sensing. In: *Proceedings of the 6th International Coral Reef Symposium*, Townsville, Australia, vol. 2, pp. 455–462.

Mumby, P.J., Gray, D.A., Gibson, J.P., and Raines, P.S. 1995. Geographic Information Systems: A tool for integrated coastal zone management in Belize. *Coastal Management* 23: 111–121.

Neelamani, S., Saif ud din, Rakha, K., Zhao, Y., Al-Salem, K., Al-Nassar, W., Al-Othman, A., and Al-Ragum, A. 2007. Coastline Evolution of Kuwait Using Remote Sensing Techniques, EC022C. Final Report, KISR, December 2007.

Ryu, J.H., Won, J.S., and Min, K.D. 2002. Waterline extraction from Landsat TM data in a tidal flat—A case study in Gomso Bay, Korea. *Remote Sensing of Environment* 83: 442–456.

Seker, D.Z., Goksel, C., Kabdasli, S., Musaoglu, N., and Kaya, S. 2003. Investigation of coastal morphological changes due to river basin characteristics by means of remote sensing and GIS techniques. *Water Science and Technology* 48 (10): 135–142.

Wu, W. 2007. Coastline evolution monitoring and estimation—A case study in the region of Nouakchott, Mauritania. *International Journal of Remote Sensing* 28 (24): 5461–5484.

Yamano, H., Shimazaki, H., Matsunaga, T., Ishoda, A., McClennen, C., Yokoki, H., Fujita, K., Osawa, Y., and Kayanne, H. 2006. Evaluation of various satellite sensors for waterline extraction in a coral reef environment: Majuro Atolll, Marshall Islands. *Geomorphology* 82: 398–411.

Zainal, A.M.J., Dalby, D.H., and Robinson, I.S. 1993. Monitoring marine ecological changes on the east coast off Bahrain with LANDSAT TM. *Photogrammetric Engineering and Remote Sensing* 59: 415–421.

Zhou, G. and Xie, M. 2009. Coastal 3-D morphological change analysis using LiDAR series data: A case study of Assateague Island national seashore. *Journal of Coastal Research: Spring* 435–447.

Zujar, O.J., Borgniet, L., Pérez Romero, A.M., and Loder, J.F. 2002. Monitoring morphological changes along the coast of Huelva (SW Spain) using soft-copy photogrammetry and GIS. *Journal of Coastal Conservation* 8 (1): 69–76.

6

Detecting Estuarine Bathymetric Changes with Historical Nautical Data and GIS

Xiaojun Yang and Tao Zhang

CONTENTS

6.1 Introduction

Bathymetry, the underwater equivalent of topography, is defined as the measurement of the depth of the lacustrine or oceanic floor from the water surface through the use of various acoustic techniques. The information on bathymetry and the change in bathymetry can be useful not only for the study of morphological evolution of seafloor but also for aquatic environmental planning and management (Jaffe et al. 1998; Cappiella et al. 1999; Gesch and Wilson 2002; Thomas et al. 2002; Buijsman et al. 2003; Cooper and Navas 2004; Bertin et al. 2005; Elias and Spek 2006).

For quite a long period of time, the only way to measure water depth was through the use of a sounding line, which was time-consuming and inaccurate. Since the early 1930s, bathymetric data have been collected by using echosounders mounted beneath or over the side of a boat, which generate and receive sounds to measure the distance to lacustrine or oceanic floor. While single-beam echosounders can measure water depth, the use of a multibeam echosounder in combination with a global positioning system nowadays has greatly improved the efficiency and accuracy of bathymetric measurements (Thurman and Trujillo 2004). Aerospace remote sensing has also been used to measure bathymetry. For example, active remote sensing has demonstrated the capability to directly measure bathymetry in shallow water areas (Culshaw 1995; Gesch and Wilson 2002; Zhou and Xie 2009) and indirectly map deep-sea topography by detecting the subtle variations of the sea surface roughness induced by the gravitational pull of undersea mountains, ridges, and other masses (Alpers et al. 2005). Optical remote sensing has also shown the potential to measure bathymetry in shallow waters (Sandidge and Holyer 1998; Stumpf et al. 2003).

Historical bathymetric data are generally available as nautical charts that date back to early 1800s (Calder 2003). When multi-year nautical charts are available for a specific study area, the change in bathymetry can be characterized (e.g., Jaffe et al. 1998; Gibbs and Gelfenbaum 1999; Foxgrover et al. 2004). However, analyzing bathymetric changes by using historical nautical chart data is not always straightforward due to errors and uncertainties associated with the data sources (Cappiella et al. 1999; Calder 2003; Johnston 2003; Foxgrover et al. 2004; Prandle 2006). Some researchers addressed these problems with a variety of methods, mostly through the use of extensive reference data (e.g., Foxgrover et al. 2004; Bertin et al. 2005). These methods, however, may not be effective for areas where sufficient reference data are not available or incomplete.

The objective of this study was to identify a method that can be used to analyze estuarine bathymetric changes from historical nautical data. The case study site covers Tampa Bay, a large, shallow estuary along the eastern Gulf of Mexico. A practical method has been developed to calibrate the bathymetric measurements derived from historical nautical data toward a common benchmark. A GIS-based approach has been identified to detect the changes between years. This chapter will discuss the technical procedures identified in this project. The following sections will provide an overview on the case study site, detail the research methodology, analyze the results, and discuss some ongoing research.

6.2 Study Site

As one of the largest estuaries in the Gulf of Mexico, Tampa Bay is the largest open-water estuary in Florida, United States (Figure 6.1). The Bay itself covers more than 1000 km² of water area and approximately 5700 km² of land area with more than 100 tributaries flowing into the receiving basin.

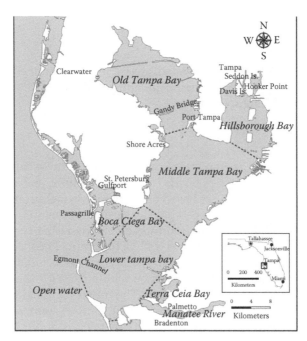

FIGURE 6.1
Location of the study area. It covers Tampa Bay, Florida. Note that the water area is shaded and the major subdivisions are labeled.

The Tampa Bay watershed is a nursery of wildlife, a metropolitan area, a hub of harbors, and a famous tourist destination. It was the home of nearly 4 million people, or 22.5% of Florida's total in 2005. Port of Tampa is the largest port in Florida.

Tampa Bay is quite shallow with average depth of approximately 4 m, but its water depths vary significantly. For accommodating deep-draft vessels, man-made ship channels have been dredged through Tampa Bay, and most of them vary from 8 to approximately 16 m in depth. In areas other than the channels, the depths are generally less than 3 m.

Physiographically, Tampa Bay is a microtidal estuary incised into tertiary platform carbonates (Brooks and Doyle 1998). It can be divided into several physiographic subdivisions with various depths (see Figure 6.1). Old Tampa Bay is a shallow and flat basin with some slopes and deep channels at the mouth. Hillsborough Bay is also shallow but it contains some dredged channels. Middle Tampa Bay and Lower Tampa Bay are relatively deeper. The other three small subdivisions are quite shallow, among which Boca Ciega Bay and Terra Ceia Bay are lagoons behind the coastal barrier islands.

6.3 Research Methodology

The research methodology included several major components: data collection, vertical datum calibration, modeling of bathymetric surfaces, and change detection (Figure 6.2).

6.3.1 Data Sources

In this study, we focused on the bathymetric data derived from two early nautical charts that were published in 1879 and 1928, respectively. The original nautical charts were based on bathymetric measurements using manual techniques. A graduated pole was used to measure soundings in areas with depth of <5 m, and lead lines and Rude-Fisher pressure tubes were used in deeper water (Gibbs and Gelfenbaum 1999). According to the instructions in Coast Survey hydrographic manual at that time (Coast Survey 1878), sounding errors would be within 0.5 ft for shallow waters (<20 ft) and 1 ft for deep waters (>20 ft). The initial digitization was completed by United States Geological Survey (USGS) Tampa Bay Study Program, and two point shape files were generated with North American datum (NAD) 27 as the horizontal datum (Table 6.1).

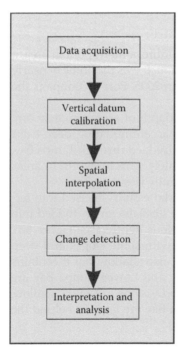

FIGURE 6.2
Flowchart of the working procedural route adopted in this study. Note that only major procedures are illustrated here.

TABLE 6.1

Summary of the Bathymetric Measurements from Two Earlier Nautical Charts

Year	Nautical Chart ID#	Chart Scale	Sounding Survey Period	Tidal Datum	Data Characteristics		
					Range (ft)	Total Points	Density[a] (pt/km²)
1879	77	1:80,000	1873–1876	Mean low water (MLW)	0.25–90	4033	2.72
1928	586/587	1:40,000	1926–1927	MLW	0.25–93	6778	5.12

[a] The calculation only considers the major subdivisions in Tampa Bay, excluding Boca Ciega Bay, Terra Ceia Bay, Manatee River, and open water.

The initial digital datasets prepared by USGS are further checked for removing possible errors introduced in digitization through a method similar to that described by Foxgrover et al. (2004). This has been an iterative process through visual inspection. In doing so, points are shaded with color according to the depth values and those showing sharp contrast in hue to their surroundings are inspected with the original nautical charts as the reference. In this way, 30–40 corrections are made, and the final datasets contain 4033 and 6778 points for 1879 and 1928, respectively.

It is noted that the geographic boundaries of Tampa Bay used in this study are largely based on the boundaries defined by National Oceanic and Atmospheric Administration (NOAA) Coastal Assessment that reflect the current condition. To accommodate the status of historical shorelines, we further modify the initial boundaries by using the 1879 and 1928 nautical charts as the reference. The final boundaries consist of all representative components of Tampa Bay, as shown in the two early nautical charts (Figure 6.1). Note that we exclude upper branches of Boca Ciega Bay and include only a small portion of open water.

6.3.2 Vertical Datum Calibration

Perhaps the biggest challenge in using nautical chart data is to deal with the variation of their vertical data used over time. For nautical charts published through U.S. Coast and Geodetic Survey, two vertical data have been used: mean lower low water (MLLW) and mean low water (MLW). MLLW is the average of the lower low water height of each tidal day observed over a 19-year National Tidal Datum Epoch (NTDE) while MLW is the average of all the low water heights observed over a certain NTDE. Apparently, conversion between MLLW and MLW can be quite complicated.

Fortunately, the two nautical charts we use were based on the same vertical datum, i.e., MLW, and there is no need to conduct vertical datum conversion. However, the data we use spanned a period of 50 years, and the change in sea level over time should be accounted for. Such an adjustment is similar to the conversion of tidal data between different NTDEs. The NTDE information for our study area has become available after 1947 when St. Petersburg Station (ID: 8726520; latitude 27°45.6′ N, longitude 82°37.6′ W) was established as part of U.S. National Water Level Observation Network (NWLON) (Zervas 2001). Because there is no NTDE information for the time span that the two nautical charts covered, the method to estimate the offset of a historical reference datum from current tidal data, as suggested by Foxgrover et al. (2004), may not be applicable to our case.

The vertical datum calibration conducted here is based on a practical method. In doing so, we acquire a complete set of monthly tidal datum records from the St. Petersburg station spanning the period of more than 59 years from January 1947 to March 2006 with a 5-month gap in 1952. Then, we model the trend of tidal datum variation from the time series dataset, and extrapolate to earlier years. Assuming that the tidal datum varies linearly, we construct a linear regression model using the 19-year moving average of MLW. A 19-year span is close to the astronomical cycle in tidal change (18.6 year) and has no seasonal bias. We use the moving average in the time series to pinpoint the general trend of tidal datum variation while suppressing the seasonal and astronomical effects on tidal data. The actual regression equation is

$$MLW_i = 0.1822953 \times i - 0.0000753$$

where
  i represents the number of months apart from January 1947
  MLW_i indicates the average MLW in that month's "Epoch" or the temporal span from the $i - 228 + 1$st month ($19 \times 12 = 228$) to the ith month

A positive i indicates a month later than January 1947 while a negative number indicates an earlier month. The MLW_i is in millimeter and is referenced to the MLLW of new NTDE (1983–2001). This linear regression model has an R^2 value of 0.913. Based on the above regression model, the MLW rose at 2.188 mm/year, which is quite consistent to the estimated sea level trend in this area (Zervas 2001).

Assuming that the referenced MLW plane in surveys of each period equals to the MLW of each "Epoch," December 1876 ($i = -841$) and December 1927 ($i = -229$) could be treated as two temporal points through which the two early MLWs can be extrapolated. Based on the regression model, the MLW values for surveys in the 1870s and 1920s are estimated at −0.161 and −0.049 m, respectively. The statistical errors for each MLW are 0.013 and 0.007 m, respectively, with the confidence level of 95%. This model also suggests that MLW increase approximate 0.112 ± 0.010 m over the entire time span.

6.3.3 Modeling Bathymetric Surfaces and Change Detection

After the raw measurements have been calibrated, we further transform the data from scattered points into continuous surfaces before a change in detection procedure is implemented. For this purpose, the triangulated irregular networks (TIN) interpolation is employed to create two continuous bathymetric surfaces. TIN has been considered as a popular method for bathymetric surface generation from point-based sounding data (Byrnes et al. 2002; Johnston 2003). The original TIN surfaces are in vector format, which has been converted into rasters (Figure 6.3). Finally, an image differencing algorithm is used to detect the changes between the two raster surfaces.

$$\text{Bathymetric change} = \text{Surface}_{1928} - \text{Surface}_{1879}$$

The resultant volumetric differenced raster (Figure 6.4) is further analyzed, which will be discussed in the next section.

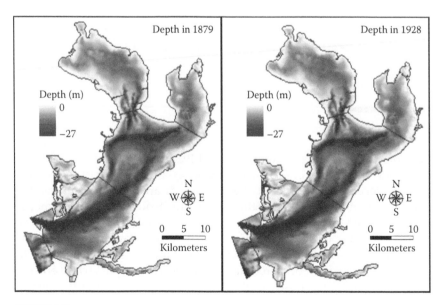

FIGURE 6.3
Continuous bathymetric surfaces created by using spatial interpolation of scattered depth points for 1879 and 1928, respectively.

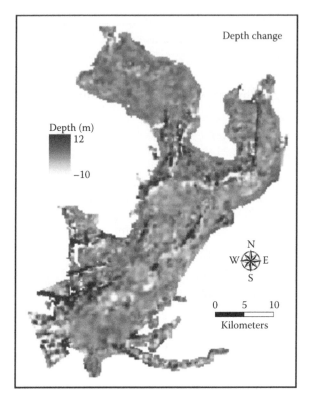

FIGURE 6.4
Volumetric changes in bathymetry in Tampa Bay during 1879–1928.

6.4 Results

The volumetric change between 1879 and 1928 can be visualized in Figure 6.4. In addition, Table 6.2 provides a statistical summary for the bathymetric differenced surface. Overall, the entire study area had an accumulation of 0.096 m in depth or 93,960,960 m^3 in volume. However, the bathymetric changes varied greatly through the bay area, ranging from −9.9 to +12.5 m. This wide range of changes indicates that Tampa Bay underwent quite different processes throughout the estuary during the time period. In general, accumulation occurred mostly in the western part while erosion (including dredging) mostly in the middle and eastern estuary. Another observation is that the bathymetry in the near-shore areas tended to aggregate, which is quite consistent with the general trend of shoreline change in this area. From 1879 to the present, the shoreline of Tampa Bay has

TABLE 6.2

Summary of the Bathymetric Changes at Different Subdivisions in Tampa Bay

Subdivisions	Area (km²)	Bathymetric Change			
		Minimum (m)	Maximum (m)	Average (m)	Standard Deviation
Old Tampa Bay	208.40	−5.345	3.635	0.121	0.452
Hillsborough Bay	109.69	−7.776	4.471	0.057	0.714
Middle Tampa Bay	270.72	−5.531	6.309	0.027	0.665
Lower Tampa Bay	242.56	−3.776	10.545	0.115	0.678
Boca Ciega Bay[a]	66.80	−6.021	10.420	0.216	0.997
Terra Ceia Bay	14.35	−1.360	2.858	0.068	0.601
Manatee River	21.61	−3.820	5.208	0.175	0.845
Open water[b]	44.63	−9.889	12.458	0.181	1.804
Total study area	978.76	−9.889	12.458	0.096	0.760

[a] Due to the limitation of sounding survey data, this unit excludes the upper branches.
[b] Only for a small area near the estuarine mouth.

been slightly shifting toward the land, mainly due to human activities. Port construction and ship channel dredging greatly modified the shoreline and affected the process of sedimentation and erosion in the study area. When overlaying the bathymetric change surface with raw nautical charts, we notice that in Hillsborough Bay, Middle Tampa Bay, and Lower Tampa Bay, most of the accumulation occurred outside the ship channels while incision largely happened inside the ship channels. Around the sites where new lands were emergent in the 1928 nautical charts, the seafloor was deepened significantly, especially in Old Tampa Bay and Hillsborough Bay. Apparently, these bathymetric change patterns are related to the dredging that began in 1880. According to Lewis (1976), the material dredged from the channels has been placed adjacent to the channels as submerged or emergent spoil areas, or used as landfill for shoreline development. The spatial pattern of bathymetric change is consistent with the known dredging practices in this estuary. For example, the maximum accumulation happened at the north side of Egmont Channel while the maximum incision occurred just inside this channel.

From Table 6.2, we can see that the subdivisions of Boca Ciega Bay and Manatee River had the largest accumulation with the highest standard deviations. In Manatee River, accumulation took place along the river banks while the incision occurred along the river centerline and around some emergent docks. In Boca Ciega Bay, accumulation was concentrated around the keys and shoals and along both sides of the dredged channels. This is partly due to the construction of several ports and the development of several cities in the areas, such as Palmetto and Bradenton along the Manatee River, and Gulfport and Pass-a-grille in Boca Ciega Bay. In addition to the localized transport of the dredged materials, there were many external landfill materials from either terrestrial or marine sources, and thus, the overall accumulation in the two subdivisions was much higher than in other units.

There were some significant changes in Tampa Bay, such as the emergence of Davis Island, Seddon Island, and Hooker Point in Hillsborough Bay and the expansion of St. Petersburg and shore acres in Middle Tampa Bay. The high standard deviations for these areas, as shown in Table 6.2, imply relatively large changes in bathymetry. The channel deepening and the adjacent dumping in Hillsborough Bay were linearly distributed, and similar patterns also appeared from Middle and Lower Tampa Bay to the estuarine mouth. The average accumulation levels in Hillsborough Bay and Middle Tampa Bay were much lower than in other subdivisions, indicating that much of the dredged materials may be used for landfill in these regions.

In general, Old Tampa Bay and Terra Ceia Bay underwent the least change among all subdivisions. In these two sites, urbanization and port construction were quite limited during the period 1879–1928. In Old Tampa Bay, the construction of Gandy Bridge and Port Tampa involved dredging and landfill activities. The area around the narrow mouth between Old Tampa Bay and Middle Tampa Bay also underwent some significant change. There was little change in bathymetry in Terra Ceia Bay where mangroves and tidal marshes dominated the shorelines. From the bathymetric change surface, some localized dredging and landfilling activities can be identified in Terra Ceia Bay.

6.5 Conclusions and Further Research

This chapter reports our research effort aiming to identify a method that can be used to detect the change in bathymetry from historical nautical charts. A critical component in this method was the development of a practical solution to calibrate the vertical datum when tidal datum records and sufficient reference points are absent. It is based on a time series of monthly tidal datum records from the St. Petersburg station spanning the period of more than 59 years from January 1947 to March 2006 with a 5-month gap in 1952. We model the trend of tidal datum variation from the time series dataset and extrapolate to earlier years. This approach is based on an assumption that the increasing of MLW in this area would be in a linear pattern in the nearest 150 years. The sea-level rise rate we model from the monthly tidal datum records is consistent to the result of an existing study conducted by Florida Geological Survey (Balsillie and Donoghue 2004).

The entire area underwent a noticeable accumulation in bathymetry between 1879 and 1928, and this change is found to be tightly linked with port construction, channel dredging, and urbanization. Bathymetric change varied across different subdivisions with the least occurred in Old Tampa Bay, the unit farthest from the open ocean water. Most of the area underwent little change, and noticeable changes were concentrated in several sites. A large portion of Hillsborough Bay, Middle Tampa Bay, and Lower Tampa Bay was deepened because of dredging activities, and much dredged materials were dumped nearby to form some long and narrow shoals that were north–south oriented. A considerable amount of the dredged materials originally from Hillsborough Bay and Middle Tampa Bay was dumped elsewhere, such as Lower Tampa Bay, where moderate accumulation occurred. Dredging activities had led to the forming of complex seafloor terrain in the estuary. On the other hand, some small subdivisions show different change patterns in bathymetry. Terra Ceia Bay had little change when comparing to Boca Ciega Bay and Manatee River. The latter underwent substantial changes, primarily due to the construction of ports and urbanization.

This study demonstrates that historical nautical chart data can be quite valuable for analyzing estuarine bathymetric change but cautions should be paid on the calibration of vertical data among the time series of raw data. Our further research will consider some more recent bathymetric measurements from modern sonar technology in order to examine the oceanic floor change since 1879 in the Tampa Bay area. In addition, we will quantify the errors and uncertainties in this type of change analysis.

Acknowledgments

We would like to thank the Florida State University for the time release in conducting this work. The research reported here has been partially supported through a major research grant from the U.S. Environmental Protection Agency's Science to Achieve Results (STAR) Estuarine and Great Lakes (EaGLe) program through funding to the CEER-GOM, U.S. EPA Agreement R829458. The authors wish to thank United States Geological Survey (USGS) Tampa Bay Study Program for sharing their bathymetric data over the Tampa Bay area.

References

Alpers, W., G. Campbell, H. Winsink, and Q. Zhang. 2005. Underwater topography. In: *Synthetic Aperture Radar Marine User's Manual*, eds. C. R. Jackson and J. R. Apel, pp. 245–262. NOAA. http://www.sarusersmanual.com/ManualPDF/ NOAASARManual_CH10_pg245-262.pdf (accessed September 20, 2010).

Balsillie, J. H. and J. F. Donoghue. 2004. A high-resolution year sea-level history for the Gulf of Mexico since the last glacial maximum (Report of Investigations No. 103). Tallahassee, FL: Florida Geological Survey.

Bertin, X., E. Chaumillon, A. Sottolichio, and R. Pedreros. 2005. Tidal inlet response to sediment infilling of the associated bay and possible implications of human activities: The Marennes-Oléron Bay and the Maumusson Inlet, France. *Continental Shelf Research* 25: 1115–1131.

Brooks, G. R. and L. J. Doyle. 1998. Recent sedimentary development of Tampa Bay, Florida: A microtidal estuary incised into Tertiary platform carbonates. *Estuaries* 21(3): 391–406.

Buijsman, M. C., C. R. Sherwood, A. E. Gibbs, G. Gelfenbaum, G. Kaminsky, P. Ruggiero, and J. Franklin. 2003. Regional sediment budget of Columbia River Littoral Cell, USA: Analysis of bathymetric- and topographic-volume change. U.S. Geological Survey Open File Report 2002-281. http://pubs. usgs.gov/of/2002/of02-281/of02-281.pdf (accessed September 20, 2010).

Byrnes, M. R., J. L. Baker, and F. Li. 2002. Quantifying potential measurement errors associated with bathymetric change analysis. Vicksburg, MS: U.S. Army Engineer Research and Development Center. ERDC/CHL CHETN-IV-50. http://chl.erdc.usace.army.mil/library/publications/chetn/pdf/chetn-iv-50.pdf (accessed September 20, 2010).

Calder, N. 2003. *How to Read a Nautical Chart: A Complete Guide to the Symbols, Abbreviations, and Data Displayed on Nautical Charts.* New York: McGraw-Hill Companies.

Cappiella, K., C. Malzone, R. Smith, and B. Jaffe. 1999. Sedimentation and bathymetric changes in Suisun Bay: 1867–1990. U.S. Geological Survey Open File Report 1999-563. http://geopubs.wr.usgs.gov/open-file/of99-563/of99-563. pdf (accessed September 20, 2010).

Coast Survey. 1878. General instructions, hydrographic surveys division. Washington, DC: U.S. Government Printing Office. http://www.thsoa.org/ pdf/hm1878/1878all.pdf (accessed on September 20, 2010).

Cooper, J. A. G. and F. Navas. 2004. Natural bathymetric change as a control on century-scale shoreline behavior. *Geology* 32: 513–516.

Culshaw, S. T. 1995. A visual interpretation of an ERS-I SAR image of the Thames estuary. *Journal of Navigation* 48: 97–104.

Elias, E. P. L. and A. J. F. Spek. 2006. Long-term morphodynamic evolution of Texel Inlet and its ebb-tidal delta (the Netherlands). *Marine Geology* 225: 5–21.

Foxgrover, A. C., S. A. Higgins, M. K. Ingraca, B. E. Jaffe, and R. E. Smith. 2004. Deposition, erosion, and bathymetric change in South San Francisco Bay: 1858–1983. U.S. Geological Survey Open File Report 2004-1192. http://pubs. usgs.gov/of/2004/1192/of2004-1192.pdf (accessed September 20, 2010).

Gesch, D. and R. Wilson. 2002. Development of a seamless multisource topographic/bathymetric elevation model of Tampa Bay. *Marine Technology Society Journal* 35(4): 58–64.

Gibbs, A. E. and G. Gelfenbaum. 1999. Bathymetric change of the Washington-Oregon Coast. In: *Proceedings of the 4th International Conference on Coastal Engineering and Coastal Sediment*, pp. 1627–1642. June 20–24, Long Island, New York.

Jaffe, B. E., R. E. Smith, and L. Z. Torresan. 1998. Sedimentation and bathymetric change in San Pablo Bay: 1856–1983. U.S. Geological Survey Open File Report 1998-759. http://geopubs.wr.usgs.gov/open-file/of98-759/of98-759. pdf (accessed September 20, 2010).

Johnston, S. 2003. Uncertainty in bathymetric surveys. Vicksburg, MS: U.S. Army Engineer Research and Development Center. ERDC/CHL CHETN-IV-59. http://chl.erdc.usace.army.mil/library/publications/chetn/pdf/chetn-iv-59.pdf (accessed September 20, 2010).

Lewis, R. R. 1976. Impact of dredging in the Tampa Bay estuary, 1876–1976. In: *Proceedings of the 2nd Annual Conference of Coastal Society*, The Coastal Society, Arlington, TX, pp. 31–55.

Prandle, D. 2006. Dynamical controls on estuarine bathymetry: Assessment against UK database. *Estuarine, Coastal and Shelf Science* 68(1–2): 282–288.

Sandidge, J. C. and R. J. Holyer.1998. Coastal bathymetry from hyperspectral observations of water radiance. *Remote Sensing of Environment* 65: 341–352.

Stumpf, R. P., K. Holderied, and M. Sinclair. 2003. Determination of water depth with high-resolution satellite imagery over variable bottom types. *Limnology and Oceanography* 48: 547–556.

Thomas, C. G., J. R. Spearman, and M. J. Turnbull. 2002. Historical morphological change in the Mersey Estuary. *Continental Shelf Research* 22: 1775–1794.

Thurman, H. V. and A. P. Trujillo. 2004. *Introductory Oceanography*, 10th edn. Upper Saddle River, NJ: Prentice Hall.

Zervas, C. 2001. Sea level variations of the United States 1854–1999. Silver Spring, MD: NOAA Technical Report NOS CO-OPS.

Zhou, G. Q. and M. Xie. 2009. Coastal 3-D morphological change analysis using LiDAR series data: A case study of Assateague Island National Seashore. *Journal of Coastal Research* 25: 435–447.

7

High-Resolution Mapping, Modeling, and Evolution of Subsurface Geomorphology Using Ground-Penetrating Radar Techniques

Victor J. Loveson and Anup R. Gjuar

CONTENTS

7.1 Introduction

Geomorphological mapping involves in deriving information on earth surface and recording them into a map/drawing with respect to two-dimensional (2D) or three-dimensional (3D) aspects. There are a number of mapping methods available including interpretation of satellite data and aerial photos in various scales.

Very often, the geomorphic signatures are totally or partially modified and get buried due to varied geological processes. These buried geomorphic features are useful and unique in terms of withholding information on the paleo-geological processes and climate variation existed during that time. In addition to the existing geomorphology, if the paleo-geomorphic details are available through buried environment, it would be appropriate to model the area, and thus, to construct the evolution of that area. Such attempt would substantiate to understand the geological processes in times and also aid to project the futuristic scenario.

The buried geomorphic features could be mapped through different geological and geophysical techniques. Again the limitations are the scale of mapping and resolution of the data retrieved by these methods. Geological methods like bore well logging, trenching, etc., are some popular methods involving collection of representative data over an area under study. The gap between sample locations is to be either simulated or manipulated through various statistical methods. Under such conditions, mapping of the area may not yield the reality of the subsurface features in between, and the resolution would depend on the interval of the sample locations. Some of the geophysical methods like seismic, geoelectric (resistivity), gravitational, etc., are good for collecting subsurface details at moderate to deeper levels and they may not yield fair results on shallower subsurface features, owing to their resolution factor. In this context, to fill this gap, the method involving ground-penetrating radar (GPR) offers an accurate, cost-effective, and nondestructive solution to map and model subsurface. It has been useful to decipher shallow geomorphic structures having various options to use different antennas for different depth penetrations (0–30 m) with higher resolution.

7.2 Principles of GPR

GPR was invented in the 1970s, originally for military purposes to locate landmines and underground military tunnels. Later, GPR application has been widely adopted for many other purposes owing to its accurate measurements and mapping shallow depth features (Mellett, 1995). GPR technique works by transmitting pulses of high-frequency (radio waves) energy into materials through a transducer (antenna), and recording the strength and the time required for the return of any reflected energy (Clough, 1976). A series of pulses over a single area makeup is called a scan. The reflection and refraction phenomena are mainly governed by the electrical properties of the ground, i.e., the dielectric constant of the media, which is the measure of the ability of the material that allows the electromagnetic energy to propagate through it. Electrical properties of any geological materials/formation are largely dependent upon volumetric water contents, sediments characteristics including mineralogy, grain size, the presence of organic matter, composition of the sediments, orientation of the grains, shape of the grains, and packing patterns of the sediments (Neal, 2004). For example, air, freshwater, seawater, unsaturated sand, silt, clay, and bedrock have relative dielectric permittivity 1, 80, 80, 2.55–7.5, 2.5–5, 15–40, and 4–6, and electromagnetic wave velocity (m/ns) of 0.3, 0.03, 0.01–0.2, 0.09–0.12, and 0.12–0.13, respectively (Neal, 2004). The dielectric properties and electrical conductivity of different layers/formations are distinctive with the different densities and moisture contents. Dielectric constants of some of the materials are listed in Table 7.1.

The relative dielectric permittivity, which is controlled by the above factors, is an essential parameter governing the reflection process and wave velocity. When a significant change in relative permittivity is encountered, part of the electromagnetic energy is reflected, the reflection being proportional to the magnitude of change. Low conductivity materials, such as unsaturated and coarse-grained sediments, cause little attenuation and, under ideal circumstances, penetration is of the order of tens of meters (Davies and Annan, 1989). Penetration depth and resolution are also influenced by the GPR frequency used for measurement. Lower antenna frequencies are suitable for greater penetration, but do not give a good resolution. Resolution is approximately a quarter of the GPR wavelength, and ranges from 0.08 m for saturated sands with 200 MHz antennas to 0.4 m for dry sands with 100 MHz antenna. Details on the various antennas best used for different depth penetration are presented in Table 7.2. However, for geological and environmental applications, 100 and 200 MHz antennas are preferred and for deeper hydrogeological and geological profiling, multilow frequency (MLF) antennas (16–80 MHz) are ideal.

A typical GPR system is essentially comprised of either a single transmitting and receiving antennae or two separate transmitting and receiving antennae, control unit, data logger, recording, and display unit (Figure 7.1). The control unit is meant for generating a short electrical pulse and transmitter/receiver are used for converting electrical pulse into an electromagnetic pulse of radio frequency and then transmitting

TABLE 7.1

Dielectric Constants of Some of the Materials

Materials	Dielectric Constant (at 100 MHz)	Materials	Dielectric Constant (at 100 MHz)
Air	1		
Water	81	Dry limestone	5.5
Seawater	81	Wet sandstone	6
Ice	4	Sandstone	6
Tills	11	Granite	4–7
Snow firrn	1.5	Dry granite	5
Frozen soil/permafrost	6	Wet granite	6.5
Frozen sand and gravel	5	Dry mineral/ sandy soils	6
Glacial ice	3.6	Shale	5–15
Water saturated sands 20% porosity	19–24	Limestone	4–8
Wet sandy soils	23.5	Bastal	8–9
Soils and sediments	4–30	PVC	3
Dry loamy/clayey soils	2.5	Concrete	4–11 (5)
Organic soils	64	Dry concrete	5.5
Peats	61.5	Travertine	8
Coal	4.5	Wet limestone	8
Wet sands	15	Wet basalt	8.5
Saturated sands	25	Wet concrete	12.5
Dry sand and gravel	5.5	Volcanic ash	13
Wet clay	27	Syenite porphyry	6
Dry clay	4	Asphalt	5
Dry salt	6	Dry bauxite	25
Dry sands	4	Potash ore	5.5

Sources: Data from http://www.geophysical.com. i) SIR System 2000 Operation Manual (2003), #MN72-338, Rev. B, Geophysical Survey System, Inc., USA., pages 27and 38. ii) RADAN for Windows, version 5.0, User's Manual (2001), MN43-162, Rev. A, Geophysical Survey System Inc., U.S.A., page 124.

it into the ground or receiving it. The transmitted signals propagate through the ground, which is then reflected from different sedimentary layers or materials having different electrical properties. These reflected signals received by the receiver of the antennae, which is then directed to a display unit via a control unit, can be interpreted in real time even in the field. They can be digitized and transferred to a personal computer, where different digital signal processing procedures can be applied to enhance the signals and apply corrections or filters for some of the distortion that is inherent to the data acquisition procedures. Then the data can be conveniently modeled through the specialized software with the support of due field ground truth information for correlation purpose.

TABLE 7.2

Best Use of Various Antennas for Achieving Different Depth Penetration

Depth Range of Interest	Best Antenna to Use	Second Choice (MHz)
0–0.5 m (0–1.5 ft)	1500 MHz	900
0–1 m (0–3 ft)	900 MHz	400
0–2.5 m (0–8 ft)	400 MHz	200
0–9 m (0–30 ft)	200 MHz	100
0–20 m (0–60 ft)	Subecho-70	100
0 >20 m (0– >60 ft)	Subecho-40	100

Source: SIR System 2000 Operation Manual (2003) #MN72-338 Rev B, Geophysical Survey System Inc., U.S.A., page 73.

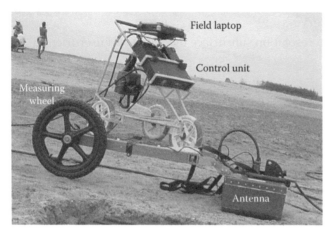

FIGURE 7.1
GPR system: laptop, control unit, antenna, and measuring wheel.

7.3 Advantages of GPR Applications

GPR is an excellent tool to scan the subsurface strata up to 30 m depth and has an edge over other conventional shallow subsurface survey methods. Some of the advantages are

- Portable, easy to use, and setup time in the field is almost immediate
- Nondestructive method and capable of working in all terrains
- Capable of working in wide range of climates and environments (glacial to desert areas)

- Manageable in the field with few staff (one or two)
- Continuous rapid data acquisition and recording (stream and point modes)
- Rapid survey speed to cover more area with less time
- High-resolution subsurface data (vertically and horizontally)
- Cost-effective and less maintenance
- Near real-time interpretation/onsite mapping and rapid 3D surveys
- Wide areas of application
- Easy detection of metallic and nonmetallic objects in the subsurface
- High sensitivity to water, enabling use of GPR for hydrogeological surveys
- High-resolution subsurface modeling and user-friendly software

7.4 Limitation of GPR Application

In spite of its huge advantages, there are a few limitations also to be considered while using GPR:

- Shallow depth of observation (up to 30 m only)
- Conductivity of the ground (signal loss at clay or saltwater intrusion areas)
- Site-specific application and needs shallow trail pits for correlation

7.5 Geological Application Using GPR

GPR can be used for many useful applications related to geological, geotechnical, and geo-environmental surveys. If not all, some of the possible applications that are established by the researcher are

- Profiling depths to bedrock, groundwater, and overburden strata
- Locating bedrock fractures and faults and measuring fracture dip
- Delineating geologic contacts and lithological changes

- Detecting buried utilities and storm drains
- Delineating sinkholes in karsts terrain
- Measuring saturated overburden thickness
- Detecting weathered and fractured bedrock
- Evaluating continuity and depth of clay strata
- Detecting mineralized zones at shallow depths
- Delineating freshwater and saltwater boundary
- Mapping sedimentary structures
- Mapping dune internal structures and their characteristics
- Delineating tsunami deposits, their extent, and depth of deposition
- Mapping shallow aquifers, water tables, and saturated zones
- Delineating paleo-lagoons and dipping of its shore features
- Constructing paleo-geomorphology by locating buried geomorphic features
- Mapping landfill boundaries and thickness and disposal extent
- Detecting contaminant plumes and estimating migration directions

7.6 Scope of the Present Case Studies and Objectives

Four case studies are discussed hereunder to understand the GPR application, especially, in the buried coastal geomorphologic studies. The first case study throws light on the depositional environments with respect to fluvial and marine regime at the south Maharashtra coast. Spit at Talashil is growing progressively and GPR survey interpretation gives an interesting insight into the depositional characteristics in the past years by means of internal structure of the spit. The second case study reveals the paleo-geomorphology of Vedaranyam coastal area and subsequent depositional environment in the recent days. The third case study deciphers the paleo-lagoon system in Velankanni coastal area, where such a lagoon was filled up with sediments and has a new look as a wide coastal plain at present. The fourth case study reveals a freshwater and saltwater boundary and saltwater wedge feature.

All these case studies indicate that GPR could be used to understand the paleo-geomorphology from subsurface profiles and also helps to construct related paleo-geological processes responsible for them. In other words, high-resolution GPR data enable to understand the buried geomorphological features accurately in respect to lateral and horizontal measurements. Dimensions and characteristics of each feature could be sketched out easily. Also, the processing software has 3D mapping facilities whereby numbers of profiles could be arranged and processed in 3D format. This facility helps to reconstruct and isolate any buried geomorphic feature in 3D perspectives and offers the same for enhancement and image processing techniques. By adapting the above techniques, the evolution of the subsurface geomorphology could be made possible by defragmenting each string of information into a full-fledged model.

7.7 GPR in Coastal Studies

Using GPR, some of the studies are reported from different parts of the world, encouraging more studies in coastal areas. Earlier studies have employed GPR in coastal settings to identify structures and facies in deltas, bars, spits, and barrier islands (Leatherman, 1987; Jol and Smith, 1991, 1992a,b,c; Meyers et al., 1994; Bridge et al., 1995; Jol et al., 1996; Sridhar and Patidar, 2005); locate buried paleo-channels (Wyatt and Temples, 1996); and to interpret the Holocene evolution of parabolic deposits along the coasts (FitzGerald et al., 1992; Hill and FitzGerald, 1992; Van Heteren et al., 1994; Daly et al., 2002) and coastal Pleistocene stratigraphy (McGeary et al., 1994; Sotsky, 1995; O'Neal, 1997). Smith and Jol (1995) have evaluated different frequency antennas and their penetration capacity in the Quaternary sediments at Utah, USA. They found a linear trend between different antenna frequencies and the maximum probable depth of penetration, suggesting that the 12.5 MHz antennas can detect strata up to 66 m deep. While mapping the deformation of Quaternary sediments of west Cumbria, UK, a number of thrust plans and folds have been clearly noticed in GPR profiles of 50 MHz antenna (Busby and Merritt, 1999). A reconstruction of bed morphology of aeolian dunes has been attempted in southern Brazil using 50 MHz antenna, achieving 10 m deep penetration (Da Silva and Scherer, 2000). Different antennas ranging from 25 to 200 MHz have been used to profile the coastal barrier spit at Long Beach, USA, where it is found that the 100 MHz provided acceptable resolution, and loss of signals was noticed at 6–8 m depth, which forms the storm wave base (Jol et al., 2002). Ground-penetrating radar profiles have been used to study late Quaternary stratigraphy along the northern margin of Delaware Bay in southern New Jersey, USA, through which six Pleistocene sea-level high stand sequences, forming two composite terraces were identified (O'Neal and McGeary, 2002). To discriminate the aquifer architecture, 100 MHz GPR profiles of about 50 km stretch have been evaluated in the coastal plain of southeast Queensland, Australia, where the shallow groundwater system has been modeling to infer the aquifer characteristics (Ezzy et al., 2003). GPR profile of 80 MHz has been used to map spit-platform fore-sets underlain by perched groundwater table and clayey fine-grained glaciolacustrine sediments in southeast Finland (Makinen and Rasanen, 2003). Beach ridge sediments have been studied with 900 MHz antenna with a penetration of 0.06 m at northern Essex coastline, UK, and a good correlation with field data on the sets of lamination and beds was found (Neal, 2004). Storm dominated sediments in barrier at Castle Neck beach, USA, has been mapped using 27 km long GPR profile of 120 MHz antenna with 10 m penetration, where garnet layers were recorded at about 1 m depth (Dougherty et al., 2004). Loveson et al. (2005) have discussed the use of GPR in heavy mineral exploration along South Indian beaches. Sedimentary records of extreme events at Maine, USA, have been correlated with GPR data and at least four storm scarps were identified (Buynevicha et al., 2004). With the advancement in the GPR system and an increase in the understanding of GPR signals with the geological materials, the use of GPR system, especially for high resolution subsurface profiling in the coastal areas, has been rapidly increasing.

7.8 Case Studies

7.8.1 Materials and Methodologies

Four different locations along the beaches in the west and east coast of India were selected for GPR subsurface profiling (Figure 7.2). The main aim of these case studies is to demonstrate the application of GPR and its utility on the identification of buried geomorphic features at various depths. Since high-resolution data were used throughout, buried landforms could be delineated clearly. For all four locations, GPR system model SIR-20, developed by GSSI, USA, was used to generate continuous profiles along with 200 MHz antenna and measuring wheel. Sometimes for confirmation, 400 MHz antenna was also used.

GPR system was initialized in the field so that the ground reality related to geoelectrical conditions of the field was considered. Either in automatic mode, the GPR system tunes itself to generate the reference dataset in respect to dielectric constant of the beach subsurface materials or parameters can be setup manually. After these initial settings, the signaled data were recorded in continuous mode by dragging the antenna along the survey line (Figure 7.3). Raw data were then processed with RADAN software, which comes along with the GPR system (GSSI, USA). Corrections related to surface and antennas were again ensured. Along the GPR

FIGURE 7.2
Location of case study areas.

FIGURE 7.3
GPR profiling in the beach.

FIGURE 7.4
Beach profiling along GPR survey line.

transect, beach profiling was also carried out using the Theodolite/other survey equipments to record elevation details (Figure 7.4), which were used during surface normalization of GPR data processing.

Case study 1: Mapping internal structure of the spit at Talashil

GPR survey was conducted at Talashil spit, located near Malvan, South Maharashtra (Figure 7.5). This spit is about 12 km length, which was built geologically in recent years and runs parallel to the coast. The Gad River flows on the eastern side and the Arabian Sea is on the western side. The width of the spit is about 1.5 km at the turning point from E-W to N-S (Tondavali village), and from there it tapers toward the southern tip, ending with a cuspate foreland. Naturally, the fluvial and coastal processes build this spit, and fluvial contribution dominates at the eastern side by way of deposition in the low-lying flood plain and marine influences and subsequent deposition on the western side.

A transect, perpendicular to the coast, was selected from the riverside to the seaside on the spit at its southern end. Survey line is about 170 m length from flood plain to intertidal zone, through dune system. Profile was generated along this transect using levels with reference to permanent bench marks. For this GPR survey, 200 MHz antenna was used with measuring wheel. Data collected were processed using RADAN software and elevation details were fed into the GPR data. The depth of penetration was achieved at 8 m depth. The results are shown in Figure 7.6.

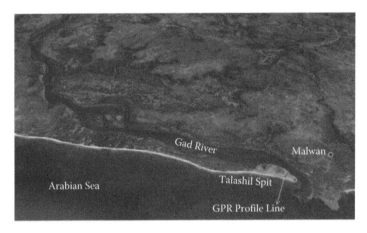

FIGURE 7.5
GPR survey line at Talashil spit.

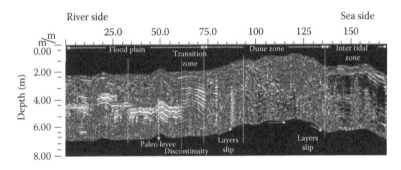

FIGURE 7.6
Subsurface structure at Talashil spit (sea and river on each side).

7.8.2 Paleo-Geomorphic Units

Various geomorphic zones under the influence of fluvial, aeolian, and marine are distinguished significantly with different subsurface structures, which substantiate to infer on geomorphic features and geological processes. Details are discussed in the following.

7.8.3 Flood Plain

More or less, the surface structures are in agreement with the morphological aspects of the present flood plain. It seems that the paleo-flood plain gradually grows into the present one by rhythmic deposition, subsequently responding to the local uplift. Paleo-levee could be marked at 2 m depth. It is also observed that below 2 m depth in the flood plain zone, consolidated sand layers are present. The features are up to 2 m depth from the surface and infer that the area was either under submergence or was lowland with frequent flooding. The transitional zone between flood plain and dune zone has regular depositional events and was always above water level in earlier times.

7.8.4 Dune System

Dune zone represents different characters of deposition compared to the flood plain. The existing topography does not correspond to subsurface topography, but infers various stages of dune-building activities in earlier times. Marine influence, by way of saline intrusion, is seen in deeper layers. Minor slumping and concurrent dipping of sand layers are observed. Another characteristic of dipping toward sea is also noticed in various depths, revealing the emergence of the beach.

7.8.5 Intertidal Zone

There is a distinctive signal change in the intertidal zone compared to the other zones. The influence of saline water dominates this zone and sand layers are saturated with salinity, thereby adsorbing the energy and resulting in loss of signals. Due to this, information on layers is not clear. Here also, up to 2 m depth, there is an indication of submergence event. However, top surface cover (nearly 1 m) is a recent feature associated with medium-grained sand deposition by sea.

Case study 2: Buried beach ridge and swale system at Vedaranyam coast

Vedaranyam coastal plain is a low-lying area with undulating topography. About 125 m length transect was selected for GPR profiling with 200 MHz antenna. The system is tuned to yield data up to 7.5 m depth. The sea is calm and wave height is at its minimum due to shallower near-shore bathymetry. The beach is about 25 m wide followed by rolling sand dunes having about 90 m width. Flat and plain land is being used, at present, for agriculture (paddy cultivation) practice. Sand dunes are low lying and highly reworked due to social forestry plantation (acacia) activities.

7.8.6 Paleo-Lagoon

GPR data show two paleo-lagoons with significant signal variability to surrounding features (Figure 7.7). The one on the left side of the figure (seaside) is about 30 m wide and nearly 1.5 m depth. This is underlain by

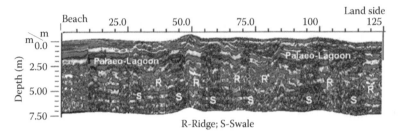

FIGURE 7.7
Buried ridge and swale features at Vedaranyam beach.

consolidated sandy layers of earlier time and overlain by recent sandy cover. The other one, toward the right side of the figure (land side), is typically a wide-basin type lagoon, having 55 m width and nearly 2.5 m depth. This lagoon formed over beach ridges but was gradually filled in by sand and clay materials. These paleo-geomorphic units, viz., lagoon, buried beach ridges, and recent sand cover, do not have similarities with present-day topography and were formed under various climatic and depositional conditions during different periods, supported by the local emergence and stability of the region. The above study implies that by observing the exiting nature of the geomorphology of the area, it may not be possible to ascertain the earlier geomorphic setup until the data like GPR subsurface information are available.

7.8.7 Buried Beach Ridges and Swale System

There are alternative beach ridges and swales buried under the sand. These buried beach ridges are seen at 2.5 m depth below surface. The width of these beach ridges ranges from 5 to 25 m. These are observed clearly from 2.5 to 5.0 m depth. Alternatively, swales (depressions) are seen in between the ridges.

7.8.8 Paleo-Environment and Processes

GPR data present a clear picture on the changes of depositional environment and paleo-geological processes related to the formation of various types of landforms. There are three different situations observed as deciphered from GPR data. Below 5 m depth indicates that the area was low lying with depressions, located parallel to the coast at the period. The entire area at that time was either partially submerged with water or flooded frequently, making the area more saturated with water. During the second situation (2–5 m depth), aeolian activity was active and had constructed series of beach ridges. Also, sea lowering must have occurred in stages. Before the third situation commenced, the beach ridges were reworked and water flooded over the area and stagnated. Due to emergence of land owing to regional uplift, these lagoons got filled up and produced the existing topography as seen today.

Case study 3: Paleo-lagoon system at Velankanni coast

Similar methodology and materials were used, as explained elsewhere in this chapter. Survey line along which the GPR was run is almost 50 m length, perpendicular to the coast. GPR data is acquired up to 7 m depth, using 200 MHz antenna. As such, the area under study is much dynamic in recent times and the coastal landforms are subjected to frequent changes as observed in the multitime satellite images. In particular, changes were significant during and after the tsunami of December 26, 2004.

7.8.9 Paleo-Lagoon

GPR profiled data signify the lateral dimension (i.e., its extent) of the paleo-lagoon (Figure 7.8). The lagoon was nearly 45 m wide and about 5 m depth. Nearly 2 m depth deposition was under rhythmic fashion, indicating no major disturbances. But, the later deposition (about 3 m) was under dynamic condition, revealing deposition under extreme condition/quick sedimentation. Due to same condition, nearly 1.5 m sand cover was deposited, filling and covering the entire lagoon area.

Paleo-lagoons shore portrays number of dipping sand layers (at right side of the Figure 7.8). Notable characteristics such as sand layers dipping toward sea and rhythmic layering pattern as the lagoon started filling up are significantly observed. The profile was taken during 2008, i.e., 4 years after the tsunami. At present, one can notice only a flat wide beach with white sands, concealing the paleo-environment, which existed a few years ago.

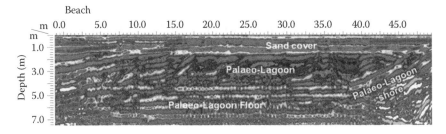

FIGURE 7.8
Buried paleo-lagoon and its shore region at Velankanni beach.

Case study 4: Hydrogeomorphic features at Sothavilai beach

GPR application in hydrogeological studies (Beres and Haeni, 1991; Harari, 1996; Nakashima et al., 2001; van Overmeeren, 2004), aquifer architecture analysis (Asprion and Aigner, 1997) hydropedological investigations (Doolittle et al., 2006) has shown appreciable results. Hereunder, an example is presented, which has been carried out in South India.

Sothavilai beach (8 06' 23" N and 77 24' 55" E), near the southern tip of India was profiled using GPR system. The beach is moderately wide (100 m) with sloping foreshore followed by two prominent elevated ridges at the backshore areas that are covered with coconut trees and other plantations. The coastline configuration trends NW–SE with rocky outcrops at surrounding beaches. At berm zone, saline water tolerant wild bushes are seen followed by a sudden change in vegetation of freshwater vegetations at backshore areas. A village (Sothavilai) is located at the elevated ridge area and the local people used to get mixed quality of groundwater at different seasons in their locality. Invariably, during the post-monsoon season, the freshwater is available at nearby dug wells. Groundwater availability at these same wells during other seasons is more saline in nature due to over pumping of groundwater from shallow aquifers and thereby results in saltwater infiltration.

A transect of nearly 52 m length perpendicular to the coastline was selected. During GPR profiling, about 2.5 m depth penetration was achieved using 200 MHz antenna. Since the beach is steep with elevated beach ridges at backshore, the survey was carried out toward seaside for practical conveniences. Based on the geoelectrical properties of the subsurface sand layers, the signals of the GPR were analyzed using various enhancement techniques and the results are presented in the following.

7.8.10 Saltwater Wedge

A saltwater wedge was significantly observed in the GPR data (Figure 7.9). In the wedge zone, which is highly saline in nature, the signals were almost lost and were indicating a strong signal loss zone. Boundary of the wedge is clearly observed. The upper limit was at 0.5 m at high tide level (HTL), which declines to 1.5 m depth at 50 m away from the HTL. Around 2 m depth, there was a baseline saline groundwater level. Below 2 m depth, it is observed that the sand layers are more compact than the wedge area and the saturation was less due to more compactness.

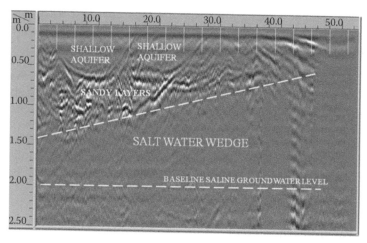

FIGURE 7.9
Saltwater wedge and shallow aquifers at Sothavilai beach.

7.8.11 Localized Shallow Aquifers

Above the wedge zone, various sand layers are observed (Figure 7.10) and their signal characters indicate that they are not much influenced by saline condition (bright white and black colors). They are dry sand layers with thin clay layers, forming localized shallow aquifers. These feature form shallow saline aquifers which was connected to the wedge at HTL zone. In between the wedge and the shallow aquifer zones, one can observe strong signals, which denote sand layers with or without

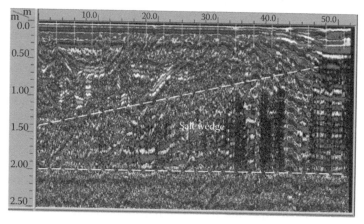

FIGURE 7.10
Fresh- and saltwater boundary line at Sothavilai beach.

freshwater content. During rainy seasons, they are filled with water and act as perched aquifers. Over them, salt-coated sand layers are observed, which have connection with the wedge at HTL area. The surface layer is totally dry and with about 20 cm thickness.

The structure of the sand layers above the wedge helps to map the shallow aquifers. There are two interconnected/compartmented shallow aquifers at 0.50 m depth with 10 m width each (Figure 7.9). It is observed that the area between 0.5 and 1 m depth, above wedge area, has coarse sand with thin layers of clay. They act as the local aquifers, which are fed entirely by the rain water. During the rainy seasons, these aquifers are filled with freshwater and for sometimes, after the rainy seasons, the local people utilize the water for domestic purpose.

7.8.12 Paleo-Beach

The GPR data display a 4 m thick subsurface vertical sand column at 42 m distance. It is also noticed that from 38 to 48 m, the upper sand layers (0.0–1.2 m depth) are dipping toward the sea. The layers are less compact with coarse sand while compared to the nearby areas. This may be the indication of paleo-beach environment. Keeping this concept of paleo-beach and sea level variations, the shallow aquifers may indicate swale systems, which are recently filled by the sand during migrating of the coastline. The shallow swales filled by the recent sands from coastal side might have brought the saline infected deposits as filled-in materials. These are observed in the GPR data (Figure 7.10) above sand layers.

7.8.13 Salt- and Freshwater Interaction Zone

The boundary of the saltwater and freshwater interaction is one of the essential factors to be assessed by the conventional methods with the help of number of dug wells, which is cumbersome. With the GPR system, the study area could be conveniently scanned rapidly and the subsurface features and, especially, the conductivity of the subsurface layers could be evaluated for further analysis. In this chapter, the interaction zone has been mapped clearly using GPR survey. Since the data are of high resolution, the boundary, especially the upper one, was clearly observed in contrast to the dry/freshwater sandy layers. This saltwater wedge is also controlled by the internal structure/arrangement of the sand layers. In the wedge zone, some areas, especially located near to HTL, have more influences with the saltwater occupation and indicate the different grain size texture (i.e., coarser and angular). The nature of the wedge disposition depicts that the wedge structure is more associated with the sand layer granulometry, and the local shallow aquifers could be infected with the overpumping.

7.9 Conclusion

GPR could be useful to derive subsurface information accurately and thus widely used in civil, geological, geotechnical, and geo-environmental applications. Considering the available technologies to map subsurface features, GPR technique stands prominent in terms of its high resolution, and nondestructive and cost-effective aspects. In this chapter, advantages and limitation of GPR techniques were presented. The usefulness of GPR application in buried coastal geomorphological mapping was discussed in detail, which is otherwise cumbersome with conventional methods. Four case studies were presented from different localities along the Indian coast. Based on the high resolution of the GPR data, buried geomorphological features were mapped accurately and the paleo-geological processes responsible for buried geomorphological features were discussed along with their evolution trend. This type of information on buried geomorphological features is another important dimension of geomorphological studies, which helps to construct paleo-environment and processes.

Acknowledgments

The authors wish to record their sincere thanks to the directors of the respective institutions for permission to prepare this chapter. Thanks are due to the Council of Scientific and Industrial Research (CSIR), New Delhi, for the financial support to procure the GPR system and to carry out the surveys at different coastal areas. We thank our field staff for their cooperation and support during data generation in the field without whom this contribution could not have been crystallized. Sincere thanks are due to Dr. S.K. Subramanian, senior scientist, National Remote Sensing Centre, Hyderabad, for having invited the authors to contribute this chapter. The opinions expressed are of authors only and not necessarily of the institutions they belong to.

References

Asprion, U. and Aigner, T. 1997. Aquifer architecture analysis using ground-penetrating radar: Triassic and Quaternary examples (S. Germany). *Environmental Geology* 31:66–75.
Beres, M. and Haeni, F.P. 1991. Application of ground-penetrating radar methods in hydrogeologic studies. *Groundwater* 29:375–386.

Bridge, J.S., Alexander, J., Collier, R.E.L.L., Gawthorpe, R.L., and Jarvis, J. 1995. Ground penetrating radar and coring used to study the large-scale structure of point-bar deposits in three dimensions. *Sedimentology* 42:839–852.

Busby, J.P. and Merritt, J.W. 1999. Quaternary deformation mapping with ground penetrating radar. *Journal of Applied Geophysics* 41:75–91.

Buynevich, I.V., FitzGerald, D.M., and van Heteren, S. 2004. Sedimentary records of intense storms in Holocene barrier sequences, Maine, USA. *Marine Geology* 210:135–148.

Clough, J.W. 1976. Electromagnetic lateral waves observed by earth-sounding radar. *Geophysics* 41:1126–1132.

Daly, J., McGeary, S., and Krantz, D.E. 2002. Ground-penetrating radar investigation of a late Holocene spit complex: Cape Henlopen, Delaware. *Journal of Coastal Research* 18:274–286.

Doolittle, J.A., Jenkinson, B., Hopkins, D., Ulmerd, M., and Tuttle, W. 2006. Hydropedological investigations with ground-penetrating radar (GPR): Estimating water-table depths and local ground-water flow pattern in areas of coarse-textured soils. *Geoderma* 131:317–329.

Davis, J.L. and Annan, A.P. 1989. Ground penetrating radar for high resolution mapping of soil and rock stratigraphy. *Geophysical Prospecting* 27:531–551.

Da Silva, F.G. and Scherer, C.M.D.S. 2000. Morphological characterization of ancient aeolian dunes using the ground-penetrating radar, Botucatu Formation, southern Brazil. *Revista Brasileira de Geociências* 30(3):531–534.

Dougherty, A.J., FitzGerald, D.M., and Buynevich, I.V. 2004. Evidences for storm-dominated early progradation of Castle Neck barrier, Massachusetts, USA. *Maine Geology* 210:123–134.

FitzGerald, D.M., Baldwin, C.T., Ibrahim, N.A., and Humphries, S.M. 1992. Sedimentologic and morphologic evolution of a beach-ridge barrier along an indented coast; Buzzards Bay, Massachusetts. In: *Quaternary Coasts of the United States: Marine and Lacustrine Systems*, eds. Fletcher C.H. III and Wehmiller, J.F., Society of Economic Paleontologists and Mineralogists Special Publication No. 48, Tulsa, OK, pp. 65–75.

Harari, Z. 1996. Ground-penetrating radar (GPR) for imaging stratigraphic features and groundwater in sand dunes. *Journal of Applied Geophysics* 36:43–52.

Hill, M.C. and FitzGerald, D.M. 1992. Evolution and Holocene stratigraphy of Plymouth, Kingston, and Duxbury Bays, Massachusetts. In: *Quaternary Coasts of the United States: Marine and Lacustrine Systems*, eds. Fletcher C.H. III and Wehmiller, J.F., Society of Economic Paleontologists and Mineralogists Special Publication No. 48, Tulsa, OK, pp. 45–56.

Jol, H.M., Lawton, D.C., and Smith, D.G. 2002. Ground penetrating radar: 2-D and 3-D subsurface imaging of a coastal barrier spit, Long Beach, WA, USA. *Geomorphology* 53:165–181.

Jol, H.M. and Smith, D.G. 1991a. Ground penetrating radar of northern lacustrine deltas. *Canadian Journal of Earth Sciences* 28:1939–1947.

Jol, H.M. and Smith, D.G. 1992b. Geometry and structure of deltas in large lakes: A ground penetrating radar overview. *Geological Survey of Finland Special Paper* 16:159–168.

Jol, H.M. and Smith, D.G. 1992c. Ground penetrating radar: Recent results. *Canadian Society of Exploration Geophysics Recorder* 27:15–20.

Jol, H.M., Smith, D.G., and Meyers, R.A. 1996. Digital ground penetrating radar (GPR): A new geophysical tool for coastal barrier research (examples from the Atlantic, Gulf and Pacific coasts, USA). *Journal of Coastal Research* 12:960–968.

Leatherman, S.P. 1987. Coastal geomorphic applications of ground penetrating radar. *Journal of Coastal Research* 3:397–399.

Loveson, V.J., Barnwal, R.P., Singh, V.K., Gujar, A.R., and Rajamanickam, G.V. 2005. Application of ground penetrating radar in placer mineral exploration for mapping subsurface sand layers: A case study. In: *Developmental Planning of Placer Minerals*, eds. Loveson, V.J., Chandrasekar, N., and Sinha, A., Allied Publishers, New Delhi, pp. 71–79.

Makinen, J. and Rasanen, M. 2003. Early Holocene regressive spit-platform and nearshore sedimentation on a glaciofluvial complex during the Yoldia Sea and the Ancylus Lake phases of the Baltic Basin, SW Finland. *Sedimentary Geology* 158:25–56.

Mellett, J.S. 1995. Ground penetrating radar applications in engineering, environmental management and geology. *Applied Geophysics* 33:157–166.

Meyers, R.A., Smith, D.G., Jol, H.M., and Hay, M.B. 1994. Internal structure of Pacific Coast barrier spits using ground penetrating radar. *Proceedings of the Fifth International Conference on GPR*, Waterloo Center for Groundwater Research, Kitchener, Canada, pp. 843–854.

Nakashima, Y., Zhou, H., and Sato, M. 2001. Estimation of groundwater level by GPR in an area with multiple ambiguous reflections. *Journal of Applied Geophysics* 47:241–249.

Neal, A. 2004. Ground penetrating radar and its use in sedimentology: Principles, problems and progress. *Earth Science Reviews* 66:261–330.

O'Neal, M.L. and McGeary, S. 2002. Late Quaternary stratigraphy and sea-level history of the northern Delaware Bay margin, southern New Jersey, USA: A ground penetrating radar analysis of composite Quaternary coastal terraces. *Quaternary Science Review* 21:929–946.

Smith, D.G. and Jol, H.M. 1995. Ground penetrating radar: Antennae frequencies and maximum probable depths in Quaternary sediments. *Journal of Applied Geophysics* 33:93–100.

Sridhar, A. and Patidar, A. 2005. Ground penetrating radar studies of a point bar in the Mahi River basin, Gujarat. *Current Science* 89:183–189.

Van Heteren, S., FitzGerald, D.M., and McKinlay, P.A. 1994. Application of ground-penetrating radar in coastal stratigraphic studies. *Proceedings of the Fifth International Conference on GPR*, Waterloo Center for Groundwater Research, Kitchener, Canada, pp. 869–881.

van Overmeeren, R.A. 1994. Georadar for hydrogeology. First Break 12(8):401–408.

Wyatt, D.E. and Temples, T.J. 1996. Ground-penetrating radar detection of small-scale channels, joints and faults in the unconsolidated sediments of the Atlantic Coastal Plain. *Environmental Geology* 27:219–225.

8

Remote Sensing in Tectonic Geomorphic
Studies: Selected Illustrations from the
Northwestern Frontal Himalaya, India

G. Philip

CONTENTS

8.1 Introduction

Man has been studying landforms of diverse origin since historic past, based on laborious field surveys, updating of topographical maps from time to time, and lately through the photogrammetrical techniques. This was arduous and time consuming with little spatial coverage in a given time frame. However, balloon aerial photography came into being in 1858 and airplane flown photography in 1909. Gradually, aerial photographs of various scales were rigorously used for landform mapping and natural resources surveys. In the 1960s, there was a technological leap when satellites were put into the orbit, yielding synoptic view of a larger area from a vantage point. This really became crucial to multiple applications, since it covered all types of terrain with ease. Thus began the era of remote sensing applications to the earth resources evaluation and mapping.

The introduction of polar orbiting geosynchronous satellites as well as dedicated space shuttle missions has brought a revolution to the observational parameters, which could overcome many limitations of conventional photography. The optomechanical scanners and lately the CCDs operational in the discrete spectral bands with high spatial and radiometric resolution have offered a variety of information, which otherwise could not be obtained from the air photos. Ever since 1972, when the first satellite of the Landsat series was launched, earth scientists have recognized the value of data recorded by these satellites in geological and geomorphological studies. Satellite image has large areal coverage, relatively low cost, and enables relatively rapid rates of mapping. Low- to high-resolution multispectral satellite imagery (e.g., Indian Remote sensing Satellite [IRS], Cartosat, IKONOS, Quick Bird, etc.) currently offers very high spatial resolutions, allowing finer details of the earth's surface to be mapped. The multidate, multispectral, and multiresolution satellite data have yielded significant information in the observation of ground reflectance over the past 40 years. Today, the technique has further progressed into digital image analysis, yielding much sharper and finer details in various dimensions than ever before. Since then, the remote sensing technology has taken a big stride in the area of mapping of landforms from space.

In general, there are three major types of landforms we observe on the earth's surface. This includes landforms that are built (depositional), landforms that are carved (erosional), and landforms that are made by the crustal movements of the earth (tectonic). Of these, our understanding of the development of tectonic landforms is considered to be very significant as they are the surface manifestation of the past and ongoing tectonic processes in the current tectonic regime. Modern tectonic geomorphology integrates many branches of earth sciences, which utilize techniques and data derived from studies of geomorphology, seismology, geochronology, structure, geodesy, and Quaternary climate change. Tectonic geomorphology, on the other hand, essentially evaluates the fundamentals of the subject, which include the nature of deformation basically in terms of faulting and folding, and also uses geomorphic markers for delineating deformation. Study of tectonic landforms, therefore, offers some basic concepts of active tectonics in terms of type of deformation and sequence and rate of tectonic movement, reconstruction of former stress field. The analysis of tectonic landforms will also help us to recognize the relation between the Quaternary tectonic movement and seismic deformation in the present and historic times.

Mapping of tectonic landforms in the hostile, rugged, and inaccessible mountainous terrain of the Himalaya has frequently been a great challenge to the geologists. It is in this context that the potential of remote sensing has been appreciated. Satellite images provide a synoptic view, which gives a regional and integrated perspective as well as interrelations between diversified land features such as lithological variation, geological structure, landform, vegetation cover, drainage pattern, etc., that can be better mapped on the image than on the ground. Easy availability of multispectral and high-resolution data and advanced capabilities of digital image processing techniques in generating enhanced and highly interpretable images have further increased the potential of satellite remote sensing in delineating the landforms, lithological contacts, and geological structures in finer details and with better accuracy. In recent years, remote sensing techniques and DEM-based studies have also been successfully employed in the field of tectonic geomorphology by many workers (Armijo et al., 1996; Philip, 1996; San'kov et al., 2000; Karakhanian et al., 2002; Kaya et al., 2004; Ganas et al., 2005; Philip and Virdi, 2006, 2007 and references therein). Besides, there are successful attempts made using interferometric radar remote sensing for deriving detailed topographic and slope information for fault scarps (Hooper et al., 2003 and references therein).

8.2 Tectonic Landforms vis-à-vis Active Faults for Seismic Hazard Evaluation in Himalaya

Seismic hazard evaluation in the tectonically active Himalaya is crucial because earthquakes pose a continual threat to the safety of the people inhabiting and adjacent to this gigantic mountain system of the world. It has now been widely accepted that active faults—faults which have moved repeatedly in the recent geological time and have potential for reactivation in the future—contribute significantly to the seismic activity (>80% seismic activity). The tectonic landforms such as displaced river terraces and alluvial fans besides fault scarps and other morphotectonic features, for instance, triangular facets, sag ponds, shutter ridges, pressure ridges and pull apart basins, deflected drainages, etc., are closely related with these faults. Geomorphic and morphotectonic analyses of tectonic landforms provide insights into rate, style, and pattern of deformation due to active tectonics. The displaced geomorphic surfaces, lineaments, river terraces, alluvial fans, drainage channels, and topographic ridges have already been cited as potential signatures of the manifestation of active tectonics (Nakata, 1972, 1989; Valdiya, 1986, 1993; Malik et al., 2003; Philip et al., 2006; Philip and Virdi, 2006, 2007; Philip, 2007).

In India, we still lack information about the distribution of active faults, their behavior, and characteristics, though information on distribution of seismicity including major (M6.5–7.5) and great (≥M8) events is available. A recent study has revealed that the October 8, 2005, Kashmir earthquake has occurred along the preexisting active fault trace, which has generated ~65 km long surface rupture with a maximum of ~5.5 m vertical displacement (Hussain et al., 2009). Identification of active faults that have moved within the current tectonic regime, i.e., during Holocene also helps in assessing whether or not tectonic movements are likely to occur and cause seismicity—which is generally associated with these faults and ascertain the seismic risk in the surroundings. In this context, remote sensing techniques have opened new vistas in studying the active faults, their distribution, and spatial extend. In the Outer Himalaya or the foothills, lying between Himalayan Frontal Thrust (HFT) in the south and the Main Boundary Thrust (MBT) in the north, numerous active faults and neotectonic features have been reported by many workers (Nakata, 1972, 1989; Virdi, 1979; Yeats et al., 1992; Philip, 1995, 1996; Philip and Sah 1999; Malik et al., 2003; Philip et al., 2006), which have generated major and great earthquakes (Bilham, 2004).

8.3 Case Studies

In an active orogenic belt like the Himalaya, expressions of active tectonics are numerous and are mainly manifested in the form of faulting and tilting of Quaternary deposits such as alluvial fans and river terraces besides the preferred stream channel migration, river capturing owing to upliftment, and development of tectonic landforms. This chapter illustrates selected case studies demonstrating the potential of remote sensing techniques in the study of active faults and associated tectonic landforms in the northwestern Frontal Himalaya (Figure 8.1), falling between the HFT and the MBT.

8.3.1 Singhauli Active Fault in the HFT Zone

Morphostructural analysis using remotely sensed data (IRS data) along with selected field investigation has helped in delineating an active fault trace, which is oblique to the HFT (Philip and Virdi, 2006) in the northwestern Frontal Himalaya (Figure 8.2a and b). The tectonic landforms and the topographic features are indicative of long-term uplift/deformation along the fault in the current tectonic regime. The cumulative slip along the fault

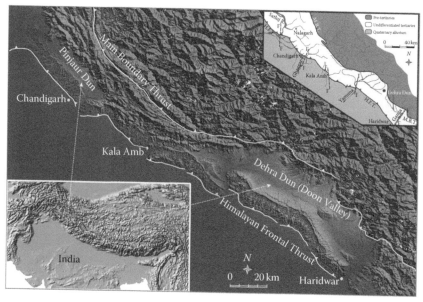

FIGURE 8.1
Regional view (SRTM image) of part of the study area. (Inset: Regional geological set up of the north western outer Himalaya.) (From Philip, G. and Virdi, N.S., *Curr. Sci.*, 90(9), 1267, 2006.)

shows manifestation of normal faulting, indicating recent tectonic activity in terms of several paleoearthquakes. Very interesting situations are available at a number of places just adjacent to the HFT, which defines the zone of convergence between the Himalaya and the Indo-Gangetic alluvial plains, where evidences of normal faulting occur. This faulting has created tectonic landforms well expressed on IRS satellite images (Figure 8.2a) and on aerial photographs. The fault observed at Singhauli near Kala Amb (Figure 8.1) is named as the *Singhauli active fault* (SAF). The SAF traverses through the middle Siwaliks (post-Miocene rocks) exposed on the hanging wall of the HFT. SAF is traceable for over 4 km, of which the western half trends ENE–WSW while the eastern half trends in E-W direction (Figure 8.2a and b). The western half merges in the piedmont alluvium in the south while the eastern segment, east of Singhauli Nala, fades away further east in the Siwaliks. The southside up behavior of this fault is in contrast to the prevalent northside up movement along the HFT. The fault steeply dips due NNW/N and has created few sag ponds to the north of the footwall zone. The fault has also caused shift of drainage channels of major and minor streams. Near its southwestern termination, north of Singhauli and Nanhari villages, another set of minor faults is also traceable for over 1 km (Figure 8.3a). These are WNW–ESE trending southside

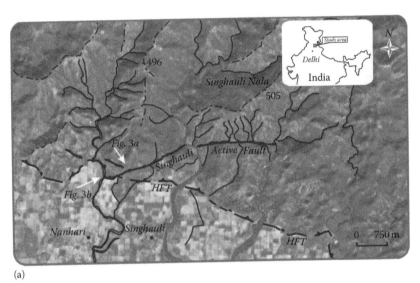

(a)

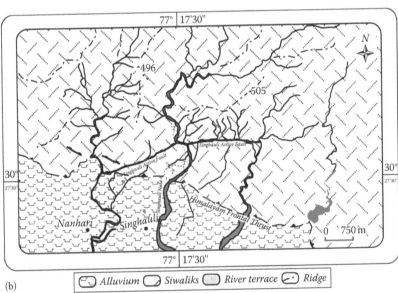

(b)

| ⬚ Alluvium | ◯ Siwaliks | ◯ River terrace | ⌒ Ridge |

FIGURE 8.2
(a) IRS-LISS-III + PAN merged image (false color composite date November 2, 2002) showing the HFT (dents on the upthrown side) and the Singhauli Active Fault zone (teeth on the downthrown side). (b) Singhauli active fault, HFT and the broad geomorphic features of the Frontal Himalaya prepared using IRS data. (After Philip, G. and Virdi, N.S., *Curr. Sci.*, 90(9), 1267, 2006. With permission.)

(a)

(b)

FIGURE 8.3

(a) Panoramic view (view toward west) showing trace of Singhauli Active Fault (F–F). Topographic high (A–B) represents the footwall of the south side up fault. A minor sympathetic fault (f–f) oblique to the Singhauli fault is also visible. (b) Singhauli fault zone showing the south side up footwall (FW). View toward NE. (After Philip, G. and Virdi, N.S., *Curr. Sci.*, 90(9), 1267, 2006. With permission.)

up normal faults dipping steeply due northeast and with well-defined fault scarps (Figure 8.3b). These faults appear to be sympathetic faults developed along with the SAF. Topographic features indicate movement along a normal fault with upthrown southern block. This extensional behavior of faulting is in contrast to the ongoing southward thrusting along the HFT, with Siwalik sediments in the hanging wall overriding the alluvial fans in the footwall. The analyses of various tectonic landforms and the relative positions of the Quaternary deposits in the area corroborate that the Frontal Himalayan region has ruptured repeatedly in the recent past.

8.3.2 Pinjaur Dun: Intramontane Valley between HFT and MBT

In the Outer Himalaya, lying between the HFT and MBT in the north, numerous active tectonic features have been reported (Nakata, 1972, 1989; Valdiya, 1992; Philip and Virdi, 2006), which have generated major and great earthquakes (Molnar and Pandey, 1989; Ambraseys and Bilham, 2000; Ambraseys and Douglas, 2004). Philip and Virdi (2007) have carried out study of these features in a very active segment of the Outer Himalaya between the Satluj and Ghaggar rivers, referred to as the Pinjaur Dun (Figures 8.1 and 8.4), lying in the meizoseismal zone of 1905 Kangra earthquake (Middlemiss, 1910). Pinjaur Dun is one of the three major Duns in the western Frontal Himalaya, viz., Soan, Pinjaur, and Dehra. The Duns are broad synclinal depressions (Figure 8.1), which develop when the growing outer ridges were constituted by the Siwalik sediments block and diverted the drainage (Nakata, 1972). The Pinjaur Dun is covered by sediments deposited over the Siwalik deposits as a series of alluvial fans brought down by tributaries from both sides to join the Sirsa River, the axial stream (Figure 8.5). In the apex region of the northern fans, the tributary streams have cut deep gorges, while in the distal portions these have wide channels with local braiding of the streams. The fan surfaces are extensively degraded due to anthropogenic activity and intense precipitation during the monsoons. A series of terraces have been carved out by erosion and deposition by both the axial stream of River Sirsa and its major tributaries.

Active tectonics in the Pinjaur Dun is reflected by the dislocation of the landforms by major and minor faults in the above Quaternary and pre-Quaternary sediments. Numerous tectonic landform features and active fault traces have been delineated on aerospace data and also verified in the field. Most of the fault traces and lineaments show a NW-SE trend, which is almost parallel to the regional trend of the MBT, Bursar, and the Nalagarh Thrusts. Uplifted and tilted Quaternary landforms and sediments indicate the recent activity along these faults. A brief account of three major active fault systems identified in the Pinjaur Dun is given in the following sections.

FIGURE 8.4

IRS-LISS-III + PAN merged image (October 4, 2002) showing the regional set up around Pinjaur Dun. The boxes (1–3) show the areas studied around active fault systems: 1, Nangal–Jhandian; 2, Bari–Batauli; and 3, Majotu. Lk, Luhund Khad; Kk, Kundlu ki Khad; Ck, Chikni Khad; Rn, Ratta Nala; Bn, Balad Nadi; Sc, Surajpur Choa; Nn, Nanakpur Nadi; Rmn, Ramnagar Nadi; Kpn, Kiratpur Nadi; Jn, Jhajra Nadi; and Kn, Koshallia Nadi. (Modified after Philip, G. and Virdi, N.S., *Curr. Sci.*, 92(4), 532, 2007.)

8.3.2.1 Nangal–Jhandian Active Fault System

Traces of parallel to subparallel active faults have been identified in Nangal–Jhandian sector to the north of Chikni Khad, a tributary of Sirsa River (Figures 8.4 and 8.6a). Dislocated fan terrace observed near the Palasra Nihla village (Figure 8.6a and b) forms a 15 m high scarp (Figure 8.7) and shows back tilting of about 10° along the stream, draining east of Palasra Nihla. The Nangal–Jhandian fault is displaced sinistrally by about 300 m along a fault, which runs parallel to the stream (Figure 8.6a and b). The fault scarp trends WNW–ESE and extends for about 6.5 km from Bahman Majra to Palasra Dittu (Figure 8.6b). A number of parallel to subparallel fault scarps are observed to the north of this fault and could be extended for about 2 km up to Jhandian, just south of Nalagarh Thrust. Well-preserved fault scarps are also observed in streams draining between two Palasras, where nearly horizontal fan gravels are thrusted

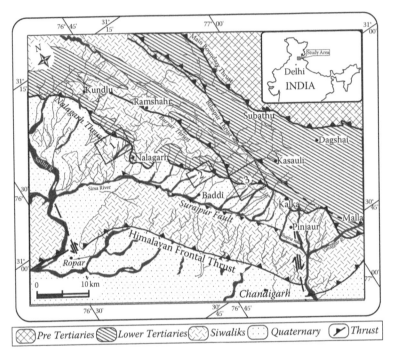

FIGURE 8.5
Regional geological map of the area. The boxes (1–3) show the areas studied around active fault systems: 1, Nangal–Jhandian; 2, Bari–Batauli; and 3, Majotu. (Modified after Raiverman, V. et al., *Petrol. Asia J.*, 6, 67, 1983.)

over the river deposits and the fan material is tilted to the northeast. The fault scarps almost perpendicular to the small tributaries of Chikni Khad originating from the lower Tertiary rocks appear to be basically due to movement along a thrust fault with north side up, though minor deflection of channels suggests a left lateral strike slip component. The diversion of NE-SW flowing Reru Nala into a NNW-SSE tributary of Chikni Khad between Dhundli and Bahman Majra is due to the movement along this fault (Figure 8.6b). The parallel to subparallel active fault systems traceable for a total of 8–9 km appear to be merging at Jhandian. A sag pond located to the northeast of Nangal corroborates the back tilting of the terraces and formation of a depression. A small dried up pond is also observed east of Bahman Majra.

8.3.2.2 Bari–Batauli Active Fault System

The Balad–Ratta Nala fan system (Figure 8.8a and b) in its apex region has been displaced by the NW-SE trending Bari–Batauli active fault. The fault traceable for about 1.5 km is a high angle reverse fault dipping

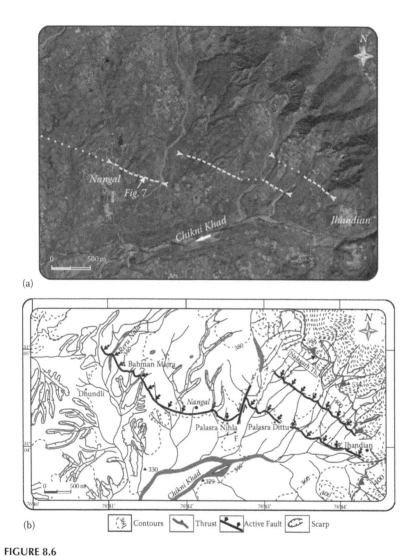

(a)

(b)   Contours   Thrust   Active Fault   Scarp

FIGURE 8.6
(a) IRS-LISS-III + PAN merged image (October 4, 2002) showing the Nangal–Jhandian active fault (Box 1 in Figures 8.4 and 8.5). The arrows show the parallel to subparallel active fault traces. (b) Nangal–Jhandian active fault system (Box 1 in Figures 8.4 and 8.5). The arrows show the upthrown northern block along the low angle thrust. A number of minor faults also occur in the hanging wall. (After Philip, G. and Virdi, N.S., *Curr. Sci.*, 92(4), 532, 2007.)

toward northeast and is marked by a distinct NW-SE scarp observed near Bauni Village at an elevation of approximately 520 m above msl. The scarp within the Quaternary terrace (Qt) is possibly generated by a major seismic event that has an average height of 20 m (Figure 8.9). It extends toward Bari–Batauli to the northwest, where it merges with the upstream terrace

FIGURE 8.7
WNW–ESE trending fault scarp (FS) near Palasra Nihla village with an average height of
15 m. (After Philip, G. and Virdi, N.S., *Curr. Sci.*, 92(4), 532, 2007.)

deposit of Ratta Nala and dies out near the Balad Nadi in the southeast
(Figure 8.8b). The associated landforms in this region also suggest that
a stream (Tatoa Nala), earlier joining the Balad Nadi to the southeast of
Bauni, was captured due to headward erosion by the Ratta Nala, which
has cut a deep gorge upstream of Bhup Nagar and meets the captured
stream at right angle at point X, indicated in Figure 8.8b. The present day
wind gap, east of Bari Batauli (Y), supplements this observation (Mukherji,
1990). An EW-trending ridge between Chhoti Batauli and Dhaulghat is
observed due to movement along the Nalagarh Thrust, where the lower
Tertiary rocks are thrusted over the fan gravels. This ridge rises to over
20 m with steep cliffs on both sides. Tilted and uplifted terraces are also
observed on the northern side along the Ratta Nala.

8.3.2.3 Majotu Active Fault System

Near the village Majotu, along a tributary of the Surajpur Choa, highly
deformed lower Tertiary sandstones ride over the terrace deposits along a
north dipping thrust (Figures 8.10 and 8.11a), which developed parallel to
and is due to the reactivation of the Bursar Thrust (Figure 8.5). The fault
scarp thus produced extends toward SE and is traceable for over 15 km
between Palakhwala and Raru, though the best expressions are between
Majotu and Koti (Figure 8.10). The fault trace is also expressed as a series
of prominent triangular facets (Figure 8.11b) and a number of ponds
aligned parallel to it. An EW trending fault has also displaced the trace
of Majotu fault north of Mandhala. Along a number of stream sections
across the fault zone, pebbles in the terrace deposit are aligned along the
fault and many of them are shattered (Figure 8.11c). Sandy horizon has
also produced sand dykes, which cut through the conglomerate.

The active faults observed in the Pinjaur Dun reflect the intermittent
tectonic impulses due to large magnitude paleoearthquakes, which produced prominent fault scarps. Since the Dun is confined between MBT,
Nalagarh, and the Bursar thrusts in the north and the Surajpur Fault

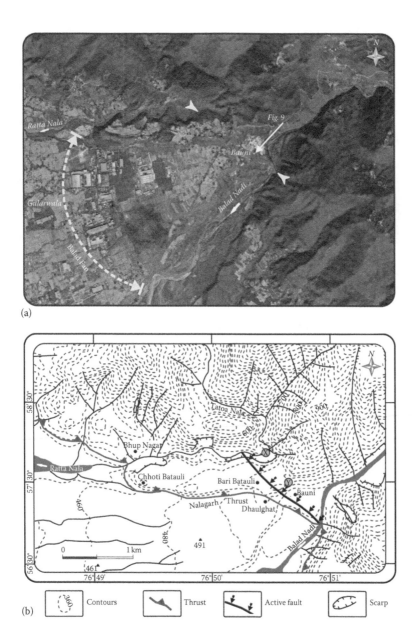

(a)

(b)

FIGURE 8.8
(a) IRS-LISS-III + PAN merged image (October 4, 2002) showing the Bari–Batauli active fault system (Box 2 in Figures 8.4 and 8.5). (b) Bari–Batauli active fault (Box 2 in Figures 8.4 and 8.5) showing a NW-SE trending fault scarp. The arrows indicate movement direction of the upthrown bock. The Tatoa Nala earlier flowing between X and Y was captured by Ratta Nala at X leaving a wind gap at Y. (After Philip, G. and Virdi, N.S., *Curr. Sci.*, 92(4), 532, 2007.)

FIGURE 8.9
Fault scarp of Bari–Batauli fault on the Quaternary fan terrace (Qt) of Balad Nadi. View due south. Arrows indicate NW-SE trace of the fault near the village Bauni. View due south. (After Philip, G. and Virdi, N.S., *Curr. Sci.*, 92(4), 532, 2007.)

and HFT in the south (Figure 8.5), the reactivation of these faults must have created fault scarps, which are almost parallel to them. The above observations further corroborate and support the findings of Malik et al. (2003), concerning a large magnitude earthquake that had struck around AD 1500 in the Frontal Himalaya. The geodetic measurements carried out by Bilham et al. (2001) suggest that the Frontal Himalaya with a recurrence interval of 500 years for great (M8) earthquakes is ready for a major event since no great (M8) earthquake has struck the region in the recorded history.

8.3.3 Kangra Valley: Near the Vicinity of the MBT Zone

Morphostructural analysis using IRS P-6 (Resourcesat) data and selected field investigations helped in delineating prominent traces of active fault systems in Kangra valley (Figure 8.12) in the vicinity of the MBT zone. The two newly recognized active faults in Kangra valley, to be referred to as the *Naddi active fault* (NAF) and the *Kareri active fault* (KAF), traverse through the lower Tertiaries and the pre-Tertiary rocks of the NW Himalaya, respectively. The NAF shows a distinct fault trace, clearly expressed on the Resourcesat (IRS-P6) satellite images and on air photos and is basically a normal fault. It is traceable for over 4 km, striking NW-SE and extending from Naddi village in the northwest to Mcleodganj in the southeast (Figure 8.13a and b). The fault dips steeply due NE and has created a prominent sag pond to the north of the footwall zone. Development of this sag pond, presently known as the Dal Lake

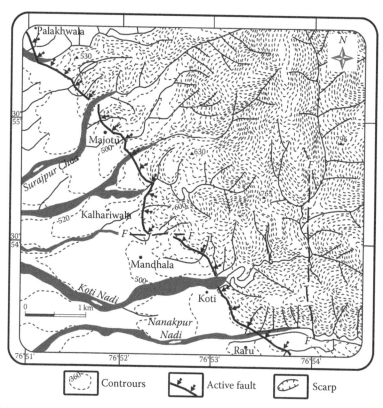

FIGURE 8.10
The Majotu active fault system (Box 3 in Figures 8.4 and 8.5) showing a prominent topographic scarp trending NW-SE direction. The arrow shows the movement of the hanging
wall of low dipping thrust fault, which also coincides with the Nalagarh Thrust in this segment. (After Philip, G. and Virdi, N.S., *Curr. Sci.*, 92(4), 532, 2007.)

(Figure 8.14a), is also partially contributed by subsequent slumping,
where Naddi fault has restricted its lower extent. The fault scarp is relatively steep in the southeastern part and linear to curvilinear along the
fault trace. The northwestern part of the fault scarp beyond Naddi is not
so clear on the satellite images because of the decreasing scarp height. On
the other hand, the southeastern part of the fault scarp is larger and also
characterized by lineaments, ponded alluvium, and drainages. The scarp
observed in this study has a maximum height of 25 m and therefore must
have been produced by multiple tectonic offsets. In particular, the south
side up behavior of the fault is in contrast to the north side up movement
along the MBT and the Chail Thrust.
   The KAF, clearly traceable on Resourcesat (IRS-P6) satellite images
and on aerial photographs, has been marked in the upper catchment of

FIGURE 8.11
(a) Field photograph showing the Majotu (Nalagarh Thrust reactivated) active fault near to the east of Kalhariwala village. Highly crushed lower Tertiary sandstones ride over the Quaternary fan gravels, view due southeast. (b) Triangular facets (TF) along the Majotu fault to the north of Kalhariwala, view due northeast. The flat ground in the foreground represents top surface of the alluvial fan. The hills in the background expose the lower Tertiary sediments. (c) Angular to subangular pebbles in the Quaternary fan conglomerate showing shattering due to southward movement along the footwall of the Majotu fault. (After Philip, G. and Virdi, N.S., *Curr. Sci.*, 92(4), 532, 2007.)

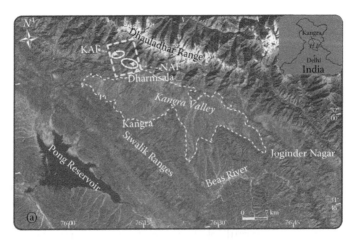

FIGURE 8.12
(a) IRS1C-LISS-III image (false color composite reproduced as grey image) showing the regional setup of the Kangra valley and the adjoining areas. The study area is marked in dotted box and Naddi active fault (NAF) and Kareri active fault (KAF) are shown in ellipses. (After Philip, G., *Int. J. Remote Sens.*, 28(21), 4745, 2007.)

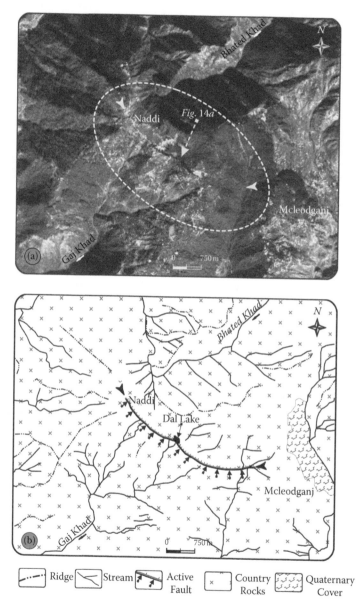

FIGURE 8.13

(a) The Resourcesat (IRS-P6) false color image reproduced as grey image (December 6, 2004) showing the expression of the Naddi active fault (also refer to NAF in Figure 8.12a). (b) The Naddi active fault as interpreted from the satellite data. The arrows indicate the NW-SE trending fault. (After Philip, G., *Int. J. Remote Sens.*, 28(21), 4745, 2007.)

FIGURE 8.14
(a) Panoramic view of the Naddi active fault exposed near Naddi village showing the north dipping normal fault with south side up foot wall (FW). The height of the fault scarp is about 25 m near to the Tibetan School. The prominent sag pond (locally known as Dal Lake) is also seen in the foreground. View due southwest. (b) Kareri active fault exposed near village Kareri showing the north dipping low angle normal fault developed in the Quaternary fan deposit with south side up foot wall (FW). The height of the scarp (FS) is about 4 m. View due southwest. (After Philip, G., *Int. J. Remote Sens.*, 28(21), 4745, 2007.)

Gaj Khad near Kareri village north of the MBT in the pre-Tertiary group of rocks (Figure 8.15a and b). KAF dips steeply to NE, where the fault scarp has been generated in the Quaternary fan deposit and the river terrace. The KAF, which is a subparallel fault to the MBT, is predominantly characterized by vertical offset with locally significant left lateral strike slip of the fan and terrace, which in turn has also laterally shifted few minor streams. The height of the fault scarp, trending nearly in a NW-SE, varies from 1 to 5 m (Figure 8.14b). The scarp is traceable for a distance of over 3 km. As in the case of NAF, the KAF also shows the south side up behavior in contrast to the north side up movement along the MBT.

The two active faults, NAF and KAF, show discrete fault scarps, which are evidently associated with two or more independent large magnitude seismic events in Kangra region. The south side up behavior of the fault scarps of the two active faults attracts more attention, since these indicate extensional conditions and may explain the normal faulting. Furthermore, the NAF and KAF are not only in close proximity of the major structural features of the Himalaya, such as the MBT, but also obliquely trending to this thrust. While considering the present-day observable length of the two fault systems, it is however difficult to infer whether these scarps are the result of a single large magnitude (M > 7) earthquake that occurred in the historical past. At the same time, generation of short length scarps

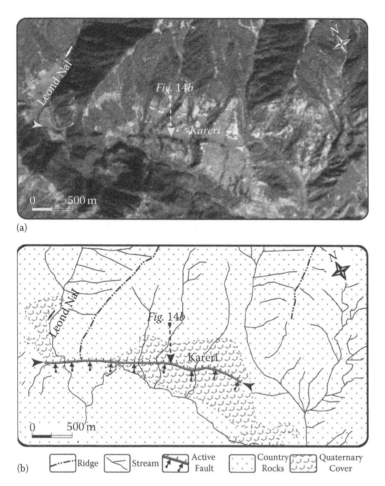

(a)

(b)

| | Ridge | | Stream | | Active Fault | | Country Rocks | | Quaternary Cover |

FIGURE 8.15

(a) The Resourcesat (IRS-P6) false color image reproduced as grey image (December 6, 2004) showing the expression of the Kareri active fault (also refer to KAF in Figure 8.12a). (b) The Kareri active fault as interpreted from the satellite data. The arrows show the NW-SE trending fault. (After Philip, G., *Int. J. Remote Sens.*, 28(21), 4745, 2007.)

due to the secondary effects of the large magnitude earthquakes in the vicinity of the MBT also cannot be ruled out. Since the fault movements are often episodic, the height of a large scarp is usually the aggregate of multiple rupture rather than generated by a single event. In either case the development of fault scarps is clearly indicative of long-term uplift/ deformation in the Holocene and cumulative slips along these faults. This, however, needs further detailed paleoseismic investigation. This study also aims to put new constraints on the rates of movement and geometry of individual faults in the northwestern part of the Kangra valley.

8.4 Conclusion

Landscape and landform mapping are basic to geomorphology. Traditionally, such mapping has been done from topographic maps, aerial photographs, and field surveys. Remotely sensed images have been used extensively in landform mapping and delineating the surface processes since the availability of early Landsat data. The above illustration has demonstrated that synergetic use of remotely sensed data can be useful for investigating the morphology of an active tectonic region. There are numerous such examples that demonstrate the potential application of remote sensing in the field of active tectonics and geomorphology. A terrain like Himalaya, which is quite rugged and more often than not inaccessible, can be better understood using remote sensing techniques while comprehending its geodynamic evolution. The study of tectonic geomorphology and active faults using multidate high-resolution satellite data will therefore fundamentally contribute to the understanding of the geodynamic processes in space and time besides contributing in the delineation of causative active fault zones, which are responsible for the future seismicity.

Acknowledgments

The author is thankful to the Director, Wadia Institute of Himalayan Geology, Dehra Dun, for facilities and permission to publish this chapter. The author is also grateful to Dr. Yuichi Sugiyama, Geological Survey of Japan, and Dr. N. S. Virdi, WIHG, for their constructive comments and suggestions for improvement of the chapter.

References

Ambraseys, N. and Bilham, R. 2000. A note on the Kangra Ms = 7.8 earthquake of 4 April 1905. *Current Science* 79(1): 45–50.

Ambraseys, N. and Douglas, J. 2004. Magnitude calibration of North Indian earthquakes. *Geophysics Journal International* 158: 1–42.

Armijo, R., Meyer, B., and King, G.C.P. 1996. Quaternary evolution of the Corinth Rift and its implications for the late Cenozoic evolution of the Aegean. *Geophysics Journal International* 126(1): 11–53.

Bilham, R., Gaur, V.K., and Molnar, P. 2001. Perspective earthquakes: Himalayan Seismic hazard. *Science* 293(5534):1442–1444.

Bilham, R. 2004. Earthquakes in India and the Himalaya: Tectonics, geodesy and history. *Annals of Geophysics* 47(2): 839–858.

Ganas, A., Pavildes, S., and Karastathis, V. 2005. DEM based morphometry of range front escarpments in Attica, Central Greece and its relation to fault slip rates. *Geomorphology* 65(3–4): 301–319.

Hooper, D.M., Bursik, M.I., and Webb, F.H. 2003. Application of high resolution, interferometric DEMs to geomorphic studies of fault scarps. Fish Lake Valley, Nevada-California, USA. *Remote Sensing Environment* 84(2): 255–267.

Hussain, A., Yeats, R.S., and Mona Lisa. 2009. Geological setting of the 8 October 2005 Kashmir earthquake. *Journal of Seismology* 13: 315–325.

Karakhanian, A., Djrbashian, R., Trifonov, V., Philip, H., Arakelian, S., and Avagian, A. 2002. Holocene-historical volcanism and active faults as natural risk factors for Armenia and adjacent countries. *Journal of Volcanology and Geothermal Research* 113(1–2): 319–344.

Kaya, S., Muftuoglu, O., and Tuysuz, O. 2004. Tracing the geometry of an active fault using remote sensing and digital elevation model: Ganos segment, North Anatolian Fault zone, Turkey. *International Journal of Remote Sensing* 25(19): 3843–3855.

Malik, J.N., Nakata, T., Philip, G., and Virdi, N.S. 2003. Preliminary observations from trench near Chandigarh, NW Himalaya and their bearing on active faulting. *Current Science* 85(12): 1793–1799.

Middlemiss, C.S. 1910. The Kangra earthquake of 4th April 1905. *Memoir Geological Survey of India* 37 (Geological Survey of India reprinted 1981): 409.

Molnar, P. and Pandey, M.R. 1989. Rupture zones of great earthquakes in the Himalayan region. Proceedings of Indian Academic Science. *Earth Planetary Science* 98(1): 61–70.

Mukherji, A.B. 1990. The Chandigarh Dun alluvial fans: An analysis of the process—Form relationship. In: *Alluvial Fans: A Field Approach* (Rachocki, A.H. and Church, M., eds.), John Wiley & Sons Ltd., New York, pp. 131–149.

Nakata, T. 1972. Geomorphic history and crustal movements of the foothills of the Himalayas. Report of Tohoku University Japan, 7th Series (Geography) 22, 39–177.

Nakata, T. 1989. Active faults of the Himalaya of India and Nepal. *Geological Society of America Special Paper* 232: 243–264.

Philip, G. 1995. Active tectonics in Doon valley. *Himalayan Geology* 6(2): 55–61.

Philip, G. 1996. Landsat Thematic Mapper data analysis for Quaternary tectonics in parts of Doon valley, NW Himalaya, India. *International Journal of Remote Sensing* 17(1): 143–153.

Philip, G. 2007. Remote sensing data analysis for mapping active faults in northwestern part of Kangra valley, NW Himalaya, India. *International Journal of Remote Sensing* 28(21): 4745–4761.

Philip, G. and Sah, M.P. 1999. Geomorphic signatures for active tectonics in the Trans-Yamuna segment of the western Doon valley, NW Himalaya. *International Journal of Applied Earth Observation and Geoinformation* 1(1): 54–63.

Philip, G., Sah, M.P., and Virdi, N.S. 2006. Morpho-structural signatures of active tectonics in parts of Kangra valley, NW Himalaya, India. *Himalayan Geology* 27(1): 15–30.

Philip, G. and Virdi, N.S. 2006. Co-existing compressional and extensional regimes along the Himalayan front vis-à-vis active faults near Singhauli, Haryana, India. *Current Science* 90(9): 1267–1271.

Philip, G. and Virdi, N.S. 2007. Active faults and neotectonic activity in the Pinjaur Dun, northwestern Frontal Himalaya, India. *Current Science* 92(4): 532–542.

Raiverman, V., Kunte, S.V., and Mukherjee, A. 1983. Basin geometry, Cenozoic sedimentation and hydrocarbon prospects in North-western Himalaya and Indo-Gangetic plains. *Petroleum Asia Journal* 6: 67–92.

San'kov, V., Deverchere, J., Gaudemer, Y. et al. 2000. Geometry and rate of faulting in the North Baikal Rift, Siberia. *Tectonics* 19(4): 707–722.

Valdiya, K.S. 1986. Neotectonic activities in the Himalayan belt. *Proceedings of the International symposium on Neotectonics in South Asia*, Survey of India, Dehradun, pp. 241–251.

Valdiya, K.S. 1992. The main boundary thrust zone of Himalaya, India. *Annals Tectonicae* 6: 54–84.

Valdiya, K.S. 1993. Uplift and geomorphic rejuvenation of the Himalaya in the Quaternary period. *Current Science* 64: 873–885.

Virdi, N.S. 1979. Status of the Chail Formation vis-à-vis Jutogh-Chail relationship in Himachal Lesser Himalaya. *Himalayan Geology* 9(Pt. I): 111–125.

Yeats, R.S., Nakata, T., Farah, A. et al. 1992. The Himalayan frontal fault system. *Annals Tectonicae* 6 (suppl.): 85–98.

9

Strain Accumulation Studies between Antarctica and India by Geodetically Tying the Two Continents with GPS Measurements

N. Ravi Kumar, E.C. Malaimani, S.V.R.R. Rao, A. Akilan, and K. Abilash

CONTENTS

9.1 Introduction

Very few studies have been conducted on the larger oceanic part of the Indian plate using space geodesy. In order to holistically determine the kinematics of the Indian Ocean Basin between Antarctica and India, the data available are very sparse, and characterization and the delineation of the plate boundaries, especially in the Indian Ocean, are poor. The study focuses on GPS geodesy to improve the understanding of the complex plate motions, diffuse and poorly located plate boundaries, and striking intraplate deformation that characterize the Indian Ocean Basin and also addresses several of these issues, as follows:

- How rigid is the Indian plate and Indian Ocean Basin?
- Does relatively high level of intraplate seismicity on the oceanic part of the Indian plate indicate internal deformation in excess of other plates?
- Is this related to the Indo-Eurasian collision and the uplift of the Himalayas?

All these lead to the new hypothesis that is possible in describing the plate kinematics of the Indian Ocean Basin relates to the unique set of forces on the boundaries. These boundary forces may have led to frequent plate boundary reorganizations in the past, and in the generation of either small plates such as the Capricorn plate or a diffuse boundary zone between India and Australia (Gordon et al., 1998), complicating kinematic interpretations. These boundary forces may also contribute to nonrigid behavior of the Indian plate. In any event, the deformation of equatorial Indian Ocean lithosphere is not ephemeral, but long-lived (Gordon, 1998). The net result of these complexities and plate kinematics of the region means that less attention has been given to these regions when compared to other regions. The issue of plate kinematics is inextricably linked to the question of plate rigidity. Despite the extensive studies on plate tectonics using geophysical investigations in the larger oceanic part of the Indian plate, major issues remain unresolved. Motions across some of the plate boundaries seem well constrained, as implied by good agreement between space geodetic and geologic models (Stein and Sella, 2004). In others, apparent discrepancies exist.

GPS-based site positions or velocity estimates may be affected by a variety of processes, including secular plate motion, elastic, and permanent deformation associated with current magmatic and tectonic event, and ongoing response to past events including rifting episodes and postglacial rebound. In order to use velocity field to investigate the secular rifting signal, the effects of these additional processes must be accurately measured or modeled per each site, and site position estimates and velocities recalculated accordingly (LaFemina et al., 2005). In addition to strain accumulation in the south of the Indian peninsula, major geological and geomorphological processes in the Indian Ocean Basin are believed to be coseismic and postseismic deformation in the Indian Ocean Basin and along the ridges.

9.2 GPS Data Acquisition

The National Geophysical Research Institute (NGRI), Hyderabad, India, has a permanent GPS station at Maitri, Antarctica, and it is operational since 1997. Having become a permanent geodetic marker, it started contributing to the Scientific Committee on Antarctic Research (SCAR) Epoch GPS campaigns since 1998.

9.3 Global Network Stations

Two global networks, namely, IND and ANT, have been chosen to geodetically connect the two continents. The IGS station at Diego Garcia (DGAR) is common to both the networks. For this purpose, 11 year data from 1997 to 2008 were used in this study. The first network ANT includes Maitri (MAIT), Davis (DAV1), Casey (CAS1) in the Antarctica plate, and several stations in the adjacent tectonic plates surrounding Antarctica, including Seychelles (SEY1), COCO, Hartebeesthoek (HARO), Yaragadee (YAR1), and Tidbinbilla (TID2). The other network, IND, includes Hyderabad (HYDE) in the Indian plate, COCO, DGAR, HRAO, IISC, IRKT, KIT3, LHAS, POL2, MALD, SEY1, WTZR, and YAR1. But for simplicity, the stations HYDE, SEY1, DGAR, and COCO are shown in Figure 9.3.

Since the baseline length between HYDE, India, and MAIT, Antarctica, is more than 10,000 km, it is mandatory to form these two different networks to improve the accuracy of the baseline measurements by GPS. This approach is to circumvent the limitation in the estimation of baseline length by GPS, maximum of 6900 km. This is due to the availability of less number of double difference observables for the GPS data analysis. Common visibility of satellites at the two chosen sites beyond 6900 km is minimum. This is the reason for the low number of double difference observables.

Kerguelen (KERG) in Antarctica plate was chosen as a reference station, since it is relatively a rigid plate site according to the plate characteristics propounded by IERS. The choice of this reference station provides a very good geographic coverage with respect to the global geometry of our network (Bouin and Vigny, 2000).

9.4 GPS Data Processing and Analysis

The optimum strategies are used for processing. The data acquired between 1997 and 2008 were processed and analyzed using GPS data processing software Bernese version 5.0. The time series of all the sites were estimated and the error bars are with one standard deviation. Time series of MAIT site coordinates till 2009 is shown in Figure 9.1. The scatter observed in the MAIT time series is due to the change of receiver from Turbo-Rogue to Ash Tech and the weighted root mean square (WRMS) values are within the acceptable limits.

The baselines between Kerguelen and all the other stations have been estimated. Table 9.1 depicts the minimal baseline shortening of all the stations MAIT, DAV1, and CAS1 in Antarctica from KERG, which reveals minimum deformation in this part of Antarctica. The lengths of the baselines of our network are between 2000 and more than 6500 km, which makes the ambiguity fixing impossible.

To eliminate the errors due to nonavailability of double difference observables, we included many stations between southern Indian peninsula and Maitri, which have longer time series of site coordinates in the IGS network so that the short baselines between the stations estimated could be tied geodetically.

Table 9.2 shows the estimated site velocities in north and east directions and their error ellipses. The velocity vector of each site has been estimated with the weighted least squares fit to the weekly solutions and with the 95% confidence error ellipse. Figure 9.2 shows the estimated velocity vector map.

Very long baselines have been estimated using ITRF 2005 from HYDE to other chosen IGS stations in and around India, including DGAR. Initially, the data were processed using ITRF 2000 and then they were updated using ITRF 2005 by including 14 transformation parameters between ITRF 2000 and 2005, published by IERS. The baseline lengths from Kerguelen to all the other stations having DGAR as a common station for both these networks and the velocity vectors of each site were also estimated. The analysis and results show increase of baseline lengths between Kerguelen in Antarctic plate and other stations and shortening of baseline lengths between HYDE in Indian plate and other common stations (Table 9.3).

By this global network analyses, the stations HYDE and MAIT are geodetically tied through DGAR. With this geodetic tie-up, as shown in Figure 9.3, having received the geodetic signatures of the geodynamical processes between India and Antarctica, continuous monitoring has enhanced the understanding of the crustal deformation processes between these two continents despite many plates, microplates, and ridges in this study region. The intercontinental networks in this study and the data analysis suggest that the high rate of movement of DGAR at the edge of Capricorn plate and COCO could be the result of excessive strain accumulation due to the Indo-Australian diffuse plate boundary forces acting upon this region. The results also conform to the genesis that the deformation of equatorial Indian Ocean lithosphere is not ephemeral, but long-lived.

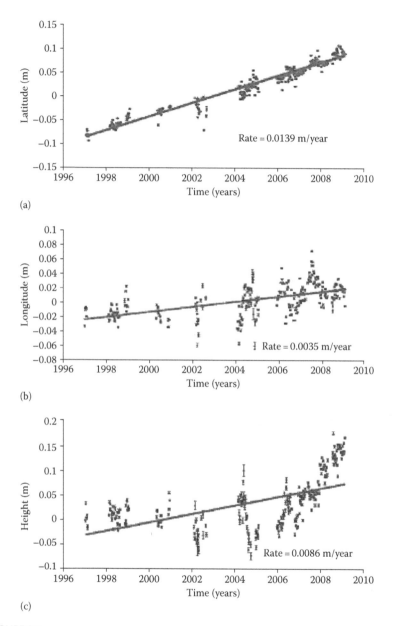

FIGURE 9.1
Time series of MAIT site coordinates: (a) time vs latitude, (b) time vs longitude, and (c) time vs height. Site velocity is given by slope of weighted least squares line fit (solid line) through the data (solid circles). Site velocity error (1 standard error) is based on weighted least squares straight line fit, accounting for white noise approximation, total time span of observation, and total number of observations.

TABLE 9.1

Estimated Baseline Lengths between Kerguelen and Other
Stations, Rate of Changes, and Their Error (1 Standard Error)
Estimates

Sl. No	Stations	Baseline Length from Kerguelen (m)	WRMS (m)	Baseline Length Change/ Year (m/year)
1.	Davis	2,172,481.3485	±0.0003	−0.0019
2.	Casey	2,933,421.2862	±0.0019	−0.0044
3.	Maitri	3,742,927.3490	±0.0006	−0.0059
4.	Yaragadee	4,323,209.3189	±0.0883	+0.0530
5.	Hartebeesthoek	4,391,931.1371	±0.0023	+0.0017
6.	Diego Garcia	4,564,601.5319	±0.0393	+0.0340
7.	Coco	4,677,608. 5252	±0.0039	+0.0553
8.	Seychelles	5,005,623.2524	±0.0270	+0.0049
9.	Tidbinbilla	6,103,757.3181	±0.0777	+0.0378

TABLE 9.2

Estimated Velocities of All the Sites and Their Error
(1 Standard Error) Estimates

SITE	N-VEL (mm/year)	E-VEL (mm/year)	N-ERROR (mm/year)	E-ERROR (mm/year)
MAIT	11.30	2.30	1.009	1.768
CAS1	12.00	7.60	2.222	4.019
DAV1	2.90	6.40	1.495	2.542
YAR1	−50.80	−27.90	5.59	2.49
TID2	−43.50	2.20	9.399	3.472
DGAR	30.28	4.50	1.821	2.267
COCO	40.30	4.80	5.696	3.694
SEY1	3.90	17.80	22.794	11.718
HRAO	−15.80	−23.60	6.087	2.703

9.5 Results and Discussion

The overall analysis is that the estimated baseline lengths between
Kerguelen and Maitri and other IGS stations, namely Casey, Davis in the
Antarctica plate, are indeed shortening at the rates of 5.9, 4.4, and 1.9 mm/
year, respectively. The rate of change in the order of mm in the baseline
lengths indicates no significant change at the one standard deviation
level and may be construed as the stations in the same plate are moving

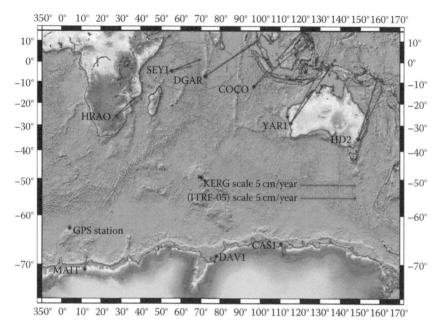

FIGURE 9.2
GPS site velocities with respect to KERG and relative to ITRF 2005 are shown. Error ellipses are two-dimensional, 95% confidence regions assuming white noise error model.

TABLE 9.3

Estimated Result of Geodetic Tie-Up between Indian and Antarctica Plate

Station	Baseline Length (m)	Baseline Change from Kerguelen (m/year)	Baseline Length (m)	Baseline Change from Hyderabad (m/year)
Seychelles	5,005,623.260381	0.0049	3,476,844.689153	−0.0117
Diego Garcia	4,564,601. 532	0.0340	2,790,790.349066	−0.005
Coco	4,677,608.665675	0.0553	3,783,980.525856	−0.0014

together. The increase in WRMS values with the increase in the baseline length confirms the baseline-dependent errors (Table 9.1), according to error relation (Lichten, 1990) $\sigma^2 = A^2 + B^2 * L^2$ with $A = 0.5$ cm and $B = 10^{-8}$.

The velocity vectors of the stations in the Australian plate indicate that they are moving away from the Indian plate conforming to the recent plate tectonic theory. The baseline lengths between Kerguelen and the stations Yaragadee (YAR2) and Tidbinbilla (TID2) in Australian plate are increasing at the rate of 5.3 and 3.8 cm/year, respectively. These estimated

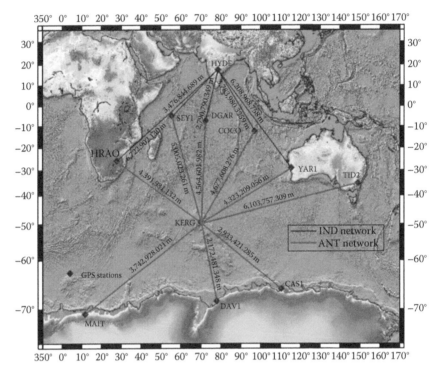

FIGURE 9.3
Map showing the geodetic tie-up of India and Antarctica through DGAR mainly and all the other GPS stations in between India and Antarctica forming two different networks with HYDE and KERG being the reference sites for two plates of India and Antarctica. The estimated baseline lengths from HYDE and KERG to all the other stations are also indicated.

rates of changes in the baseline lengths of YAR2 and TID2, which are original contribution in this study, are in support of the theory that India and Australia lie in two different plates. The velocity vectors of these two stations, which corroborate this theory, are discussed later in this chapter. Table 9.1 reveals that the distance between Kerguelen (KERG) and the islands in the Indian Ocean, Seychelles, and COCO are lengthening at the rates of 4.9 and 5.5 cm/year, respectively. The station Hartebeesthoek (HRAO) in South African plate shows a trend of minimal shortening at the rate of 1.7 mm/year.

The estimation of the velocity vectors of the stations in and around Antarctica does conform to the estimated velocity vectors of SCAR GPS campaigns, which are consistent with rigid plate motion (Bouin and Vigny, 2000). The major results provide new information on the overall direction and rate of Indian plate motion and on some tectonic processes in the intraplate seismicity of the Indian Ocean, including the elastic strain accumulation. The effects of elastic strain are assessed on measured site velocities.

9.5.1 Elastic Strain Accumulation

The strain accumulation using GPS geodesy are computed in our global network using the following formula:

$$\text{Strain accumulation (year}^{-1}) = \frac{\text{change in baseline length } (\Delta L)}{\text{baseline length (L)}}$$

Very significant results are emerging. Normally, in the intraplate region, the strain accumulation is around (Kreemer et al. 2003) 10^{-8} to 10^{-10} year^{-1}. The results of the intraplate strain accumulation within Antarctica plate, covering three sites MAIT, CAS1, and DAV1 are 1.8×10^{-9}, 1.6×10^{-9}, and 1.1×10^{-9} year^{-1}, respectively. Similarly, the estimates of interplate strain accumulation between Antarctica and other plates such as Somalia (SEY1), Africa (HARO), Australia (YAR1), and diffuse plate boundary between India and Australia (COCO) are found to be 1.1×10^{-9}, 1.0×10^{-10}, 1.27×10^{-8}, and 1.18×10^{-8} year^{-1}, respectively (Table 9.4).

The rate of change of COCO makes an interesting observation. The high rate of movement of COCO relative to all the other sites agrees with the global strain rate in the Indian Ocean near COCO, proposed by Gordon et al. (1998) and Kreemer et al. (2003).

The increase in baseline length between KERG and SEY1 at the rate of 5.6 mm/year clearly indicates that SEY1 is moving away from KERG. The velocity vector of SEY1 shows the movement toward the Indian peninsula and this also agrees with the plate model (Gordon and DeMets, 1989). If this rate of movement is verified upon, this may result in the increase in the strain accumulation in the southern Indian peninsula.

TABLE 9.4

Estimated Strain Accumulation of All the Chosen Sites

Site	Strain Accumulation
MAIT	1.8×10^{-9} year^{-1}
CAS 1	1.6×10^{-9} year^{-1}
DAV1	1.1×10^{-9} year^{-1}
SEY1	1.1×10^{-9} year^{-1}
HARO	1.0×10^{-10} year^{-1}
YAR1	1.27×10^{-8} year^{-1}
COCO	1.18×10^{-8} year^{-1}

It is evident from plate kinematic studies that the strain rates are higher up to a factor of 25 in the weak diffuse plate boundaries than in the strong plate interiors, particularly in the region between 75°E and 100°E longitude in the Indian Ocean (Gordon, 1998). This is also corroborated by IERS in their estimation of plate sites characteristics by ITRF 2000 reference frame by declaring that the region between about 75°E and 100°E longitude is a deforming zone.

9.6 Conclusion

High precision space geodetic data from 1997 to 2008 have been analyzed to investigate the tectonic activity, plate boundary organizations, crustal deformation in the southern Indian peninsula, the driving mechanisms, and the response of the Indian Ocean lithosphere.

By the two global networks (IND and ANT) that have been chosen and analyzed, the stations HYDE and MAIT are geodetically tied through DGAR. With this geodetic tie-up having received the geodetic signatures of the geodynamical processes between India and Antarctica, continuous monitoring had enhanced the understanding of the crustal deformation processes between these two continents. GPS data from three sites—MAIT, CAS1, and DAV1—within Antarctica plate result in the accumulated strain of the order of 10^{-9} year^{-1}. Similarly, an order of 10^{-8} to 10^{-10} year^{-1} for accumulated strain is the result between Antarctica and other plates, where the stations SEY1, HARO, YAR1 and the diffuse plate boundary exist. GPS data at COCO suggest the high rate of movement that could be the result of excessive strain accumulation due to the Indo-Australian diffuse plate boundary forms acting upon this region. The GPS analysis confirms the emergence of diffuse plate boundary between India and Australia and relates to the late Miocene Himalayan uplift. The calculated stress field in the west of the Indian peninsula has a roughly N-S directed tensional and E-W oriented compressional character (Bendick and Bilham, 1999; Sella et al. 2002), and the velocity vectors of all other sites throw a significant insight into the plausible causes of the strain accumulation processes in the Indian Ocean and the northward movement of Indian plate.

Acknowledgments

We record due thanks to Dr. H.K. Gupta, former secretary, Department of Ocean Development, Government of India, for launching this scientific program and constant support. We would also like to thank Dr. V.P. Dimri, Director, NGRI, Hyderabad, for permitting to publish this chapter. Thanks are also due to all the IGS stations for contributing the data.

References

Bendick, R. and R. Bilham, 1999. Search for buckling of the southwest Indian coast related to Himalayan collision, *Geological Society of America Special Paper* 328: 313–321.

Bouin, M.F. and C. Vigny, 2000. New constraints on Antarctic plate motion and deformation from GPS data, *Journal of Geophysical Research* 105 (B12): 28,279–28,293.

Gordon, R.G., 1998. The plate tectonic approximation: Plate non-rigidity, diffuse boundaries, and global plate reconstructions, *Annual Review of Earth and Planetary Sciences* 26: 615–642.

Gordon, R.G. and C. DeMets, 1989. Present day motion along the Owen fracture zone and Dalrymple trough in the Arabian Sea, *Journal of Geophysical Research* 94: 5560–5570.

Gordon, R.G., C. DeMets, and J.-Y. Royer, 1998. Evidence of for long-term diffuse deformation of the lithosphere of the equatorial Indian Ocean, *Nature* 395: 370–374.

Kreemer, C., W.E. Holt, and A.J. Haines, 2003. An integrated global model of present-day plate motions and plate boundary deformation, *Geophysical Journal International* 154: 8–34.

LaFemina, P.C., T.H. Dixon, R. Malservisi, T. Árnadóttir, E. Sturkell, F. Sigmundsson, and P. Einarsson, 2005. Geodetic GPS measurements in South Iceland: Strain accumulation and partitioning in a propagating ridge system, *Journal of Geophysical Research* 110 (B11405), doi 1029/2005/JB003675.

Lichten, S.M., 1990. High accuracy global positioning system orbit determination: Progress and prospects, in *Global Positioning System: An overview, IAG Symposium No. 102*, eds. Y. Bock and N. Leppard, pp. 146–164. Springer, New York.

Sella, G.F., T.H. Dixon, and A. Mao, 2002. REVEL: A model for recent plate velocities from space geodesy, *Journal of Geophysical Research* 107 (B4), 10.1029/2000JB000033.

Stein, S. and G. Sella, 2004. Investigation of Indian Ocean intraplate deformation and plate boundary evolution using GPS, NSF proposal number 0440346, (2004). pp. 6–12.

10

Indian Ocean Basin Deformation Studies by Episodic GPS Campaigns in the Islands Surrounding India

E.C. Malaimani, N. Ravi Kumar, A. Akilan, and K. Abilash

CONTENTS

10.1 Introduction

GPS geodesy studies were initiated to improve the understanding of the complex plate motions, diffuse and poorly located plate boundaries, and striking intraplate deformation that characterize the Indian Ocean basin. Although much has been learned using marine geophysical and seismological data and described by geologic plate motion models based on magnetic anomalies, transform azimuths, and earthquake slip vectors, the kinematics are poorly characterized compared to simpler regions, which inhibits understanding of the regional dynamics. The geographic distribution of the rigid Indian plate comprises the 2000 km long chain of islands atop the Chagos-Lakshadweep ridge. These include Lakshadweep islands, part of India, Maldives islands, and the islands that are located on the Indian plate proper (the southernmost part of the chain includes Diego Garcia, which may lie on the Capricorn plate). This unique geographic feature lies on the oceanic lithosphere, in the stable interior of the Indian plate. In effect, it constitutes a 1200 km long "strain gauge," optimally oriented in the N-S direction, which is capable of measuring possible deformation of this oceanic part of the Indian plate, for comparison with deformation of its continental part.

It is anticipated that significant advances in knowledge of the kinematics of the complex plate interactions and intraplate deformation in the Indian Ocean basin, which is the type example showing that oceanic, as well as continental, plates can deviate dramatically from the ideal rigid behavior assumed in classic plate tectonics. The Indian Ocean basin, roughly speaking the area surrounding the triple junction where the Central Indian (CIR), Southwest Indian (SWIR), and Southeast Indian ridges (SEIR) meet (Figure 10.1), strikingly illustrates the complexities of applying ideal rigid plate tectonics to oceanic plates.

The Central Indian basin thus plays a major role in evolving ideas about when to regard a region as a plate and how to characterize its boundaries. Plate tectonic ideas have evolved from viewing plates as rigid and divided by narrow boundaries to accepting that plates can deform internally and be separated by broad boundary zones (Gordon and Stein, 1992; Gordon, 1998; Gordon et al., 1998; Stein and Sella, 2002). Such zones are common on land, but rarer in the oceans, making the Indian Ocean of

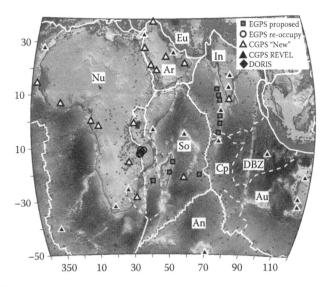

FIGURE 10.1
Proposed and existing geodetic site locations for the Indian Ocean basin, plate boundaries. White lines are plate boundaries, dashed where diffuse or uncertain. EGPS sites (circles) shown across the rift and on Madagascar will be reoccupied. Squares denote new island EGPS sites. Triangles are existing CGPS sites; solid were used in REVEL model and open have become available since 2000. Plate abbreviations An, Antarctica; Ar, Arabia; Au, Australia; Cp, Capricorn; Eu, Eurasia; In, India; Nu, Nubia; So, Somalia.

special interest. Because the concepts of plates and plate boundaries are kinematic, kinematic data provide rigorous means of examining them. Hence, India, Australia, and Capricorn are viewed as distinct plates and the earthquakes between them as plate boundary earthquakes, rather than as intraplate deformation within a single Indo-Australian plate, because three rigid plates fit the rates and directions of motion recorded by magnetic anomalies and transform fault azimuths better than would be expected purely by chance due to the additional free parameters (Stein and Gordon, 1984).

10.2 GPS Data Acquisition and Analysis

The Lakshadweep island map and the chosen sites are given in Figure 10.2. Continuous GPS + GLONASS measurements were carried out in Kavaratti, Chetlat, and Minicoy islands. These state-of-the-art GNSS receivers could track 30 GPS and 11 GLONASS satellites with 5° elevation mask. The data were processed in a global network solution using Bernese software version 5.0. The strategy used in the data processing and analysis is given in Table 10.1. The acquired data were processed in ITRF 2005 Reference Frame.

FIGURE 10.2
The map showing Lakshadweep islands and the sites chosen for episodic GPS campaigns. The triangle indicates the sites Kavaratti, Chetlat, and Minicoy chosen for GPS campaigns.

TABLE 10.1

Strategies Adopted for GPS Data Processing

Parameters	Description
GPS processing software	Bernese version 5.0
Sessions	24 h
Earth orientation parameters	IERS bulletin B
Elevation angle cut-off	5°
Ambiguity resolution	Free-not resolved
Tropospheric dry delay model	Saastamoinen
Tropospheric wet delay	Estimated every 4 h
A priori station positions	ITRF 2005
Station position constraints	Free network approach

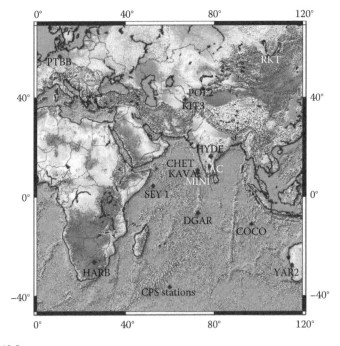

FIGURE 10.3
The global network of stations chosen to estimate the site coordinates and the velocity vectors.

The global network of stations chosen to estimate the site coordinates and the velocity vectors is shown in Figure 10.3.

It is expected that when all the proposed sites for episodic GPS (EGPS) are reoccupied and velocity vectors are determined, it would yield very significant results.

It is suggested to extend this study using continuous GPS (CGPS) data, including new sites that have become available since 2000, and episodic (survey, campaign, or EGPS) measurements can be made. Some of the EGPS sites have been already established and others are planned to be established. As shown in Figure 10.1, CGPS sites resolve the horizontal velocities used for plate motions to about 1 mm/year in 5 years. EGPS data are less precise due to errors in setting up equipment and because they are occupied only for a few days every few years. It is found that the former can be reduced using fixed antenna heights, making the resulting velocities only about two times less precise, because the uncertainty is dominated by the time series length. Because EGPS is cheaper, it is a cost-effective way of addressing tectonic problems. These results are combined with those from marine geophysical and seismological studies, which have the complementary advantages of not being restricted to island sites. In addition, geologic data have the ability to sample over time. It will be assessed how well the earlier models fit the GPS data, explore possible temporal changes in plate motion, and compare differing estimates of plate rigidity and intraplate deformation. Further referring to Figure 10.1, it will be augmented by others such as the ones in Australia and Antarctica. The sites are divided into several groups, such as the following:

- Existing CGPS sites of the International GPS Service (IGS) network, which provide publicly available data. Such data were used by Sella et al. (2002) in REVEL model and comparable studies by others. Site velocities from that study will be better constrained, owing to the longer time series (REVEL used data from 1993 to 2000, and precision increases roughly as 1/length of data). However, additional sites will give better estimates of plate motion.

- Three CGPS sites operated by NGRI at Hyderabad and Mahendragiri, India, and Maitri, Antarctica.

- About 11 new island EGPS sites: Europa and Grande Comore in the Mozambique Channel, Mauritius and Rodriguez in the Central Indian Ocean, Socotra in the Gulf of Aden, and six in the Laccadive-Maldives islands. These will give significantly better coverage of the Somali and Indian plates, which improve Euler vector estimates. In addition, the Laccadive-Maldives sites span about 11° of latitude and provide unique sampling of the Indian plate, permitting "strain gauge" measurements of possible deformation.

- Nine reoccupied EGPS sites: Seven in Tanzania/Malawi spanning the East African rift and two in Madagascar, established by NGRI. The new EGPS sites have been chosen both in terms of scientific objectives and operational considerations, including ease

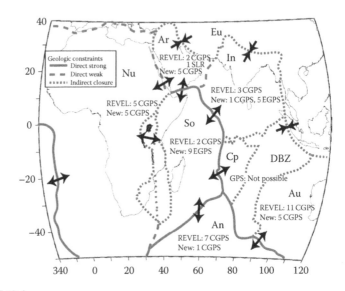

FIGURE 10.4
Boundaries in study area showing relative knowledge of plate motion from geologic models. Our study, using GPS data beyond those used in REVEL model as shown, together with longer time series at REVEL sites will improve Euler vector estimates for all plates listed except Capricorn, which does not have islands. See Figure 10.1 for abbreviations.

and cost of access. Despite the extensive studies to date, major issues remain unresolved. As illustrated in Figure 10.4, motions across some of the plate boundaries seem well-constrained, as implied by good agreement between space geodetic and geologic models. In others, apparent discrepancies exist.

10.3 Results and Discussion

The estimated site coordinates of Kavaratti, Chetlet, and Minicoy and the baseline lengths between Hyderabad and all these islands are shown below (Malaimani et al., 2009).

Estimated site coordinates of Kavaratti

Latitude 10° 33′ 22.2200″ ± 0.0199

Longitude 72° 37′ 54.1650″ ± 0.0174

$Height_{(ellip)} = -82.0332 \pm 0.0199\,m$

Orthometric height = 11.567 m (height above msl)

Estimated site coordinates of Chetlat

Latitude 11° 41' 23.6471" ± 0.0207

Longitude 72° 42' 38.3379" ± 0.0228

$Height_{(ellip)} = -83.3895 ± 0.0654\,m$

Orthometric height = 8.8880 m (height above msl)

Estimated coordinates of Minicoy

Lat = 08° 17' 02.536284" ± 0.0115

Lon = 73° 03' 23.986176" ± 0.0117

$Height_{(ellip)} = -83.6667 ± 0.0151$

Orthometric height = 13.155 m (height above msl)

Estimated baseline length in ITRF 2005 in the global network solution between

Hyderabad and Kavaratti is 991,303.3067 ± 0.0082 m

Hyderabad and Chetlat is 892,216.5594 ± 0.0040 m

Hyderabad and Minicoy is 1171,071.8777 ± 0.0065 m

The estimated velocity vectors in ITRF-2005 Reference Frame are as shown in Figure 10.5. Table 10.2 shows the estimated site velocities.

10.3.1 Indian Plate Motion

It is anticipated to improve estimates for Indian plate motion that are due to both longer time series at sites used by Sella et al. (2002), the new NGRI site, and the EGPS sites. Improving the velocity estimate is important, because previous GPS studies yield estimates of Indian-Eurasian motion 14%–25% slower than predicted by NUVEL-1A (Holt et al., 2000; Shen et al., 2000; Paul et al., 2001; Sella et al., 2002). At this level it is not clear whether it reflects systematic errors in NUVEL-1A (Gordon et al., 1999), true deceleration, or a combination. The issue is important because plate motions have been evolving as a result of India–Eurasia convergence.

10.3.2 Indian Plate Rigidity

The Indian plate deviates significantly from an ideal rigid plate, even given that much of the oceanic seismicity and deformation near the equator is now recognized to be part of the plate's diffuse southern boundary. Its continental boundaries to the north and west are also broad deformation zones. In addition, earthquakes occur within peninsular India, south of 15°N (Bilham and Gaur, 2000) and the adjacent oceanic lithosphere north

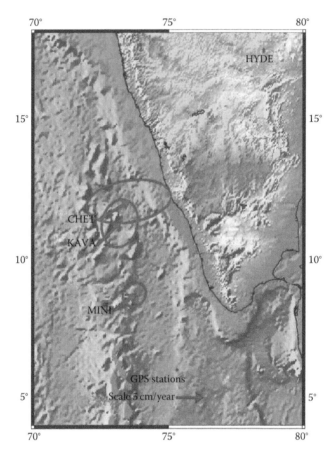

FIGURE 10.5
The estimated velocity vectors in ITRF-2005 Reference Frame with error ellipses.

TABLE 10.2

Estimated Site Velocities with Their Corresponding Sigma Values

Site	N-VEL (Month/Year)	E-VEL (Month/Year)	N-ERROR (m)	E-ERROR (m)
Kavaratti	0.0404	0.0432	±0.0465	±0.0358
Chetlat	0.0353	0.0587	±0.0421	±0.0841
Minicoy	0.0286	0.0375	±0.0232	±0.0230

of the diffuse boundary zone, and continental crust along the west coast appears to be flexed (Subrahmanya, 1996; Bendick and Bilham, 1999), as shown in Figure 10.6. The earthquakes, such as the 1967 Koyna (Ms 6.3) and 1993 Latur (Ms 6.4) events, can be very destructive (the latter caused over 11,000 fatalities). Although the area seems geodetically relatively

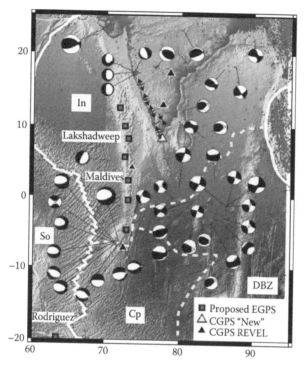

FIGURE 10.6

Site locations for GPS study of deformation in oceanic part of Indian plate illustrated by seismicity (cluster of dark dots) and focal mechanism data. Anticlines and synclines show N-S compression on land (Bilham and Bendick, 1999). Squares denote new island EGPS sites that form a "strain gauge." Triangles are the existing CGPS sites; solid were used in REVEL model and open have become available since 2000. White continuous lines are plate boundaries, dashed ones are uncertain plate boundaries. Cp, Capricorn; DBZ, diffuse boundary zone; In, India; So, Somalia.

rigid, the deformation may be significant (Malaimani et al., 2000). Paul et al. (2001) estimate rigidity at the level of 7×10^{-9} 1/year permits 4 mm/year over a 500 km baseline, enough for large earthquakes every few hundred years. These earthquakes may be due to internal deformation of the Indian plate, since they are well south of the seismic zone, where the 2001 Bhuj (Mw 7.7) earthquake occurred, which may be part of the diffuse India–Eurasia boundary (Li et al., 2002; Stein et al., 2002).

10.4 Conclusion

The episodic GPS measurements at Lakshadweep islands and future reoccupation of these sites would lead to understand the inferred continental flexure in the west and south of India. These studies would also help investigate the nonrigidity of this oceanic portion of the plate. The west of southern India constitute a 1200-km-long "strain gauge" optimally oriented almost parallel to compression seen on land, perhaps due to the effects from the Himalayan collision, or extension of the Capricorn–India diffuse boundary that extends this far north (Henstock and Minshull, 2004). These data can be combined with data from sites in continental India to compare the rigidity of the continental and oceanic regions. Thus, the use of GNSS geodesy would improve the understanding of the complex plate motions, diffuse and poorly located plate boundaries, and striking intraplate deformation that characterize the Indian Ocean basin. Although much has been learned using marine geophysical and seismological data and described by geologic plate motion models based on magnetic anomalies, transform azimuths, and earthquake slip vectors, the kinematics are poorly characterized compared to simpler regions, which inhibits understanding of the regional dynamics. This study would resolve many of the above issues.

Acknowledgments

We record due thanks to Dr. V.P. Dimri, Director, NGRI, Hyderabad, for his constant support and according permission to publish this chapter. We would also like to thank the National Centre for Antarctica and Ocean Research (NCAOR), Goa, Department of Ocean Development under Ministry of Earth Science, Govt. of India, for funding this project. Thanks are also due to all the IGS stations for contributing data. The authors are also grateful to Seth A. Stein and Giovanni F. Sella for the NSF proposal for Collaborative Research: Investigation of Indian Ocean intraplate deformation and plate boundary evolution using GPS (NSF PROPOSAL NUMBER 0440346) from where most of the material for this chapter is obtained.

References

Bendick, R. and Bilham, R., 1999. Search for buckling of the southwest Indian coast related to Himalayan collision, In *Himalaya and Tibet: Mountain Roots to Mountain Tops*, eds. A. Mcfarlane and R. B. Sorkhabi, *Journal of Quade* 313–321. Geological Society of America Special paper 328.
Bilham, R. and Gaur, V., 2000. Geodetic contributions to the study of seismotectonics in India. *Current Science* 79: 1259–1269.

Gordon, R., 1998. The plate tectonic approximation: Plate non-rigidity, diffuse plate boundaries, and global reconstructions. *Annual Review of Earth and Planetary Science*, 26: 615–642.

Gordon, R. G., Argus, D. F., and Heflin, M. G., 1999. Revised estimate of the angular velocity of India relative to Eurasia, Eos Trans. *American Geophysical Union*, 80: F273.

Gordon, R. G., DeMets C., and Royer J.-Y., 1998. Evidence for long-term diffuse deformation of the lithosphere of the equatorial Indian Ocean. *Nature*, 395: 370–372.

Gordon, R. and Stein, S., 1992. Global tectonics and space geodesy. *Science*, 256: 333–342.

Henstock, T. J. and Minshull, T. A., 2004. Localized rifting at Chagos Bank in the India–Capricorn plate boundary zone. *Geology*, 32: 237–240.

Holt, W., Chamot-Rooke, N., Le Pichon, X., Haines, A. J., Shen-Tu, B., and Ren, J., 2000. The velocity field in Asia inferred from Quaternary fault slip rates and Global Positioning System observations. *Journal of Geophysical Research*, 105: 19,185–19,209.

Li, Q., Liu, M., and Yang, Y., 2002. The 01/26/2001 Bhuj earthquake: Intraplate or interplate? In *Plate Boundary Zones*, eds. S. Stein and J. Freymueller, American Geophysical Union, Washington, DC.

Malaimani, E. C., Cambell, J., Gorres, B., Kotthoff, H., and Smaritschnik, S., 2000. Indian plate kinematics studies by GPS-geodesy. *Earth, Planets, Space*, 52: 741–745.

Malaimani, E. C., Ravi Kumar, N., Akilan, A., and Abilash, K., 2009. Episodic GPS campaigns at Lakshadweep islands along the Chagos-Laccadive ridge to investigate the inferred continental flexure in the west of India and the nonrigidity of the oceanic part of the Indian Plate. *Indian Geophysical Union*, 13(1): 1–7.

Paul, J., Burgmann, R., Gaur, V. K. et al., 2001. The motion and active deformation of India. *Geophysical Research Letters*, 28: 647–650.

Sella, G. F., Dixon, T. H., and Mao, A., 2002. REVEL: A model for recent plate velocities from space geodesy. *Journal of Geophysical Research*, 107: 10.1029/2000JB000033.

Shen, Z.-K., Zhao, C., Yin, A. et al., 2000. Contemporary crustal deformation in east Asia constrained by Global Positioning measurements. *Journal of Geophysical Research*, 105: 5721–5734.

Stein, S. and Gordon, R. G., 1984. Statistical tests of additional plate boundaries from plate motion inversions. *Earth Planetary Science Letters*, 69: 401–412.

Stein, S. and Sella, G. F., 2002. Plate boundary zones: Concept and approaches, In *Plate Boundary Zones*, Geodynamics Series 30, ed. S. Stein, *Journal of Freymueller* 1–26, AGU, Washington, DC.

Stein, S., Sella, G. F., and Okal, E. A., 2002. The January 26, 2001 Bhuj earthquake and the diffuse western boundary of the Indian plate, In *Plate Boundary Zones*, Geodynamics Series 30, eds. S. Stein and J. Freymueller, pp. 243–254, American Geophysical Union, Washington, DC.

Subrahmanya, K. R., 1996. Active intraplate deformation in south India. *Tectonophysics*, 262: 231–241.

11

Remote Sensing and GIS in Groundwater Evaluation in Hilly Terrain of Jammu and Kashmir

G.S. Reddy, S.K. Subramanian, and P.K. Srivastava

CONTENTS

11.1 Introduction

Groundwater is one of the components of the water circulatory system of the hydrogeological cycle. It comes into existence with the process of infiltration at the surface. Geomorphology, in terms of relief variations, plays an important role in understanding the surface runoff (SR) and infiltration. It also influences the groundwater flow and quality to a certain extent. Then, it percolates into the ground, consisting of different rock formations having different hydrogeological properties (Rao et al., 2001; Sankar, 2002; Bahuguna et al., 2003). They are the primary hosts for the storage of groundwater. The storage capacity depends on the porosity of the rock. In the rock formation, the water moves from areas of recharge to areas of discharge under the influence of hydraulic gradients, depending on the permeability. Therefore, at a given location, the occurrence of groundwater depends not only on the storage capacity and the rate of transmission but also on the landform. The identification of some geomorphic features in fluvial environment, such as palaeochannels, alluvial plains have direct bearing on groundwater prospects. The groundwater reacts to the force of gravity and begins to flow in the direction of the steepest inclination of the water table. This direction is the hydraulic gradient, which is analogous to topographic or stream gradients, except that it refers explicitly to the water table surface.

As water in the groundwater system drains away from high relief areas, the water table surface will become a subdued replica of the topographic surface. If the region undergoes a long period of drought, the water table will eventually become flat. The groundwater moving down the hydraulic gradient eventually encounters a location where the water table is elevated above the land surface. These locations are termed as ponds, swamps, lakes, and streams. In the hilly terrain, occupying a sizeable part of India, where most of the water goes as runoff and recharge is minimal, delineation of groundwater potential zones is a highly difficult task. However, the study and analysis of geomorphic features using remote sensing techniques lead to understanding of the groundwater regime to a greater extent with acceptable accuracy. In this chapter, the efficacy of remote sensing data in studying and analyzing landform characteristic features in hilly terrains and its implication in evaluating the groundwater condition have been attempted. A part of hilly terrain from Jammu and Kashmir, India, has been taken as type area for carrying out the detailed study.

11.2 Advantage of Satellite Data in Groundwater Study

The satellite images acquired using remote sensing technology provide diagnostic signatures on the parameters, i.e., rock types, landforms, geological structures, and recharge conditions, which control the occurrence and distribution of groundwater. Using these signatures, all these parameters can be studied and mapped accurately avoiding a detailed field survey, which is time-consuming and expensive. Particularly in hilly terrains, where the area is inaccessible, the satellite data are highly useful (Mather, 1987; Lillesand and Kiefer, 2004). The study of groundwater demands a systematic inventory of all the details of the parameters for drawing meaningful conclusions. It is difficult to bring out all the details pertaining to the parameters by fieldwork, using conventional methods.

Further, the occurrence and distribution of groundwater at a given location cannot be assessed just based on the consideration of parameters and have to be considered in their totality. The synoptic view of the image, derived due to large area coverage, provides information about the parameters on a regional scale thereby facilitating the understanding of the groundwater regime as a whole. The groundwater regime is a dynamic system and, for the proper understanding of the system, the role of each parameter in forming the aquifer and the degree of its influence on the groundwater prospects need to be evaluated. The satellite data provide information about all the parameters in an integrated form, so that the role and influence of each parameter can be studied with respect to the other parameters, and thereby a better evaluation of groundwater conditions can be made. The three-dimensional view provided by the satellite data facilitates to study the terrain conditions and geomorphology more effectively.

11.3 Study Area

The study area is located between north latitude 33° 12′ 20″–33° 12′ 30″ and the east longitude 74° 26′ 40″–74° 26′ 50″. The elevation varies from 670 to 1200 m above the mean sea level. It is covered in the Survey of India topomap no. 43 K/8 (Figure 11.1). Regionally, the study area forms part of major Himalayan mountain belt. Locally, it is a severely dissected terrain comprising hills, steep slopes, and narrow-deep valleys formed by the weathering and erosion of the Tawi river system flowing in the west and the Nihari Tawi river systems flowing in the east. The annual rainfall in the study area is about 1700 mm.

11.4 Data Used

High-resolution Indian remote sensing (IRS) satellite—P6 LISS-IV MX (5.8 m resolution) data of January 8, 2005, were used for inventory of landforms and also other parameters, namely, rock types, geological structures, and recharge conditions. IRS-CARTOSAT-1 data were used for

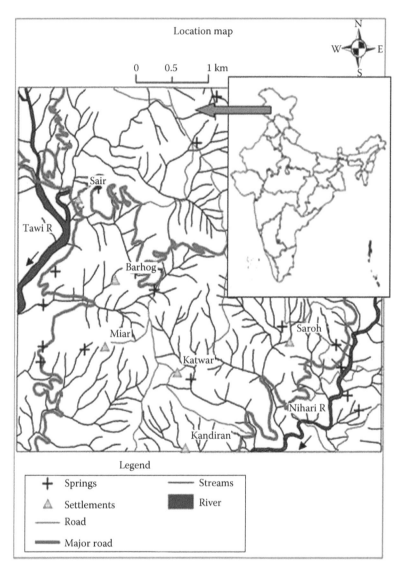

FIGURE 11.1
Location map of the part of Tawi river system, Jammu and Kashmir.

generating digital terrain model (DTM). Contour data available from topomaps on 1:25,000 scale and altitude data obtained from GPS survey have been used for correcting DTM data (Jenson and Domingue, 1988; Mark, 1988). Ground truth information on water table and springs collected during pre-monsoon and post-monsoon periods have been used for validating the results.

11.5 Methodology

Photointerpretation keys like tone, texture, shape, pattern, size, and terrain elements aspects have been used as criteria for interpretation (RGNDWM, 2008). The water available for the recharge is estimated from the analysis of monthly rainfall data. Digital elevation model (DEM) for the study area has been generated using CARTOSAT-1 data. The DEM has been generated from GPS survey and contour data of 1:25,000 scale toposheet. Considering the spring locations collected from the field and their altitudes from DEM, the virtual water table (Ghoneim and El-Baz, 2007) is created for monsoon season. Recharge from the available source is estimated for the given terrain using SCS curve matching technique from daily rainfall data. The discharges from the springs were taken into account while estimating the baseline water table and also the perched water table condition. All the data have been subjected to modeling (Solomon and Quiel, 2006), using empirical formula to derive the ground-water regime in the hilly terrain.

11.6 Analyses of Factors Controlling Groundwater

11.6.1 Geomorphology/Landforms

In order to understand the topographic variations and to bring out the geomorphic units clearly, the satellite data have been integrated with the contour using image processing techniques and a DTM for the study area (Figure 11.2).

The DEM reveals that the terrain exhibits a steplike topography inclined toward southeast. It is developed due to differential weathering and erosion of the hard and compact sandstone beds and the soft and impervious shale/mudstone, which dip in the southeast direction. Narrow and deep valleys are formed on different steps at different levels due to headward erosion along the fractures and dissected the steplike terrain resulting in the evolution of high relief structural hills capped with hogback ridges surrounded by steep slopes. The attitude of the beds and the fracture network are the controlling factors in the evolution of the present-day topography.

The study of the DEM indicates that the relief varies from 700 to 1200 m above the mean sea level in an area of about 25 km². There is a drop of 500 m elevation within such a small area. Based on the slope and altitude, the major constituents of hilly terrain, the study area is demarcated into four geomorphic units (Figure 11.3), namely, river/valley bottom,

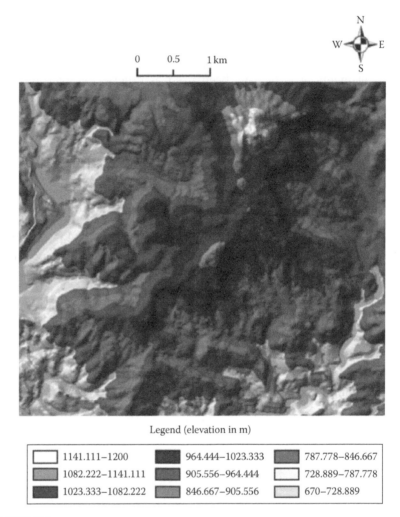

FIGURE 11.2

Digital elevation model of the study area.

plateau/hill tops, steep slope around plateau/hill tops, and escarpments around plateau/hill tops.

11.6.2 Rock Types

The area is occupied by the sedimentary rocks of Precambrian to Oligocene age. Upper part consists of sandstone and mudstone sequence of Upper Eocene–Oligocene, middle part consists of shale and siltstone of Late Paleocene–Middle Eocene, and lower part consist of limestone of Precambrian age (Figure 11.4).

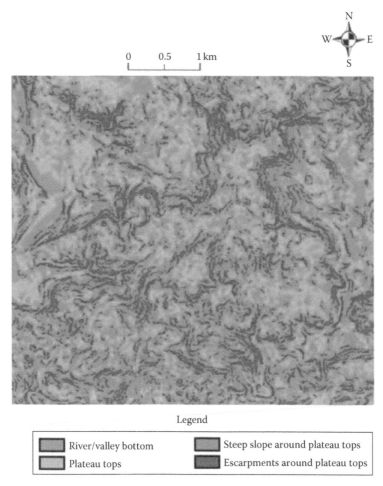

FIGURE 11.3
Geomorphology/landform map.

The rocks occur as a thick pile of beds dipping toward southeast with an angle of 10°–15°. The sandstone and the mudstone sequence belong to the lower Murree group and occupy mostly the elevated part of the terrain. The sandstone is fine grained, compact, and brittle. Closely spaced fractures are noticed in the sandstone beds. Though the sandstone is fine textured, the fracture network and the bedding planes make the litho unit porous and permeable. The mudstone is soft and clayey and it is impervious and the infiltration capacity is very poor. It acts as a barrier for the vertical movement of groundwater. Due to collapsing nature, it poses a serious problem in construction of roads along slopes of the hills in this region. The shale and siltstone, belonging to the Subathu group, occur

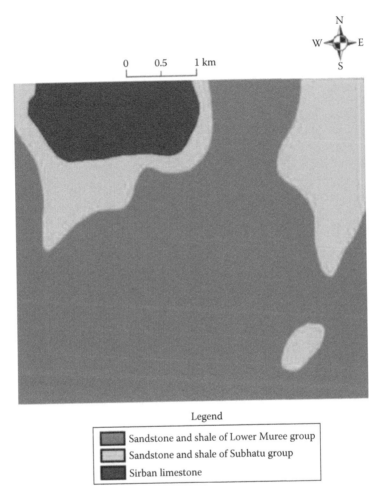

FIGURE 11.4
Lithology map of part of Tawi river basin, Jammu and Kashmir.

below the Murree group of rocks and are exposed along the escarpments and steep slopes, occurring on the way to Kalakot and Matka villages. The porosity and permeability of these rocks are negligible. At the bottom, the area is occupied by Sirban limestone. The detailed stratigraphy of the study area is given in Table 11.1.

11.6.3 Geological Structures

The area forms a part of an anticlinal and synclinal fold belt. A small habitation, on the highest elevation, is located on the axis of an anticline

TABLE 11.1

Stratigraphic Sequence of the Study Area

Top	Sandstone and mudstone sequence	Lower Murree group (Upper Eocene–Oligocene)
Middle	Shale and siltstone Shale with coal seams	Subathu group Late Paleocene–Middle Eocene
Bottom	Limestone	Sirban group (Precambrian)

running in east–west direction. The fold acts as a guiding factor in the movement of groundwater in the area. The surface and subsurface water moves away from the axis of the fold. The area is crisscrossed by two sets of fracture systems. One set trends in east–west direction and cuts the beds obliquely. The second set of fractures is perpendicular to the first and trends in NNW-SSE direction (Figure 11.5). The east–west trending fracture system is dominant in the area and the major stream courses flow either along the bedding planes of the rock formation or along the fracture zones. The fractures facilitated in the development of secondary porosity and permeability in the harder components of the rock formations thereby becoming the locales for the groundwater. They act as conduits for the movement of the groundwater.

11.6.4 Recharge Condition

Rainfall is the main source of recharge. The average annual rainfall is about 1700 mm. There are no other sources of recharge-like water bodies and perennial stream courses. The recharge from the rainfall is estimated using water balance approach based on the following equation (Kalinski et al., 1993; Pradeep et al., 2009):

$$Q = P - ET_a - SR \tag{11.1}$$

where
  Q is the net recharge
  P is the precipitation
  ET_a is the actual evapotranspiration
  SR is the surface runoff

In the first step, the unknown parameters, i.e., SR and ETa are calculated.
  The SR is calculated as per the modified equation for Indian condition (*Handbook of Hydrology*, 1972)

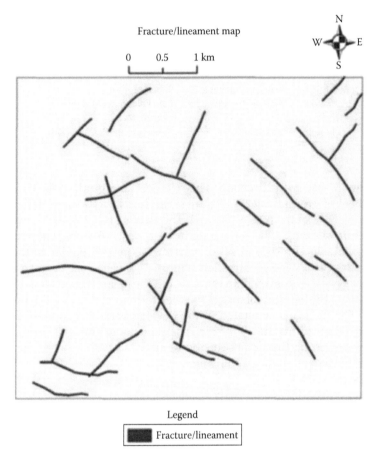

FIGURE 11.5
Geological structure map of part of Tawi river system, Jammu and Kashmir.

$$SR = \frac{(P - 0.35S)^2}{(P + 0.7S)} \quad \text{(if } P - 0.35S \text{ is negative, } SR = 0) \qquad (11.2)$$

where
    P is the precipitation
    S is surface potential retention
    S is calculated as

$$CN = \frac{25{,}400}{(254 + S)}$$

where CN is curve number.

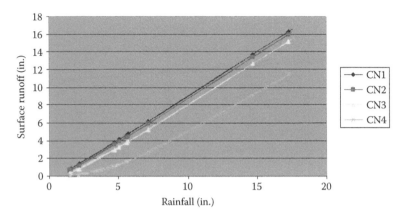

FIGURE 11.6
Rainfall vs. surface runoff in different geomorphic units.

The CN is calculated for each recharge zone based on the land cover category, land use practices, soil type, and hydrologic condition. The study area is classified into four categories as CN1, CN2, CN3, and CN4.

Accordingly, the SR has been estimated for each CN, i.e., in each geomorphic unit (Figure 11.6).

The ET_a is obtained from potential evapotranspiration (ET_p) and crop water stress index (CWSI). The CWSI for the area is found to be 0.35 using NDVI, surface temperature, and climatic data. The ET_a can be calculated from CWSI and ET_p (recorded at meteorological observatories) as follows:

$$CWSI = 1 - \left(\frac{ET_a}{ET_p} \right) \qquad (11.3)$$

In the second step, the net recharge is calculated from Equation 11.1. Precipitation of the 5 year average has been taken for calculation for net recharge in the study area. There are four recharge zones corresponding to four geomorphic units. The recharge in different geomorphic unit is escarpment = 89 mm; steep slope around plateau/hill tops = 153 mm; plateau/hill tops = 305; river courses/valley bottoms = 808 mm per unit area (Figure 11.7).

11.7 Ranking and Weightage Assignment to Parameters

A conceptual model has been developed for the study area keeping in view of the terrain condition. It is assumed that irrespective of the

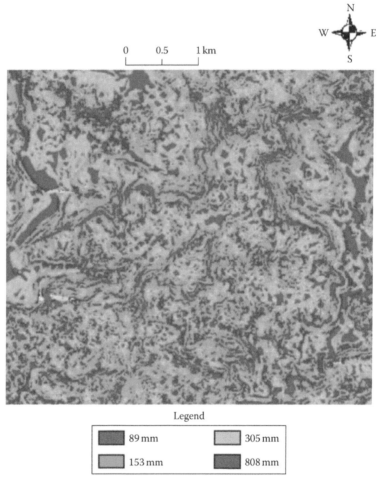

FIGURE 11.7
Recharge condition in four different landforms of Tawi river system, Jammu and Kashmir.

hydrogeological character of the individual parameter (Murthy and Mamo, 2009), the landform plays a dominant role in the occurrence of groundwater in hilly terrain. Accordingly, the landform is given highest weightage, followed by recharge condition, geological structure, and rock type.

A well-established procedure for assigning weights to a set of factors (Franssen et al., 2009) that correspond to a factor is Saaty's analytical hierarchy process (AHP). In this method, a matrix of pair-wise comparisons (ratios) between the factors is built. This matrix is constructed by eliciting values of relative importance on a scale of 1 to 9; e.g., a ratio value of 1 between a pair of factors indicates equal importance of the factors, while ratio value more than 1 indicates one factor is more important than the other. The process of allocating weights is a subjective one and can be done in participatory mode, in which a group of decision makers may be encouraged to reach a consensus of opinion about the relative importance of the factors. Based on the Saaty's AHP approach, all the four factors were modeled in GIS to make groundwater potential map (Barillec and Cornford, 2009).

In groundwater studies, many of the factors that are considered are qualitative. As a result, a problem arises in modeling the parameters. Therefore, the parameters pertaining to all four factors have been quantified by assigning with suitable ranks, which represents its hydrogeological character and the role it plays in a particular terrain, ranging 1 to 9. The number "1" indicates unfavorable hydrogeological character of a parameter and "9" indicates favorable hydrogeological character of a parameter. Similarly, "1" indicates lowest influence of the factor, while "9" indicates highest influence of the factor. The ranks and weightages assigned to each parameter and corresponding factor are given in Table 11.2.

There are a number of springs occurring in the study area (Figure 11.8). Springs, in terms of their elevation and discharge, reveal that the recharge that is taking place in the geomorphic units corresponding to $0°–5°$ slopes occurring at higher elevation is getting discharged within no time and not reaching the groundwater table. Therefore, the recharge components from these zones are not considered in the statistical analysis.

All the parameters have been adopted in GIS environment resulting in the creation of a groundwater potential index (GWPI) map. The following equation is used for integrating the parameters:

$$GWPI = \frac{\left(L_w L_r + G_w G_r + RC_w RC_r + FL_w FL_r\right)}{\Sigma w} \tag{11.4}$$

where
  G represents geomorphology
  L represents lithology
  FL represents fracture/lineament
  RC represents recharge condition
  r represents the rank of a parameter
  w represents the weight of a factor

GWPI is a dimensionless quantity that helps in indexing the probable groundwater potential zones of an area. The probable occurrence and distribution of groundwater in the study area are shown in Figure 11.9.

TABLE 11.2

Ranks and Weightages Assigned to Each Parameter and Corresponding Factor

Parameter	Rank	Normalized Rank	Factor	Weightage	Normalized Weights
Sandstone and shale of Lower Murree group	1	0.07	Rock types	1	0.05
Sandstone and shale of Subatu group	5	0.33			
Sirban limestone	9	0.60			
Area without any fracture	1	0.10	Geological structure	3	0.16
Fracture/lineament	9	0.90			
Escarpment zones around plateau tops	1	0.05	Geomorphology/landform	9	0.47
Steep slopes around plateau tops	3	0.16			
Plateau tops	6	0.32			
River banks/valley bottoms	9	0.47			
Zone 1: 89 mm per unit area	1	0.05	Recharge condition	6	032
Zone 2: 153 mm per unit area	3	0.16			
Zone 3: 305 mm per unit area	6	0.32			
Zone 4: 808 mm per unit area	9	0.47			

11.8 Validation

Well observation data from four shallow hand pump wells occurring in the higher altitudes and three wells in the valley bottom zone have been assessed for validating the study.

Out of the four wells drilled in the higher altitudes, three are unsuccessful. The remaining one is successful; however, it is dry in most part of the year. Two shallow hand pump wells located in the valley bottom zone are highly successful and yield 100–150 lpm water even during the summer season. The third well, located in the fracture zone meeting the Tawi river, yields more than 250 lpm round the year.

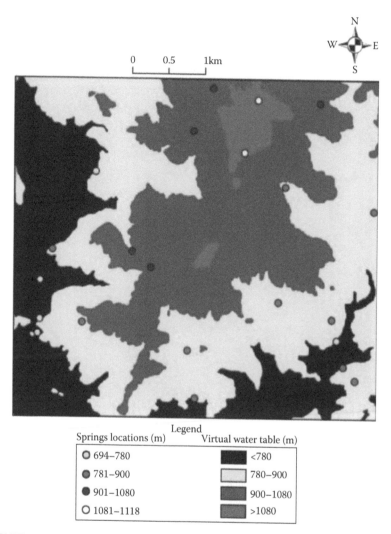

FIGURE 11.8
Location map of the springs occurring at various altitudes and depth of virtual groundwater table.

11.9 Conclusions

The study reveals that geomorphology plays a dominant role in the assessment of groundwater prospects in the hilly terrain. Though the rainfall is high, the recharge is limited due to high relief, as infiltration is greatly influenced by the slope. Hence, the higher altitudes act mainly as runoff

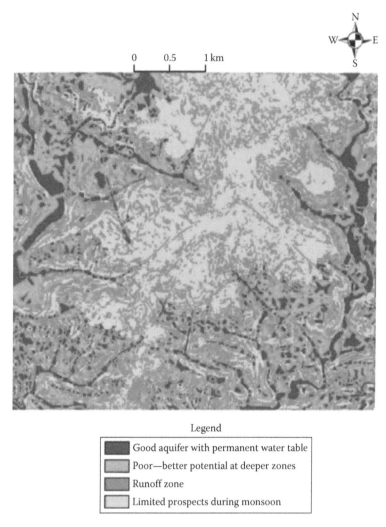

FIGURE 11.9
Groundwater prospects map for part of Tawi river basin, Jammu and Kashmir.

zone. All the limited local recharge that takes place through the narrow valleys and depressions occurring within the hills gets discharged as springs immediately. The ephemeral nature of the springs indicates that the recharge is seasonal. Therefore, the groundwater does not occur under water table condition, particularly in the higher altitudes of the terrain. As the aerial extent of the terrain increases at lower altitudes, the terrain gets regional recharge in addition to the local recharge. At these elevations, the scope for the formation of water table increases. Therefore, the main emphasis of the groundwater study in hilly terrain should be on locating the regional water table. The depth of the well in a given area depends on the depth to the water table in that area. This procedure can be adopted in the water-starved northeastern Himalayan terrain while implementing the drinking water supply schemes, especially for the habitations situated on the hill tops.

Acknowledgments

We are thankful to the Department of Drinking Water Supply (DDWS), Ministry of Rural Development (MoRD), Government of India, for financial support to the Rajiv Gandhi National Drinking Mission (RGNDWM) project under which the study was carried out. We are also thankful to the director, NRSC, and dy. director, RS&GIS-AA, NRSC, for their encouragement in carrying out this work. The help provided by colleagues from the Hydrogeology Division of NRSC in preparing this chapter is gratefully acknowledged.

References

Bahuguna, I. M., Nayak, S., Tamilarsan, V., and Moses, J. 2003. Groundwater prospective zones in Basaltic terrain using remote sensing. *Journal of the Indian Society of Remote Sensing* 31(2): 107–118.

Barillec, R. and Cornford, D. 2009. Data assimilation for precipitation now casting using Bayesian inference. *Advances in Water Resources* 32(7): 1050–1065.

Franssen, H. J., Alcolea A., Riva, M., Bakr, M., Wiel, N., van der Stauffer, F., and Guadagnini, A. 2009. A comparison of seven methods for the inverse modelling of groundwater flow. Application to the characterization of well catchments. *Advances in Water Resources* 32(6): 851–872.

Ghoneim, E. and El-Baz, F. 2007. DEM-optical-radar data integration for palaeohydrological mapping in the northern Darfur, Sudan: Implication for groundwater exploration. *International Journal of Remote Sensing* 28(22): 5001–5018.

Handbook of Hydrology. 1972. Soil Conservation Department, Ministry of Agriculture, New Delhi.

Jenson, S. K. and Domingue, J. O. 1988. Extracting topographic structure from digital elevation data for geographic information system analysis. *Photogrammetric Engineering and Remote Sensing* 54(11): 1593–1600.

Kalinski, R. J., Kelly, W. E., and Bogardi, S. 1993. Combined use of geoelectrical sounding and profiling to quantify aquifer protection properties. *Ground Water* 31: 538–544.

Lillesand, T. M. and Kiefer, R. W. 2004. *Remote Sensing and Image Interpretation*, 4th edn. John Wiley & Sons, New York, 725 pp.

Mark, D. M. 1988. *Network Models in Geomorphology, Modelling in Geomorphological Systems*. John Wiley, New York.

Mather, P. M. 1987. *Computer Processing of Remotely Sensed Images: An Introduction*. John Wiley & Sons, New York, 352 pp.

Murthy, K. S. R. and Mamo, A. G. 2009. Multi-criteria decision evaluation in groundwater zones identification in Moyale-Teltele subbasin, south Ethiopia. *International Journal of Remote Sensing* 30(11): 2729–2740.

Pradeep, V. M., Krajewski W. F., Mantilla, R., and Gupta, V. K. 2009. Dissecting the effect of rainfall variability on the statistical structure of peak flows. *Advances in Water Resources* 32(10): 1508–1525.

Rao, N. S., Chakradhar, G. K. J., and Srinivas, V. 2001. Identification of groundwater potential zones using remote sensing techniques in and around Guntur town, A.P. *Journal of the Indian Society of Remote Sensing* 29(1–2): 69–78.

RGNDWM. 2008. *Ground Water Prospects Mapping Manual for Rajiv Gandhi National Drinking Water Mission*. National Remote Sensing Centre, ISRO, Department of Space, Hyderabad, India, 256 pp.

Sankar, K. 2002. Evaluation of groundwater potential zones using remote sensing data in upper Vaigai river basin, Tamilnadu, India. *Journal of Indian Society of Remote Sensing* 30(3): 119–129.

Solomon, S. and Quiel F. 2006. Groundwater study using remote sensing and geographic information systems (GIS) in the central highlands of Eritrea. *Hydrogeology* 14(6): 1029–1041.

12

Remote Sensing in Delineating Deep Fractured Aquifer Zones

Siddan Anbazhagan, Balamurugan Guru, and T.K. Biswal

CONTENTS

12.1 Introduction

The World Health Organization (WHO) estimates that over a billion people worldwide do not have access to clean drinking water. The excessive use and continuous mismanagement of water resources and the increasing water demands of proliferated users have led to water shortages, pollution of freshwater resources, and degraded ecosystems worldwide (Clarke, 1991; Falkenmark and Lundqvist, 1997; De Villiers, 2000; Tsakiris, 2004). Under these circumstances, the identification of groundwater potential zones and the proper utilization of water resources in hard rock terrain are important, particularly in drought-prone areas. The state of Tamil Nadu has a similar hard rock terrain located in the southern part of India. About two-thirds of the area is covered by a hard crystalline terrain comprising metamorphic, meta-sedimentary, and igneous rocks. Remote sensing is an excellent tool for geologists and water resource engineers for groundwater exploration, and several researchers have adopted this technique to delineate groundwater potential zones in hard crystalline terrains (Farnsworth et al. 1984; Gupta et al. 1989; Ramasamy et al., 1989; Waters et al., 1990; Boeckh, 1992; Anbazhagan, 1993; Greenbaum et al., 1993; Gustaffson, 1993; Ramasamy and Anbazhagan, 1994; Ramasamy et al., 1996; Saraf and Chaudhary, 1998; Srinivasa Rao et al., 2000; Madan Jha et al., 2007). Groundwater exploration in hard rock always requires an integrated approach owing to the heterogeneity of the aquifer systems (Nag, 1998; Pratap et al., 2000; Srivastava and Bhattacharya, 2000; Sarkar et al., 2001; Subba Rao et al., 2001). Sometimes, detailed hydrogeological investigation, geophysical resistivity survey, and aquifer characteristic studies, etc., have not yielded satisfactory results in identifying groundwater targets.

Remote sensing techniques are effectively utilized for the demarcation of fractures and for locating potential aquifers in the hard rock terrain (Raju et al., 1979; Roy, 1981; Satyanarayana Rao, 1983; Raju et al., 1985; Ramasamy et al., 1989; Bobba et al., 1992; Anbazhagan, 1993; Meijerink, 2000; Anbazhagan et al., 2001; Ramasamy et al., 2001; Kumanan and Ramasamy, 2003; Balamurugan et al., 2009). The application of linear features called "lineaments" to locate fractures and potential groundwater zones was first adopted by Lattman and Parizek (1964) in the folded and faulted carbonate terrain.

In recent years, the hydrogeology of deep fractured aquifers has received more attention owing to an interest in finding new sources of groundwater for use in geothermal activities and artificial recharge and in locating sites for the disposal of liquid waste. Deep aquifers with 1–5 km depths were studied in Jurassic and Paleozoic rocks in the Negev Desert, Israel (Native et al., 1987), and such aquifers were delineated with the help of the electromagnetic (EM) method (Tournerie and Choutean, 1998). Fractured aquifer studies in hard rock terrain are limited because of the heterogeneous and consolidated nature of the crystalline terrain. In the case of boreholes, various methods are being adopted to determine fractured aquifer properties (Ballukraya et al., 1983; Muralidharan, 1996; Subramanyam et al., 2000). For example, a downhole televiewer can map fracture traces on the borehole wall and determine their density and orientations (Acworth, 1987; Lee et al., 1991). However, it is well known that not all of these fracture traces correspond to water conducting fractures. This chapter, which is a case study from the Hosur–Rayakottai region in the state of Tamil Nadu, located in the southern part of India, demonstrates how the remote sensing technique was utilized to delineate deep fracture aquifer zones. The major lineaments were interpreted from satellite imagery integrated with a geophysical resistivity survey and field investigation data.

12.2 Study Area

The area selected for the investigation of the deep fractured aquifer zone is located east of the Hosur region, in the northwestern part of Krishnagiri district of the Tamil Nadu state, and covers 1670 km². It is bounded by 12°30' and 12°53' northern latitudes and 77°50' and 78°12' eastern longitudes (Figure 12.1). Normally, a subtropical climate prevails over the study area without any sharp variation. The temperature rises slowly to a maximum of 40°C during summer (May) and goes down to a minimum of 20°C in winter (January). The average annual rainfall of the study area is 850 mm per year (WRO Report, 2004). The area receives maximum precipitation (55%) during the northeast monsoon (October–December), medium precipitation (25%) from the southwest monsoon (June–September), lesser precipitation (18%) during summer (March–May), and very less rainfall, about 2%, during the winter months (January–February).

The maximum elevation of the region is 991 m above mean sea level (MSL) around Rayakottai and 960 m above MSL near Hosur. The major part of the area is covered by an undulating terrain condition with massive rock shoots and fractures. The Ponnaiyar river is the main river flowing in

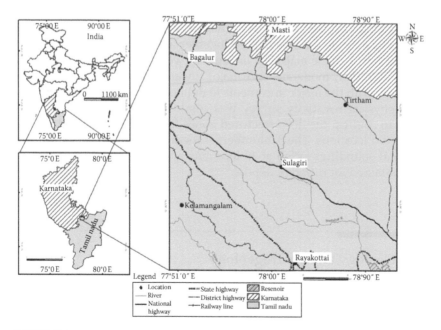

FIGURE 12.1
Location map indicates the region of study in the southern part of India.

the study area from the northwest to the southeast. The important irrigation project in the district is the Krishnagiri dam constructed across the Ponnaiyar river near Periya Muthur in the southeastern corner, and the Chinnar river, a tributary of the Cauvery river flowing in the southwestern part of the study area.

12.3 Materials and Methods

The integrated remote sensing study has proved its credentials in groundwater exploration, development, assessment, and management. Similarly, in this chapter, for locating and characterizing the deep fractured aquifer system, moderate resolution satellite data were utilized and integrated with data obtained from field investigation and a geophysical resistivity survey. IRS 1D LISS III satellite imagery collected from the National Remote Sensing Centre (NRSC), Hyderabad, was used in this study. The LISS III sensor has four band data with a 23 m resolution for the first three bands and the fourth SWIR band has a 70 m resolution. The digitally processed satellite outputs, such as edge enhancement, filtering, principle component analysis (PCA), and ratioed images, were utilized for the interpretation of lineaments. The lineaments interpreted from different processed outputs were combined together and all the lineaments were brought into a single map. The lineaments were classified based on their length as minor, moderate, and major lineaments. Further studies proceed only with the major lineaments to delineate deep fractured aquifers. A geophysical resistivity survey was conducted along the major lineaments to confirm the deep fractured zones for depths from 140 to 460 m at selected locations. The Schlumberger method was adopted for vertical electrical soundings (VES) at 19 locations and the apparent resistivities were obtained for different depths.

12.4 Geology and Hydrogeology

The major rock types in the study area are broadly classified as follows in the order of increasing age: alkaline carbonatites, basic dykes, granites (Krishnagiri), peninsular gneissic complex, high-grade schists of the Satyamangalam group, migmatites, the charnockitic group, and older metasedimentary rocks (Subramaniam and Selvan, 2001). The charnockitic suite of rocks occupied in the middle region, while hornblende biotite gneiss and garnetiferous quartzo-felspathic gneiss covered the west. The peninsular gneissic complex consisting of pink migmatite and granitoid gneiss occupied a major portion of the study area. The quaternary formation was restricted to narrow stretches as fluviatile sediments along the valley in the form of buried channels and river terraces. The geology map for the region of study (Figure 12.2) was compiled from the geological map published by the Geological Survey of India (GSI, 1995) and updated with field data.

Hydrogeology deals with the distribution and movement of groundwater in the soils and rocks of the earth's crust. Fractured rock aquifers are difficult to characterize because of their heterogeneous nature. The understanding of the hydraulic properties of fracture aquifers is difficult and time consuming. The field testing techniques for determining the location and connectivity of fractures are limited. In the present study area, the main source of groundwater was yielded from the fractured aquifers. Joints and fractures are the secondary porosity, in part of fractured aquifers, and are developed after the geological formations. The thickness of the highly weathered zone, the weathered zone, the fractured zone, and the less fractured zone were measured during the field–well inventory survey. The interpreted lineaments and associated fractures were carefully correlated in the well sections as their expression on the surface topography. The subsurface lithology is not evenly distributed and varies from place to place in the following manner: soil thickness from 0 to 20 m, a weathered zone from 2 to 38 m, and the jointed zone varies from 3 to 82 m.

12.5 Interpretation of Lineaments

Satellite imagery, a low altitude radar image, and aerial photographs are the remote sensing data readily available to the geologist for locating fractures (Dinger et al., 2002). Roads, railway lines, power lines, and pipelines represent the linear features in the remote sensing data. However, our interest is to interpret lineaments that are linear or curvilinear in nature, a structurally controlled feature particularly for fracture and faults. IRS 1D LISS III satellite data were used for extracting lineaments in the region of study (Figure 12.3). Four band data were imported to the ERDAS 8.7 Imagine image processing software and digital image processing was carried out. The purpose of image processing and spectral enhancement is to accentuate lineaments and to be able to discern and select fractures better. In order to extract all the lineaments from LISS III satellite data, various standard image processing techniques, including filtering, contrast stretching, and PCA, were performed. Lineaments interpreted from

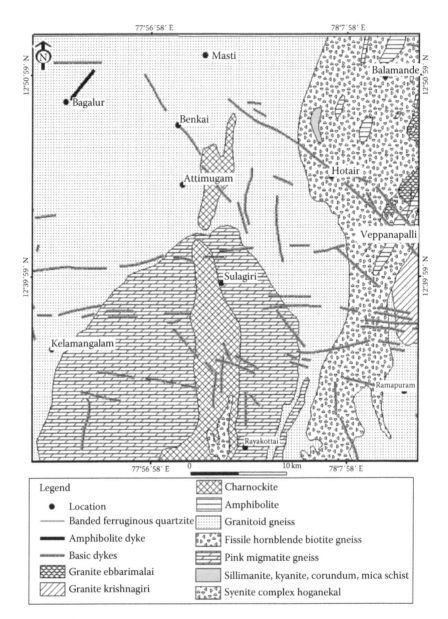

FIGURE 12.2
The geology map of the study area.

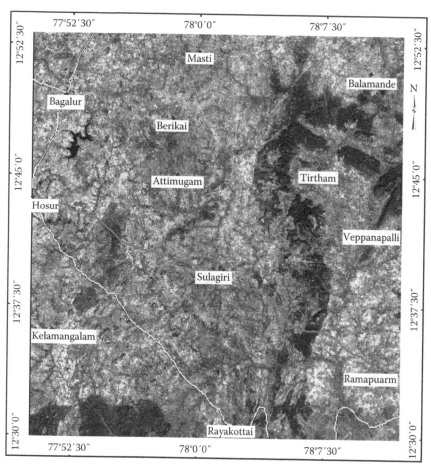

FIGURE 12.3
IRS-1D False Color Composite image of the study area.

different processed images often appeared to be separate and offset when plotted in a single map (Dinger et al., 2002). The lineaments interpreted from every individual processed output were brought into the GIS environment through spatial integration as a single lineament map of the study area. During field investigations, artificial linear features, like roads, railway lines, and communication corridors, were removed. The integrated lineament map shows all major, medium, and minor lineaments in the study area (Figure 12.4). The lineaments correlated straight topographical features, such as valleys with vegetation, barren valleys, fractures with vegetation, ridgelines, saddles, and first-order to third-order streams. The length of the lineaments is an important controlling parameter in groundwater movement and accumulation (Singhal and Gupta, 1999). The longer

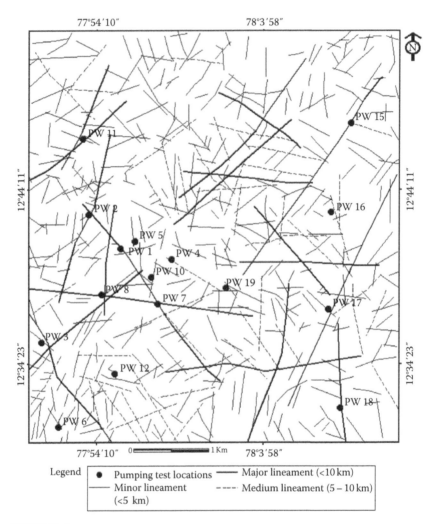

FIGURE 12.4
Classified lineament map of the study area.

the lineaments, the greater the role they play in groundwater movement in a larger area.

During field investigations, the geological, topographical, and structural aspects like joints, fractures, and their orientations were studied (Table 12.1). Most of the lineaments were oriented in the NNE–SSW and NE–SW directions. Lineaments were found to correlate with systematic and non-systematic joints and fractures. Systematic fracture/joint sets in the NNE–SSW and EW directions were located at Bagalur (Figure 12.6a and b) and in the Sulagiri NNE–SSW direction. At Uddanapalli, lineaments were

TABLE 12.1

Locations and Fracture Geometry

Location Name	Lat	Long	Lithology Orientation	Fracture/Joint
Hosur to Mattur	12°45'27.3"	77°50'0.32"	Quartz: N75°E, dip 43°W	Fracture: N15°E-dip vertical
Sittanapalli to Mattur	12°46'24.5"	77°50'03.1"	Gneiss: N15°E, dip vertical	
Bairasandiram	12°47'46.8"	77°51'34.3"	Foliation: N32°E, dip vertical	Fracture: N53°W-dip vertical, N42°E-Dip 50°SE
Thalapalli to Sivanapuram	12°51'57.8"	77°51'42.7"		Fracture: N18°W-dip vertical, N48°E-dip vertical
Thalapalli to Sivanapuram	12°52'21.8"	77°51'39.9"		Fault: N30°E and N38°W, fracture: N55°W-dip vertical, N8°E-dip 45°SE
Nagdenhalli to Dinnehalli	12°52'13.4"	78°02'12.1"	Foliation: N25°E, dip vertical	Fracture: N70°E-dip vertical
Masti to Ajjiapanhalli	12°52'51.1"	78°03'52.1"	Foliation: N12°E, dip vertical	
Masti to Ajjiapanhalli	12°52'36.3"	78°04'09.5"	Foliation: N20°W, dip vertical	Fracture: EW-dip vertical
Balamande	12°51'21.0"	78°09'54.1"	Foliation: N42°W, dip vertical	Fracture: N60°E-dip vertical
Karapalli to Chinnamutthali	12°45'40.7"	77°54'02.5"	Foliation: N10°E, dip vertical	
Peddamuthali to Attimugam	12°46'04.7"	77°53'03.8"	Foliation: N10°E, dip vertical	Fracture: N10°E-dip vertical, N75°E-dip vertical
Navathi to Gurupathi	12°41'45.4"	77°49'14.9"	Foliation: N8°E, dip 70°30E	
Achichatipalli	12°40'46.8"	77°49'34.8"	Foliation: N20°E, dip vertical	Fault: N10°W-dip vertical
Near Kelamangalam	12°39'02.0"	77°50'41.0"	Foliation: N10°E, dip 78°E	
Near Hosur	12°41'43.8"	77°49'05.8"	Dyke: N10°E	
Hosur	12°43'37.2"	77°50'11.5"	Foliation: N10°E	
Hosur to Rayakottai (4 km from Hosur)	12°42'39.3"	77°50'31.3"	Foliation: N20°W, dip 23°NW	
Karapalli colony (Hosur to R. kottai)	12°41'45.9"	77°51'08.6"	Foliation: N20°E, dip vertical	Joint: N20°E
Onnalvadi (1/2 km toward R. kottai)	12°40'29.1"	77°51'42.3"	Foliation: N15°E	

(continued)

TABLE 12.1 (continued)

Locations and Fracture Geometry

Location Name	Lat	Long	Lithology Orientation	Fracture/Joint
Gudisaganapalli	12°40′00.4″	77°52′35.0″		Fracture: N50°E-dip vertical
Nayakanapalli (Biresetipalli to R. kottai)	12°38′18.4″	77°54′44.4″	Dyke-N10°E, foliation: N10°E	Joint: N45°E and N44°W-dip vertical Joint: N20°W and N73°E-dip vertical
Uddanapalli	12°36′26.6″	77°55′57.9″	Foliation: N10°E	
Uddanapalli to Kelamangalam	12°36′31.7″	77°55′28.5″	Foliation: N35°W	Fault: N15°W
Kolegopasandiram	12°32′52.7″	77°48′58.3″	Foliation: N8°E, dip vertical	
Girisettipalli to Bevanatam	12°32′33.1″	77°51′08.7″	Foliation: N8°E, dip vertical	Fault: N9°E and N75°E-dip vertical
In between Girisettipalli and Bevantam	12°32′16.0″	77°51′23.4″	Foliation: N25°W	Fracture: N63°E
Anusonai to Kelamangalam	12°35′03.8″	77°53′08.0″	Foliation: N20°W	Fault: N50°E-dip vertical
Anusonai to R. kottai	12°33′57.3″	77°54′29.5″	Foliation: N10°E, dip 46°E	
Nallarapalli to Alesibam	12°34′43.3″	77°57′58.8″	Foliation: N25°E, dip 23°E	Fold
Alesibam well	12°33′57.6″	77°56′38.3″		Fracture: N73°E
Uddanapalli	12°36′47.5″	77°56′29.4″	Foliation: N25°E, dip 78°E	Fracture: N23°E and N75°E-dip vertical
Perandapalli to Bukkanapalli, Punagaram	12°43′05.8″	77°56′09.5″	Foliation: N10°W, dip vertical	Fault: N9°W-dip vertical Fracture: N65°E-dip vertical
Usdanapalli	12°42′22.5″	77°55′54.8″		Fracture: N70°W-dip 78°S
Chempatti or Sempatti well	12°43′40.7″	77°56′41.0″		Fracture: N80°W-dip vertical
Biraipalayam to Ulagam	12°39′09.9″	78°00′46.8″	Foliation: N15°W, dip 44°E	
Gettur to Ulagam	12°35′43.9″	78°01′14.6″		Fold: N13°W, then N75°W, then N15°W
Hosur to Krishnagiri	12°43′38.5″	77°50′53.4″		Fracture: N25°W-dip 78°NE, N40W-dip 85°NE, N65°E-dip 80°SSW
Rayakottai to Krishnagiri	12°30′38.5″	78°02′59.3″	Foliation: N10°W, dip 63°E	Fracture: N10°W-dip 63°E, EW-dip vertical

			Bed orientation / Foliation	Fracture / Fault details
Sikaripura or Sikarimedu	12°35'15.6"	78°08'50.8"	Bed orientation: N65°W	Fault: N20°W-dip 68°W; Fracture: N46°W-dip vertical; Fracture: N55°W-dip 58°NE, N70°E-dip vertical
Attigunda (K. giri to Vappanapalli)	12°40'12.3"	78°11'21.3"		
Basavanakoil	12°46'52.6"	78°05'15.8"		Fracture: N60°E-dip vertical, N47°E-dip 46°NE
Bikkanapalli (Sulagiri to K. giri)	12°36'24.4"	78°06'01.5"	Dyke: N30°W, dip 78°SW	Fracture: N40°E-dip 80°NW, N65°W-dip 45°NE, N75°E,N15°E-dip vertical, N40°W-dip 85°SW
Hosur to Bagalur (minor)	12.75169°	77.83437°	Foliation: N12°E	Fracture: N38°E and N3°E dip: Both vertical
Bagalur to Berigai (major)	12.82217°	77.89707°		Fracture: N27°E and N70°E dip:Both vertical
Permalpatti to Devisettipalli (major)	12.80350°	77.91513°		Fracture: N40°E (vertical dip), N75°W dip: 44°50'
Bagalur to Berigai	12.81283°	77.93416°	N72°W,N15°E	Fracture: N72°W dip: vertical
	12.79493°	77.97646°	Foliation: N15°E	Fracture: N10°E, N58°E, dip: Both vertical
Attimugam to Sulagiri	12.73624°	77.98140°	Foliation: N14°E	Fracture: N 33°W, N70°E, dip: Both vertical
Attimugam to Sulagiri	12.71990°	77.99039°	Foliation: N36°E	Fracture: N68°E, N44°W, dip: Both vertical fault: N30°E, foliation: N12°E, N15°E
Perandapalli (Ponnaiyar river)	12.71476°	77.89136°		Fracture: N83°E-dip 43°NW, N18°W-dip vertical joint: N59°E and N73°W.
Perandapalli to Attimugam	12.71773°	77.89532	Foliation: N8°E	
Hosur to Sulagi	12.69677°	77.91452°		Fracture: N23°W, dip 70°NE, N83°W dip vertical fold limb only: 65° NW23°, 86°15' NW63°
Hosur to Sulagi	12.68700°	77.92813°		Fracture: N65°W dip vertical
Kottarpalli (medium)	12.67226°	77.97695°		NW-SE tanks are aligned
Near Sundagiri (major)	12.64960°	78.04170°		Fracture: N31°W dip-56°, N53°W dip vertical
Chinnar (major)	12.64013°	78.06229°		Fracture: N-S dip 45°E, joint: N80°E and N10°E
Hosur to Krishnagiri (along NH-7)	12.63860°	78.06590°		Fracture: N75°E, N23°W dip:Both vertical , joint: N20°E dip 28° NW50°

(continued)

TABLE 12.1 (continued)

Locations and Fracture Geometry

Location Name	Lat	Long	Lithology Orientation	Fracture/Joint
Hosur to Krishnagiri	12.61653°	78.08655°	Foliation: N12°E	Fracture: N12°E, N73°E, dip vertical, third fracture is horizontal
Along Ponnaiyar river (Kurubarapalli)	12.59716°	78.13406°	Foliation: Qtz ridge is N18°W	Fracture: N25°E and horizontal fracture
Kurubarapalli to Kupuchiparai (Gundapalli)	12.63914°	78.13776°		Fracture: N65°W dip 60° NE20°, N20°E dip vertical
Nedusalai to Veppanapalli	12.66174°	78.14702°	Foliation: N23°E dip 79° NW53°	Fault plane: N72°W, fracture: N6°W, joint: N30°W, fracture in trap rock: N12°E, N39°W, dip vertical both
Thirtham	12.76372°	78.11584°	Foliation: N64°W dip	Fracture: N35°E,N53°W, dip: Both vertical
Thirtham	12.78755°	78.06772°	Foliation: N20°E	Fracture: N40°E, N40°W dip: Both vertical, western side of the river (fracture): N50°E dip 52°NE joint: N30°W dip 45° NE73°
Ramandodi to Sulagiri	12.76488°	78.04446°	Foliation: N23°E	Fracture: N23°E, N30°W
Masti to Dinnahalli	12.87033°	78.03645°	Foliation: N15°E	Fracture: N15°E
Nagenhalli to Kaderahalli	12.87203°	78.13695°	Dyke: N87°W	Fold: N10°E, N80°E; fracture: N26°W, N82°W
Yarkol to Thirtham	12.79766°	78.16199°		Fracture: N57°Edip vertical, N63°W dip-58°SW35°
Ramandodi to Sulagiri	12.67667°	78.01748		
Bevanattam to Anusonai	12.54530°	77.88066°	Dyke: N-S Foliation: N22°W	Fracture: N72°E, N23°W, dip: both vertical
Sulagiri to Ullagam	12.59550°	78.02085°	Vein: N-S	Fold: N10°W—EW—N12°W fracture: horizontal

Location	Latitude	Longitude	Foliation	Fracture
Sulagiri to Rayakottai (Ponnaiyar)	12.57557°	78.01909°		Fracture: N37°W dip 66°30′ SW46°, fracture: N23°E dip 64° NW50° fracture: N71°W dip 68° SW25°, N55°E dip vertical
Rayakottai to Krishnagiri (major)	12.53127°	78.06052°	Foliation: N10°E	Fracture: N10°E, N10°W dip: Both vertical fracture: E-W dip 73°20′ NE10°
Rayakottai to Krishnagiri	12.50914°	78.13565°		Fault plane: N15°E, bed N35°W,N35°W
Krishnagiri to Vappanapalli	12.57336°	78.18033°	Foliation: N16°E	Fracture: N81°E, dip vertical
Krishnagiri to Vappanapalli	12.57716°	78.18314°		Fracture: N80°E, N15°W dip: Both vertical
Krishnagiri to Vappanapalli	12.62992°	78.18731°	Foliation: Dyke N60°E	Fracture in Dyke: N51°E dip84°NE, N10°E dip68°SW, fracture in rock: N10°W, N36°E dip: Both vertical
Vappanapalli	12.70267°	78.19739°	Foliation: Dyke N28°W	
Vappanapalli to Trtam	12.70901°	78.16644°		Fracture: N80°W
Vappanapalli to Chinakuttur (river)	12.71894°	78.13271°		Fracture1: N87°W dip: 50° to 54° Fracture2: N55°E dip: 72°SW
Chinakuttur to Kurubarapalli	12.69589°	78.13417°		Fracture1: N45°W dip: vertical fracture2: N84°W dip: 69°SW
Chinakuttur to Kurubarapalli	12.66660°	78.14205°	Foliation: N20°E	Fracture: N65°W dip:28°NE
Rayakottai to Hosur	12.53793°	78.00816°		Fracture1: N13°E, N68°W dip: 36° NE
Rayakottai to Hosur	12.55624°	77.98943°		Fracture1: N13°E, vertical dip joint: N20°W, and N58°E
Rayakottai to Hosur	12.56467°	77.98135°	Foliation: N20°E	Fracture1: N20°E dip: vertical fracture2 and 3: N50°W dip:58°, N47°E dip:38°
Rayakottai to Hosur	12.58325°	77.96670°		Fracture1 and 2: N62°E, E-W dip:Both vertical fracture3: N26°W dip:62°
Rayakottai to Hosur (small drainage)	12.59548°	77.95493°		Fracture: N51°E dip:68° SE
Rayakottai to Hosur	12.61355°	77.93147°		Fracture: N4°E dip:34° E joint: N84°E, N4°E

(continued)

Geoinformatics in Applied Geomorphology

TABLE 12.1 (continued)

Locations and Fracture Geometry

Location Name	Lat	Long	Lithology Orientation	Fracture/Joint
Uddanapalli to Hosur	12.63077°	77.92007°	Foliation: N34°E	Fracture1: N4°E, N34°E dip: both vertical
Uddanapalli to Hosur	12.63787°	77.91538°	Foliation: N14°E	Fracture1to4: N14°E, N80°E, N42°E, N65°W
Uddanapalli to Hosur	12.64920°	77.89410°	Foliation: N48°E	Fracture1 and 2: N30°W dip:75°, N46°E dip: vertical
Uddanapalli to Hosur	12.65003°	77.86785°		Joint1to4: N68°W, N20°W dip:79°, N44°W, N35°E
Rayakottai to Hosur	12.57362°	77.86272°	Foliation: N24°E	Fracture1 and 2: N62°W dip:79°, N80°W dip:78° fracture3 and 4: N44°E dip:50°, N80°E dip:80°
Rayakottai to Hosur	12.54438°	77.83935°		Fracture1 and 2: N55°W, N50°E dip: vertical both joint: N4°E
Rayakottai to Hosur	12.53783°	77.85564°	Foliation: N10°W	Joint1to3: N79°W, N8°W, N64°E dip:79°
Rayakottai to Hosur	12.51363°	77.86684°	Foliation: N18°E	Fracture1 and 2: N18°E dip:58°, E–W dip 75° Joint1 and 2: N44°W and N28°E
Rayakottai to Hosur	12.53414°	77.87981°	Foliation: N–S (up to 10° variation)	Joint1to3: N80°W, N50°W, and N80°E
Rayakottai to Hosur	12.56921°	77.90456°	Foliation: N20°E	Fracture1: N5°E dip:43° joint1 and 2: N80°W and N40°E dip: vertical both
Rayakottai to Hosur	12.56475°	77.95133°		Joint1to3: N60°W, N30°E Dip:53°, E–W dip:62°
Rayakottai to Hosur (Kelamangalam)	12.54932°	77.97183°	Foliation: N18°E	Joint1to3: N20°E dip:54°, N35°W dip:68°, E–W joint4: N62°W
Krishnagiri to Rayakottai	12.52420°	78.15711°		Fracture1 and 2: N63°W dip:83°, N22°W dip:65° fracture3 and 4: N12°E dip:61°, horizontal fracture
Krishnagiri to Rayakottai	12.53297°	78.14888°		Fracture1 and 2: N16°W dip:72°, N76°E dip:88°
Krishnagiri to Rayakottai	12.52167°	78.13591°		

Location	Latitude	Longitude	Foliation	Fracture/Joint
Krishnagiri to Rayakottai	12.49963°	78.13567°		Fracture1to3:N45°W dip:84°,N69°E dip:86°, N42°E joint1 and 2: N15°E dip:80, N80°W dip:80°
Krishnagiri to Rayakottai	12.49436°	78.12501°	Foliation: N18°E	
Krishnagiri to Rayakottai	12.49880°	78.09224°		Fracture1 and 2: N10°W dip:58°, N72°W dip: vertical fracture3 and 4: N10°E dip:79°, N22°E dip: vertical joint1 and 2: N-S dip:52°, N8°E dip:40°
Krishnagiri to Rayakottai	12.50226°	78.08812°	Foliation: N4°E	Fracture1 and 2: N21°W dip:81°, N30°E dip:48° fracture3 and 4: N-S, N10°W dip:Both vertical
Krishnagiri to Rayakottai	12.53426°	78.06101°	Foliation: N4°E	Fracture1 and 2: N12°W dip:69°, N60°W dip:80°
Krishnagiri to Rayakottai	12.51290°	78.05005°	Foliation: N4°E	Fracture1 and 2: N16°E dip:40°, N10°E dip:72° joint1 and 2: N4°E, N18°E
Rayakottai to Sulagiri	12.52686°	78.02767°	Foliation: N13°E	Fracture1 and 2: N58°W dip: vertical, N4°E dip:53° joint: N76°E
Rayakottai to Sulagiri	12.56951°	78.01560°		Fracture1 and 2: N60°W dip:78°, N44°E dip: vertical joint1 and 2: N3°E and horizontal joint
Rayakottai to Sulagiri (near Ulagam)	12.58820°	78.02173°	Foliation fault1: N40°W,N62°W and fault plane: N48°E. fault2: N42°W,N80°E,N52°W and fault plane: N28°E	Fracture1 and 2: N30°E dip: vertical, N20°E dip:80° joint: N16°W dip:78°
Ulagam to Samalpallam	12.58290°	78.04717°		Joint1 and 2: N63°E, N78°E
Kuttur to Samalpallam (river)	12.57527°	78.07478°	Foliation: N-S dip:48°	Fracture1 and 2: N40°W dip:80°, N35°E dip:78° fracture3 and 4: N82°W dip:88°, N42°E dip:76° fracture5 and 6: N58°E dip:79°, N60°W dip: 62° fracture7: N15°E dip:86° joint1 and 2: N8°W dip:60, N56°W dip:80° joint3 and 4: N27°W dip: vertical, gentle horizontal

observed as systematic NE–SW trending fractures, and in other locations, it was emplaced as nonsystematic (irregular) lineaments. In the region of study, more than 230 lineaments were verified at 130 locations during the field reconnaissance survey. During field investigations, it was observed that some of the lineaments were obliterated due to subsequent weathering and soil cover. The structural data collected during field investigations were plotted as a rose diagram. Similarly, 472 lineament-interpreted satellite images were plotted in a separate rose diagram (Figure 12.5). The lineaments show two prominent trends in the NNE–SSW and E–W directions. The less dominant fracture sets are oriented in the NE–SW and the NW–SE directions. However, the major fracture-controlled lineaments are oriented in the NE–SW direction. Figure 12.6 shows field photographs of

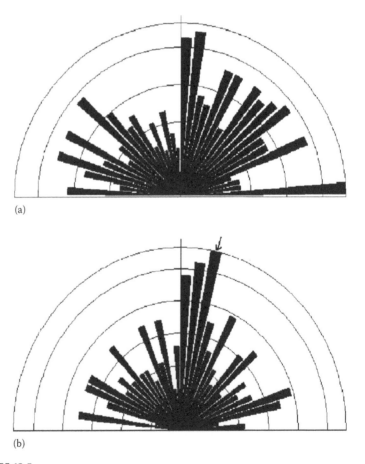

(a)

(b)

FIGURE 12.5
Rose diagrams show (a) orientation lineaments interpreted from a satellite image and (b) orientation of the field fractures.

FIGURE 12.6
The photographs (a), (b), (c), and (d) show the lineaments controlled by the NNE–SSW and E–W fractures, and the NNE–SSW fractures, the NE–SW fractures, and the NNE–SSW fractures, respectively, at different locations in the study area.

the structural features of systematic fractures and joints, and these fractures have been transmitting surface water to subsurface fractured aquifer zones.

12.6 Geophysical Resistivity Survey

The electrical resistivity method consists of the injection of electrical current into the subsurface by current electrodes and a measurement of the change in voltage by potential electrodes (Todd, 1980). The change in voltage measured by the potential electrodes allows for the resistivity of the subsurface materials to be calculated. The resistivity of a material is simply the inverse of the conductivity. Geologic materials, like sands, gravels, and hard rock, have relatively higher resistivity. On the other hand, water-filled voids, clays, and silts have relatively low resistivity. Vertical Electrical Sounding (VES) was carried out at 19 locations along major lineament zones in the study area. The principle of VES was established during 1920's (Gish and Rooney, 1925). The Schlumberger configuration was adapted to the maximum depth penetration of 460 m. The SSR-MP-AT resistivity meter of the IGIS product was utilized for the geophysical resistivity survey. This instrument directly measured the resistance and apparent resistivity of a formation at a particular depth. Field data acquisition was generally carried out by moving two or four of the electrodes between each measurement. The average measurement of the apparent resistivity value was taken at every stage of the investigation. The resistivity surveys were conducted along and across the major lineaments to measure their depth persistence and fracture condition. The depth of investigation ranged from 130 m at the Agaram village to 460 m near the Sittanapalli village. There was a constraint involved in conducting a constant depth of investigation at all locations due to the uneven terrain conditions, the slopes, and the hindrances due to the presence of vegetation/crop land. The apparent resistivity values were interpreted to a particular depth by using inverse slope techniques proposed by Sankarnaryana and Ramanujachary (1967). The inverse slope is a commonly adopted technique in India to measure resistivity values.

12.7 Results and Discussion

The general subsurface configuration of the study area is soil, followed by a highly weathered zone, a weathered formation, a jointed/fractured zone, a less fractured zone, and a massive rock zone. The thickness of the various subsurface layers varies depending on the intensity of the weathering and fractured conditions. The maximum soil thickness observed in the study area is 20 m. The thickness of the weathered zone varies from 2 to 38 m and the jointed zone varies from 3 to 82 m. In total, 472 lineaments were interpreted from processed satellite data. The lineaments were further classified, based on their length, into three categories, as major (>10 km), medium (5–10 km), and minor lineaments (<5 km) (Figure 12.4). Nineteen major lineaments were interpreted in the study area and considered for further field investigations and a geophysical resistivity survey. During field investigations, the structural details like the joints/fractures and their geometry in the study area were measured. The lineaments controlled by fractures have a systematic pattern (e.g., Uddanapalli) at most of the locations, and in a few places there is a nonsystematic (irregular) pattern. More than 200 lineaments were thoroughly verified in the field and the geometry was measured at 130 locations. It was observed that some of the lineaments were obliterated due to weathering and the development of soil formation. Overall in the study area, the lineaments show two prominent trends in the NNE–SSW and E–W directions and less dominant fracture sets in the NE–SW and the NW–SE directions.

The main purpose of conducting VES along major lineaments is to confirm the deep fractured zones at various depth and to explore the possibility of deep aquifers (>100 m depth) for safe groundwater supply for villages and towns located nearby. A resistivity survey was carried out at 19 selected locations in the study area and the apparent resistivity was measured with the help of the inverse slope method. Deep fractured aquifer zones were inferred at different depths from 100 to 400 m below ground level (bgl). Fractured zones were identified at 100 m depth in Agaram-VES2 and 400 m depth at Sittanapalli-VES1. Similar deep fracture zones were located at different depth intervals in the following locations: Kirnapalli-VES1, Birjapalli-VES1, Birjapalli-VES2, Enusonai-VES, Chinnamuthalli-VES1, Marasandiram-VES1, Torapalli-VES1, and Sittanapalli-VES1. From detailed investigations, it was indicated that out of 19 VESs, 12 VES locations could act as deep fractured aquifer zones, where potential groundwater was explored below the depth of 200 m (i.e., >650 ft). The possible locations of deep fractured aquifer zones are listed in Table 12.2.

In order to validate the findings, bore well litho–log data for the study area were collected from the Central Groundwater Board (CGWB). CGWB has carried out exploratory drilling at selected locations for depths between 135 and 300 m bgl. Six bore well locations fall within the study

TABLE 12.2

Probable Deep Fractured Aquifer Zones Interpreted from the Resistivity Survey

Location No.	Village Name	Recommendation for Deep Bore Well (m bgl)	Fractured Zone Depth (m bgl)
1	Birjapalli no.1	250	165–200
2	Birjapalli no.2	250	165–200
5	Uddanapalli	200	80–160
6	Kirnapalli no.1	300	140–230
			260–300
8	Enusonai	270	195–220
10	Chinamuthalli	360	250–320
13	Marasandiram no.1	400	340–360
14	Marasandiram no.2	270	200–220
15	Torapalli no.1	350	260–300
16	Torapalli no.2	250	192–220
19	Permalpalli	360	268–300

TABLE 12.3

Details of Depth of Fractured Zones and Yield of the Bore Wells (Litho–log data)

Sl. No.	Name of the Location	Depth of Drilling (m bgl)	Lithology	Fractured Zone (m bgl)	Yield of Bore Well (lps)
1	Kelamangalam	250	Granitic gneiss	109–180	6.2
			Pink granite	180–250	
2	Bagalur	300	Granitic gneiss	107–229	3.9
3	Thorapalli	300	Granitic gneiss	175–245	2.6
4	Hosur	135	Granitic gneiss	088–113	5.5
			Biotite gneiss	113–135	
5	Elumichchanhalli	170	Granitic gneiss	075–132	6.2
			Granitic gneiss and charnockite	132–170	
6	Billakottai	210	Granitic gneiss	068–131	12.1
			Charnockite with Pink granite	131–210	

Source: CGWB, Chennai.

area. Lithology data have shown the depth of drilling, the lithology, and the depth of the fractured zones along with the discharge (Table 12.3). Most of these bore wells are located either in the major or medium lineaments zone, which was interpreted from satellite data. Overall, CGWB bore well litho–log data have indicated the presence of deep fractured aquifer zones at depths between 68 m and 250 m. In order to pictorially represent the deep aquifers in the study area, a 250 m buffer zone was created surrounding the major lineaments, and the locations of the deep fractures interpreted from the geophysical resistivity survey were demarcated (Figure 12.7).

12.8 Conclusion

The groundwater condition in parts of Hosur region is mainly controlled by secondary porosities like fractures, joints, and faults, and invariably considered as the fractured aquifer system. The geophysical resistivity survey conducted up to the depth of 460 m along the major lineaments confirms the presence of deep fractures at the depth of >200 m. Bore well litho–log data collected from the CGWB have validated the study. Most of the major lineaments have deep depth persistence and hence could be tapped as deep fractured aquifer zones for safe drinking water supply as well as for aquifer replenishment through injection wells at selected locations. The systematic approach adopted in this chapter, such as the

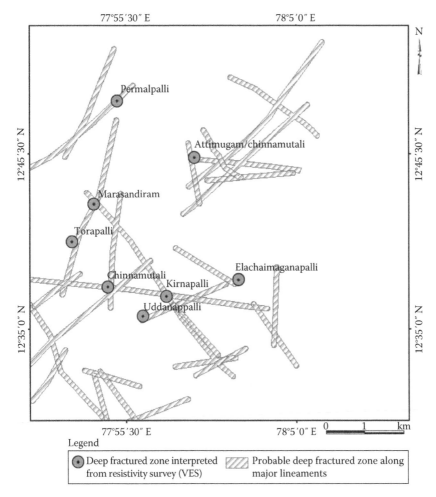

FIGURE 12.7
Favorable zones for deep fractured aquifers in parts of the Hosur–Rayakottai region, Tamil Nadu.

interpretation of major lineaments from satellite data, the characterization of fractures/lineaments through field geometry measurements, and the geophysical resistivity survey, provided convincing results. Similar procedures could be adopted to study the fractured aquifers in other parts of the hard rock terrain. In the present scenario, groundwater development in the region of study is at a satisfactory level (70%). However, groundwater development could also include the use of deep fractured aquifers for domestic, agricultural, and aquifer management purposes.

Acknowledgments

The authors acknowledge the state Groundwater Department and the Central Groundwater Board for providing necessary data. The second author acknowledges the AICTE for providing the National Doctoral Fellowship.

References

Acworth, R.I., 1987. The development of crystalline basement aquifers in a tropical environment. *Quarterly Journal of Engineering Geology* 20:265–272.

Anbazhagan, S., 1993. Integrated groundwater study in drought prone Pennagaram, Dharampuri district, Tamilnadu. *Bulletin of the Indian Geologists Association* 26(2):117–123.

Anbazhagan, S., Aschenbrenner, F., and Knoblich, K., 2001. Comparison of aquifer parameters with lineaments derived from remotely sensed data in Kinzig basin, Germany. XXXI. IAH congress: New approaches to characterising groundwater flow, Munich, 10–14 September, Vol. 2, pp. 883–886.

Balamurugan, G., Anbazhagan, S., Biswal, T.K., and Kusuma, K.N., 2009. Surface temperature mapping using ETM—TIR images in parts of Precambrian terrain, India. In: *Exploration Geology and Geoinformatics*, eds. Anbazhagan, S. et al., MacMillan India, New Delhi, pp. 237–244.

Ballukraya, P.N., Sakthivadivel, R., and Baratan, R., 1983. Breaks in resistivity sounding curves as indicators of hard rock aquifers. *Nordic Hydrology* 14:33–40.

Bobba, A.G., Bukata, R.P., and Jerome, J.H., 1992. Digitally processed satellite data as a tool in detecting potential groundwater flow systems. *Journal of Hydrology* 131(1–4):25–62.

Boeckh, E., 1992. An exploration strategy for higher-yield boreholes in the West African crystalline basement. In: *The Hydrogeology of Crystalline Basement Aquifers in Africa*, eds. Wright, E.P. and Burgess, W.G., The Geological Society of London, London, U.K., pp. 87–100.

Clarke, R., 1991. *Water: The International Crisis*, Earthscan Publications Ltd., London, U.K.

De Villiers, M., 2000. *Water: The Fate of Our Most Precious Resource*, Mariner Books, Houghton, Mifflin, Boston, MA.

Dinger, J.S., Andrews, R.E., Wunsch, D.R., and Dunno, G.A., 2002. Remote sensing and field techniques to locate fracture zones for high-yield water wells in the Appalachian Plateau, Kentucky. In: *Proceedings of the National Ground Water Association Fractured-Rock Aquifer 2002, Conference*, March 13–15, 2002, Denver, CO, pp. 195–199.

Falkenmark, M. and Lundqvist, J., 1997. *World Freshwater Problems—Call for a New Realism*, UN/SEI, New York/Stockholm.

Farnsworth, R.K., Barret, E.C., and Dhanju, M.S., 1984. *Application of Remote Sensing to Hydrology Including Ground Water*, IHP-II Project A. 1.5, UNESCO, Paris.

GSI, 1995. Geological and mineral map of Tamil Nadu and Pondicherry, Geological Survey of India.

Gish, O.H. and Rooney, W.J., 1925. Measurement of resistivity of large masses of undisturbed earth. *Terrestrial Magnetism and Atmospheric Electricity* 30:161–188.

Greenbaum, D., Carruthers, R.M., Peart, R.J. et al., 1993. Groundwater exploration in southeast Zimbabwe using remote sensing and ground geophysical techniques. BGS Technical Report WC/93/26, p. 10.

Groundwater Report of Dharmapuri and Krishnagiri Districts, 2004. Water Resource Organisation (WRO).

Gupta, A.K., Ganesh Raj, K., and Yogarajan, N., 1989. Use of digital techniques of remotely sensed data for geological applications—A review. *International Conference on Image Processing*, Singapore.

Gustaffson, P., 1993. High resolution satellite data and GIS as a tool for assessment of groundwater potential in semi arid area. In: *Proceedings of 9th Thematic Conference on Geological Remote Sensing*, Pasadena, CA, p. 123.

Kumanan, C.J. and Ramasamy, S.M., 2003. Fractures and the transmissivity behaviour of the hard rock aquifer systems in parts of Western Ghats, Tamil Nadu, India. *Escap Water Resources Journal, Bangkok* (June):53–59.

Lattman, L.H. and Parizek, R.R., 1964. Relationship between fracture traces and the occurrence of groundwater in carbonate rocks. *Journal of Hydrology* 2:273–291.

Lee, D.R., Milton, G.M., Cornett, R.J., and Welch, S.J., 1991. Location and assessment of groundwater discharge. In: *Proceedings of 2nd Annual International Conference on High Level Radioactive Waste Management*, Vol. 2. American Nuclear Society, LaGrange Park, I 11, pp. 1276–1283.

Madan Jha, K., Chowdhury, A., Chowdary, V.M., and Peiffer, S., 2007. Groundwater management and development by integrated remote sensing and geographic information systems: Prospects and constraints. *Water Resource Management* 21:427–467.

Meijerink, A.M.J., 2000. Groundwater. In: *Remote Sensing in Hydrology and Water Management*, eds. Schultz, G.A. and Engman, E.T., Springer, Berlin, pp. 305–325.

Muralidharan, D., 1996. A semi-quantitative approach to detect aquifers in hard rocks from apparent resistivity data. *Journal of Geological Society of India* 47:237–242.

Nag, S.K., 1998. Morphometric analysis using remote sensing techniques in the Chaka sub-basin, Purulia district, West Bengal, India. *Journal of Indian Society of Remote Sensing* 26(1&2):69–76.

Native, R., Bachmat, Y., and Issar, A., 1987. Potential use of deep aquifer in the Nagav Desert, Israel—A conceptual model. *Journal of Hydrology* 94:237–265.

Pratap, K., Ravindran, K.V., and Prabakaran, B., 2000. Groundwater prospect zoning using remote sensing and geographical information system: A case study in Dala-Renukoot area, Sonbhadra District, Uttar Pradesh. *Photonirvachak Journal of Indian Society of Remote Sensing* 28(4):249–263.

Raju, K.C.B., Kareemuddin, Md., and Prabhakara Rao, P. 1979. *Operation Anantapur*. Geological Survey of India, Miscellaneous Publications 17:58.

Raju, K.C.B., Rao, P.N., Rao, G.V.K., and Kumar, B.J., 1985. Analytical aspects of remote sensing techniques for groundwater prospection in hard rocks. In: *Proceedings of 6th Asian Conference on Remote Sensing*, NRSA, Hydrabad, pp. 127–132.

Ramasamy, S.M. and Anbazhagan, S., 1994. *Remote Sensing for Artificial Recharge of Groundwater*. NNRMS Bulletin, B18, India, pp. 35–37.

Ramasamy, S.M., Anbazhagan, S., and Moses Edwin, J., 1996. Control of fracture systems in artificial recharge with special reference to crystalline aquifer system in Tamil Nadu. In: *Trends in Geological Remote Sensing*, ed. Ramasamy, S.M., Rawat Publ, Jaipur, pp. 274–280.

Ramasamy, S.M., Nagappan, N., and Selvakumar, R., 2001. Fracture pattern modeling and ground water hydrology in hard rock aquifer system, central Tamil Nadu, India. In: *Spec. Vol. of the ISRS on Spatial Technology for Natural Hazards Management*, eds. Singh, R.P. and Vinod, T., pp. 280–291.

Ramasamy, S.M., Thillaigovindarajan, S., and Balasubramanian, T., 1989. Remote sensing based appropriate methodology for groundwater exploration—A case study. In: *Precambrians of South India*, ed. Gupta, C.P., NGRI, Oxford Press, pp. 327–332.

Roy, A.K., 1981. Case studies in groundwater. In: *Proceedings of In-Regional Seminar on Groundwater in Hard rocks, Coimbatore, India*, UNESCO, Paris, pp. 85–88.

Sankarnaryana, P.V. and Ramanujachary, K.R., 1967. Inverse slope method for determining absolute resistivities geophysics. *Geophysics* 32(6):1036–1040.

Saraf, A.K. and Chaudhary, P.R., 1998. Integrated remote sensing and GIS for groundwater exploration and identification of artificial recharges sites. *International Journal of Remote Sensing* 19(10):1825–1841.

Sarkar, B.C., Deota, B.S., Raju, P.L.N., and Jugran, D.K., 2001. A geographic information system approach to evaluation of groundwater potentiality of shamri micro-watershed in the Shimla taluk, H.P. *Journal of Indian Society of Remote Sensing* 29(3):151–163.

Satyanarayana Rao, R., 1983. Application of integrated deformation model to groundwater targeting in peninsular gneissic complex through remote sensing studies. In: *Proceedings of National Symposium on Remote Sensing in Development and Management of Water Resources*, New Delhi, pp. 249–254.

Singhal, B.B.S. and Gupta, R.B., 1999. *Applied Hydrogeology of Fractured Rocks*, Kluwar Academic Publication, Dordrecht, the Netherlands, p. 400.

Srinivasa Rao, Y., Reddy, T.V.K., and Nayudu, P.T., 2000. Groundwater targeting in a hard-rock terrain using fracture-pattern modeling, Niva River basin, Andra Pradesh, India. *Hydrogeology Journal* 8:494–502.

Srivastava, P.K. and Bhattacharya, A.K., 2000. Delineation of ground water potential zones in a hard rock terrain of Bargarh district, Orissa using IRS data. *Journal of Indian Society of Remote Sensing* 28(2&3):129–140.

Subba Rao, N., Chakradhar, G.K.J., and Srinivas, V., 2001. Identification of groundwater potential zone using remote sensing technique in and around Guntur town, A.P., India. *Journal Indian Society of Remote Sensing* 29(1&2):69–77.

Subramaniam, K.S. and Selvan, T.A., 2001. Geology of Tamil Nadu and Pondicherry. *Journal of Geological Society of India*. 7–19.

Subramanyam, K., Ahmed, S., and Dhar, R.L., 2000. Geological and hydrogeological investigations in the Maheshwaram watershed, Ranga Reddy District, Andhra Pradeh, India. Tech report no. NGRI-2000-GW-292.

Todd, D.K., 1980. *Groundwater Hydrology*, John Wiley & Sons, New York.

Tournerie, B. and Choutean, N., 1998. Delineation of deep aquifer in the Senegal Basin using EM methods. In: *SAGEEP, 98 and CGU*, Quebec, Mai, 1998, pp. 1–3.

Tsakiris, G., 2004. Water resources management trends, prospects and limitations. In: *Proceedings of the EWRA Symposium on Water Resources Management: Risks and Challenges for the 21st Century*, September 2–4, 2004, Izmir, pp. 1–6.

Waters, P., Greenbaum, P., Smart, L., and Osmaston, H., 1990. Applications of remote sensing to groundwater hydrology. *Remote Sensing Review* 4:223–264.

13

Remote Sensing and GIS for Locating
Artificial Recharge Structures for
Groundwater Sustainability

S.K. Subramanian and G.S. Reddy

CONTENTS

13.1 Introduction

Average annual groundwater resource in India is estimated as 396 billion cubic meter. As per the norms, this replenishable resource is only to be used to maintain the balance between annual recharge and annual draft. However, to meet the increasing demands for various activities, groundwater is being mined from below static levels without taking natural recharge into consideration. Because of this, the water levels are declining in many parts of India resulting in the failure of existing wells. This is not only adversely affecting agricultural production but is also leading to a shortage of drinking water.

Augmenting groundwater reserve through artificial recharge helps to a great extent in improving the sustainability of existing groundwater sources (Karanth 1987; Rao et al. 1996). In many parts of the world, efforts are being made to recharge groundwater by constructing recharge structures (Kulbhushan et al. 2008), and it is reported that there is a significant rise in groundwater levels. Now, it has become an established practice to recharge groundwater through artificial methods (Shah 2008).

The total reserve of groundwater is the result of rain and snowmelt water percolating through various layers of soil and rocks. But, the amount of percolation varies greatly within the same region depending upon climatic factors like temperature and humidity, terrain conditions, the amount and pattern of rainfall, and the characteristics of soils and rocks (*Handbook of Hydrology* 1972). These issues need to be addressed for effective recharge and to achieve the targeted benefits. Very often, recharge structures are constructed on administrative grounds resulting in the ineffectiveness of the structures and wastage of money.

Much work has been done on all of these aspects, and information is available for specific locations. As far as the construction of recharge structures on the ground on operational levels is concerned, there are a few gaps that need to be addressed. There are two major issues: (1) availability of data for the study and analysis and (2) a study and analysis of the data related to artificial recharge for the selection of site-specific recharge structures.

This chapter demonstrates the efficacy of space technology for groundwater sustainability through artificial recharge. The advantage of satellite data as an integrated input database for the selection of site-specific recharge structures on an operational basis has already been brought out (RGNDWM 2008). A remote sensing and geographic information system (GIS)-based methodology for the identification of suitable locations that require the study and analysis of factors influencing recharge in the given area and the selection of suitable recharge structures for the given conditions has been developed.

The technical guidelines and the methodology developed in this chapter help in constructing terrain-specific recharge structures for better precipitation, thereby improving the sustainability of groundwater sources.

Groundwater comes into existence with the process of infiltration. The storage capacity of rock formations depends on the porosity of the rock. In rock formations, water moves from areas of recharge to areas of discharge under the influence of hydraulic gradients depending on the hydraulic conductivity or permeability. Therefore, the hydrogeological framework within which groundwater occurs is the same for groundwater recharging (Figure 13.1).

The hydrogeological properties depend on the deformation and the geomorphic processes that the original rock formation has undergone.

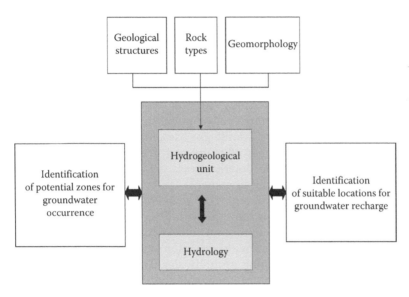

FIGURE 13.1
Factors influencing groundwater occurrence and recharge.

These modifications bring significant variations in the hydrogeological properties within the rock type, thereby changing groundwater storage and the transmitting abilities both horizontally and vertically. Therefore, the framework that governs groundwater occurrence as well as groundwater recharge is as varied as that of rock types, as intricate as their structural deformation and geomorphic history, and as complex as that of the balance among the lithologic, structural, and geomorphic parameters.

13.2 Satellite Image: Input Database

The hydrogeological properties of the units that act as a geological framework for groundwater recharge are manifested in the form of rock types, geological structures, and landforms. These processes and the resultant changes are manifested on the surface. Satellite imagery is one of the best tools for providing information pertaining to all these parameters and also available in an integrated environment. Based on the interpretation of satellite imagery, in conjunction with limited ground truth information, the extraction and mapping of the spatial distribution of rock formations, landforms, the structural network, and hydrological conditions can be done accurately. They can be better studied and understood in association with each other. This is not possible through conventional ground surveys. Apart from this, it takes a lot of time and energy, thereby making groundwater survey costly. An estimation of groundwater resources is generally based on geology, in particular on lithological characteristics. Particularly, data on landforms, geological structures, and recharge conditions are not feasible through conventional surveys.

If an area is occupied by granite gneisses, intruded by dolerite dykes, and cut across by a number of faults and lineaments, it is possible to draw conclusions that the dolerite dykes act as barrier for the movement of water, whereas the lineaments/faults that cut across them act as conduits for water movement. The weathered zones within the granite gneisses are feasible for limited quantities of groundwater recharge. Surface water bodies, like ponds and tanks, which are seen on the images as black patches, not only provide irrigation facilities in the area but also contribute toward recharging groundwater. Thus, by providing appropriate hydrogeological information, satellite data facilitate the proper identification and mapping of prospective groundwater recharge zones. Satellite data, by providing a spatial distribution of the irrigated crop land as bright red patches in a false color composite or green in the true color composite, are not only useful in calculating where and how much of groundwater is being tapped for irrigation, but also in classifying the entire area into overdeveloped, underdeveloped, optimally developed, and undeveloped zones, indicating the status of groundwater development. An analysis of multispectral high-resolution data clearly depicts minor faults and lineaments indicated by offsets and gaps in the dyke ridges. These lineaments act as conduits for the movement of water below the ground and form prospective groundwater recharge zones. In addition, some minor fractures originating from these major lineaments, and passing through water bodies, also form potential zones for groundwater recharge.

13.3 Hydrogeological Units and Recharge Conditions

Hydrogeological units are derived based on the integration of lithology, landform, structure, and recharge conditions. Each unit is unique with respect to the presence of the controlling factors. These units are considered as three-dimensional homogenous entities with respect to the hydrogeological properties and the recharge potential. The properties for groundwater recharge are expected to be uniform in each unit. However, some amount of heterogeneity may exist at the microlevel and it can be brought out only by using high-resolution images on the scale of mapping. The degree of heterogeneity and the resultant variations in groundwater recharge need to be accounted for depending on the scale of the study.

The hydrogeological unit is considered as the unit area of the geological framework that governs groundwater recharge. The geological framework is considered as a three-dimensional entity that includes the slope, the soil thickness, the weathered zone, and the basement rock. The infiltration capacity of the soil is an important factor that governs the rate of saturation of the vadose zone and thereby the efficacy or otherwise of a recharge structure. Impervious soil horizons act as barriers for recharge. Similarly, the weathered zones with additional porosity and permeability facilitate the recharge and form shallow aquifers. Buried zones, formed due to the deposition of transported material that is clayey, act as impervious horizons for groundwater movement. The basement rock, along with the weathered zone, acts as aquifer material. Storage coefficient, availability of storage space, and permeability vary significantly for different geological materials. Very high permeability of the geological material results in the loss of recharged water due to subsurface drainage, whereas low permeability of the geological material reduces the recharge rate. Moderate permeability facilitates a good recharge rate and retains the recharged water for a sufficient amount of time during the lean period. Older alluvium, buried channels, alluvial fans, dune sands, glacial outwash, etc., are the favorable landforms for recharge. In hard rock areas, fractured, weathered, and cavernous zones are capable of allowing a high intake of water. Basaltic rocks, comprising of lava flows, usually have large local pockets that can take recharge water.

13.3.1 Study and Analysis of Factors

In order to derive the hydrogeological units, the rock types, geological structures, and geomorphic units are mapped based on the interpretation of satellite imagery (Figure 13.2). The existing maps and literatures are helpful in understanding the geological setting of the area and the different rocks types that occur or are likely to occur in the area. With this "a priori knowledge," satellite imagery is studied to correlate the image characteristics with different rock types. The contrasting rock types and their boundaries can be seen very clearly on the satellite imagery with different colors and tones or land use. In other cases, complementary evidences have to be considered to demarcate the boundaries between different units.

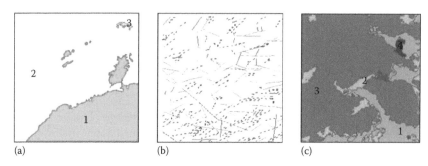

(a)   (b)   (c)

FIGURE 13.2
Map showing the distribution of (a) rock types 1–3, (b) geological structures, and (c) geomorphic units 1–4.

Faults, fractures, dykes, veins, etc., which act as conduits and barriers for the movement of groundwater, and bedding, schistosity, foliation, folds, etc., which contribute toward the creation of primary porosity and permeability, are considered for mapping.

While demarcating the geomorphic units, the topomaps were consulted to comprehend the relief variations and other topographic features. A digital elevation model can also be generated using stereo data, which in turn is used for deriving the slope of the area (Jenson and Domingue 1988). The geomorphologic map, showing the assemblage of different landforms, is prepared based on the lithological map so that each rock type is classified into different geomorphic units. Sometimes, one lithologic unit may be classified into two or more geomorphic units and vice versa. The geomorphic units that have been classified into shallow, moderate, and deep categories, based on their depth of weathering, the thickness of the deposited material, etc., have to be verified on the ground by observing the stream cuttings, well sections, etc. However, the contacts between the shallow, moderate, and deep categories are gradational.

13.3.2 Derivation of Hydrogeological Units

In the second step, the lithological, geomorphological, and structural maps have been subjected to an overlay analysis in the GIS environment. As a result of the integration, the areas with unique lithology, landform, and structure are delineated, thereby taking into account the primary porosity and permeability of the rock formations, the secondary porosity and permeability developed due to structural deformation, the geomorphic process, and landform genesis. These integrated lithological–structural–geomorphic units (Figure 13.3) are treated as homogenous areas with respect to the hydrogeological properties.

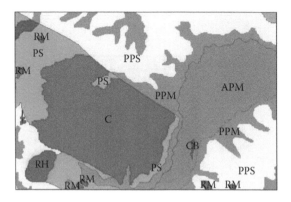

FIGURE 13.3
Delineation of aquifers integrated with lithology and geomorphology: C, Cuesta; RM, residual mound; RH, residual hill; I, inselberg; PS, piedmont slope; PPS, pediplain shallow weathered; PPM, pediplain moderately weathered; CB, channel bar.

13.4 Prioritization of Hydrogeological Units

The first step in planning a recharge structure is to demarcate the area for recharge. Such an area should, as far as possible, be a hydrogeomorphic unit where the hydrogeological properties are unique. However, localized structures can also be taken up for the benefit of a single hamlet or a village. In either case, the demarcation of the area has to be based on certain broad criteria.

13.4.1 Criteria for Prioritization

The following criteria have been adopted for the prioritization of the area, i.e., the hydrogeomorphic units for constructing recharge structures:

- Presence of villages with drinking water scarcity (mainly due to the decline in water table)
- Status of groundwater development
- Areas where groundwater levels are fast declining due to overexploitation
- Areas where a water-quality problem exists
- Areas where recharge is poor/limited due to unfavorable hydrogeological conditions

Satellite data, in conjunction with limited ground truth (Figure 13.4), forms an ideal dataset for deriving information pertaining to these parameters.

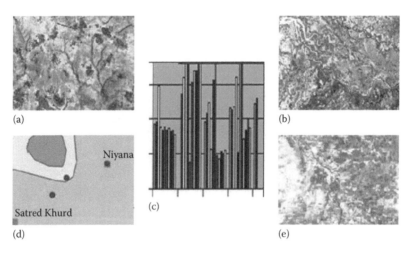

(a)

Niyana

Satred Khurd

(c)

(b)

(d)   (e)

FIGURE 13.4
IRS satellite images: (a) presence of villages with drinking water scarcity, (b) status of groundwater development, (c) areas where groundwater levels are declining fast due to overexploitation, (d) areas where water quality problems exist, and (e) areas where recharge is poor/limited due to unfavorable hydrogeological conditions.

The following five priority categories are identified and considered for prioritization:

1. Very high priority
2. High priority
3. Moderate priority
4. Low priority
5. No priority

Hydrogeomorphic units, where drinking water sources have dried up, water levels are declining fast, or a larger number of drinking water scarcity villages are located or where the percentage of the groundwater irrigated area is very high or a quality problem is reported (which can be improved by dilution through recharge), are delineated "very high priority" zones. Similarly, units that are mainly covered under forests, are inhabited, or have shallow water tables, with good-to-excellent recharge from canals, surface water bodies, rivers, etc., are designated "no priority" zones. The remaining units are given "high priority"/"moderate priority"/"low priority" status, depending on the usage of groundwater, the stress on the aquifer, and water-quality problems.

Well inventory data include type of well, depth-to-water table, water table fluctuation (i.e., pre- and post-monsoon water tables), yield, total depth of well, and type of subsurface formations collected from all the lithologic–landform combinations. This integrated information needs to be used for evaluating the hydrogeological units in different prioritization categories.

The priority with which the hydrogeological unit is to be considered for artificial recharge is based on the hydrogeological condition of a suitable location and a cost–benefit analysis.

13.5 Estimation of Surface Water Available for Recharge

Before undertaking the construction of a recharge structure, it is important to assess the adequacy of water available for artificial recharge in the hydrogeological unit or a part of the hydrogeological unit, as the case may be (Sang-Ki et al. 2004). The sources of recharge water, the form in which surface runoff occurs, and the amount of water available for a particular recharge structure are the issues involved in it.

Precipitation is the main source of recharge. Large reservoirs with a canal network also form sources. The surplus water can be diverted for recharge, without violating the rights of other users. Rainfall data need to be analyzed for understanding its pattern and evaporation losses in order to determine the amount of water available from given catchments. The main aspects to be considered are

- Minimum annual rainfall during the previous 10 years
- Number of rainy spells in a rainy season and duration of each spell
- Amount of rainfall in each rainy spell
- Rainfall intensity (maximum), 3 hourly, 6 hourly, etc., as may be relevant for a region

The surface water available for recharge is considered to occur in the form of overland flow, channel flow, and subsurface flow (Frot and van Wesemael 2009). Surface runoff is the water flow that occurs when soil is infiltrated to full capacity and excess water flows in the forms of over-land/sheet flow, channel/stream flow, and subsurface flow. The prediction of surface runoff is one of the most useful hydrological capabilities of GIS. In this chapter, the soil conservation services (SCS) model is used for estimating runoff. Input parameters, including the slope required for estimating runoff, were generated from satellite imagery.

Surface runoff (SR) is calculated as per the modified equation for the Indian condition (*Handbook of Hydrology* 1972):

$$SR = \frac{(P-0.35S)^2}{(P+0.7S)} \quad \text{(If } P - 0.35S \text{ is negative, } SR = 0\text{)}$$

where
  P is the precipitation
  S is the potential of maximum soil moisture retention after runoff begins
  S is calculated as $S = 25{,}400/CN - 254$

where CN is the curve number.

The CN is calculated for each recharge zone based on the land use category, land use practices, soil type, and hydrologic conditions.

Rainfall data available from district statistical reports and the meteorological department were used. Data on the infiltration capacity of the soils, together with maps showing infiltration rates, have been taken from the district agriculture officer and the central and state groundwater boards. The stream network and distribution of existing surface water bodies have been generated from satellite data. Using the SCS model, water available for artificial recharge is estimated (Dages et al. 2009).

13.6 Selection of Site-Specific Recharge Structures

Hydrogeological units, as per priority, need to be taken up for the identification of sites for artificial recharge, vis-à-vis selection of suitable recharge structures. The sites are to be mainly identified based on the adequacy of surface water (Saraf and Choudhury 1998), and the recharge structures are to be selected based on the given hydrogeological properties as well as the terrain condition.

There are many types of practiced recharge structures with different engineering designs and of a nomenclature-particular location and terrain. However, as far as the form of surface flow and its harvesting is concerned, the following seven types of recharge structures are considered (Table 13.1). The structure can be designed suitably while constructing these recharge structures.

Considering the surface flow available for harvesting and the hydrogeological properties of the given location, a suitable recharge structure, listed in Table 13.1, is to be selected for different forms in the hydrogeological unit (Sekar and Randhir 2007). In general, the locations for recharge structures are identified about 200–300 m upstream of the habitations so

TABLE 13.1

Types of Recharge Structures Suitable for
Harvesting Different Forms of Surface Flow

Type of Recharge Structure	Forms of Surface Flow
Percolation tank Check dam	Overland flow
Nala bund Desilting of tank	Stream flow
Invert well (recharge well) Recharge pit	Interstream divides
Subsurface dyke	Base flow

that drinking water sources in the habitations also become sustainable. The locations identified for harvesting overland and stream flow should be mainly on first- to third-order streams, and at the most, up to the initial stages of fourth-order stream (Horton 1945). No recharge structure should be located on major streams or rivers occupying large areas. The criteria used for the identification of sites for the construction of the seven types of recharge structures are as follows.

Recharge pit: Recharge pits are to be located around the habitations where drainage does not exist, e.g., water divide areas, hill/plateau tops, etc. Such structures can be constructed for harvesting rain water/limited surface flow available in the interstream divide areas where the retention capacity of the naturally recharged water is low due to slopes (Figure 13.5). In order to improve the sustainability of drinking water sources, such structures may be preferred in the existing tanks.

FIGURE 13.5
IRS P6 LISS IV image of part of Mysore District, Karnataka, India, showing a village situated on the interstream divide area.

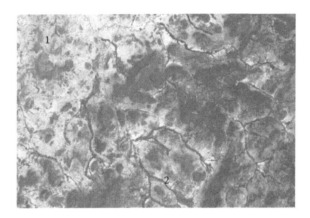

FIGURE 13.6
IRS 1D LISS III image of parts of Latur District, Maharashtra, India, showing a lateritic upland with dry channels.

Subsurface dyke: Sites for constructing subsurface dykes are to be located across the streams flowing in lateritic and alluvial terrains. Specific retention is very poor due to the psolitic and cavernous texture in the laterites and the high percentage of void space in alluvium (Figure 13.6). The subsurface dykes act as a subsurface barrier to retard the base flow and store water upstream below the ground surface.

Check dam: The check dams are to be identified mainly along the foothill zones where the slope is between 0% and 5%. The locations can be the first- and second-order streams where the overland flow from hills and uplands gets accumulated (Figure 13.7).

Percolation tank: Percolation tanks are to be identified in weathered and fractured zones of hard rock areas where secondary porosity and permeability exist. The overland surface flow from plains and valleys, accumulated in the first- to third-order streams, can be selected as the locations for constructing structures with a suitable engineering design (Figure 13.8).

Nala bund: A nala bund is a structure that is to be constructed for harvesting the stream flow in the river bed where acquisition of land for constructing other recharge structures is not possible. Limited water can be stored in the river bed for some time, which can increase recharge in the surrounding areas. Higher order (third to fourth) streams flowing through the plains and valleys can be suitable locations for constructing these structures (Figure 13.9).

Invert well/recharge well: These are basically well-type structures that are to be constructed in the areas where the transmissivity of the upper strata

FIGURE 13.7
IRS P6 LISS IV image of part of Nalgonda District, Andhra Pradesh, India, showing foot hill slope zone traversed by first- and second-order streams.

FIGURE 13.8
IRS P6 LISS IV image of parts of Rangareddy District, Andhra Pradesh, India, showing valleys formed by fracture zones. Crops concentration along fracture zones indicates over-exploitation of groundwater.

is poor. The surface flow is expected to recharge the underlying porous and permeable formation. Impervious shaly horizons underlain by sandstones, buried pediplains formed by transported clayey material with low permeability, Deccan traps where vesicular basalt is overlain by massive basalt or thick black cotton soil, or an impervious zone are the suitable areas for the construction of recharge wells (Figure 13.10).

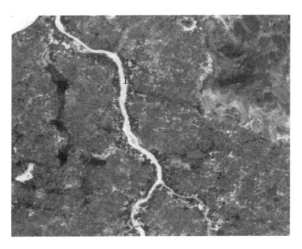

FIGURE 13.9
IRS 1D LISS III image of part of Karimnagar District, Andhra Pradesh, India, showing an ephemeral stream of higher order flowing in the pediplain.

FIGURE 13.10
IRS P6 LISS IV image of part of Nalgonda District, Andhra Pradesh, India, showing weathered gneiss overlain with transported black soil.

Desiltion of tanks: Silt and clay act as an impervious layer and barrier for groundwater recharge. Desilting facilitates normal recharge from the water impounded in the tank. Normally, desilting is recommended in small tanks that are partially silted up (Figure 13.11). Siltation in the tanks can be assessed based on the study of the multitemporal satellite images in conjunction with ground data.

Accordingly, in each hydrogeomorphic unit, locations for the suitable type of recharge structures are identified and shown in Figure 13.12.

FIGURE 13.11
IRS P6 LISS IV image of parts of Mysore District, Karnataka, India, showing silted water bodies/tanks.

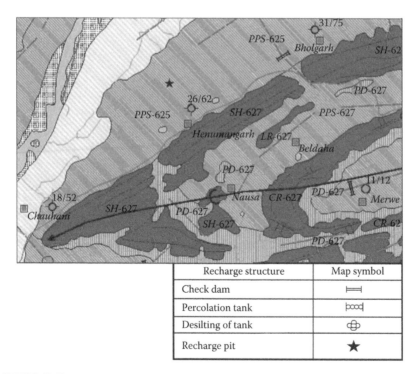

Recharge structure	Map symbol
Check dam	⊨⊨
Percolation tank	⋈⋈
Desilting of tank	⊕
Recharge pit	★

FIGURE 13.12
Groundwater recharge map of part of Thikamghad District of Madhya Pradesh, India, showing hydrogeomorphological units with recharge structures.

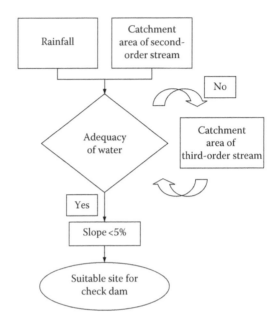

FIGURE 13.13
Decision support system for selecting sites for check dams.

13.7 Development of Decision Support System

A semi automated decision support system (DSS) to map the locations for various artificial recharge structures has been developed. The system is developed on a GIS platform based on logical expression. The criterion used for locating recharge structures manually is considered as the input for formulating the DSS. Figure 13.13 depicts, as an example, the DSS developed for locating sites for check dams.

Similarly, based on the criteria discussed against each structure, the DSSs have been developed for percolation tanks, nala bunds, invert wells, the desilting of tanks, recharge pits, and subsurface dykes. Using the DSS, the recharge structures can be located automatically in a given area and it acts as a cost- and time-effective tool.

13.8 Conclusion

Systematic planning with scientific rationale is essential in the development of groundwater resources and sustainability. The factors that control the groundwater and sustainability regime have been grouped into lithology, geological structures, geomorphology, and recharge conditions. As described in this chapter, by using these inputs derived from remote sensing data blended with ground information and integrated with the GIS domain, it is possible to know more precisely where to construct suitable types of water harvesting structures for the sustainability of groundwater.

The demand for groundwater in a hard rock area is more than the resources available. Unless the augmentation of the resources by way of constructing the recharge structures is planned scientifically and in time, most of the area may turn into overexploited or dark zones.

Favorable sites with the required ground conditions for different types of water recharging structures are explained above with a few examples. Though the augmentation of the resource is possible everywhere, except in a few hydrogeomorphologically unfavorable zones, implementing it in a larger area all at the same time during a given period may not be feasible. In this endeavor, a prioritization of the area and the time- and cost-effective use of satellite and GIS techniques have proven to be strong tools in the achievement of the goal. Recent and archival data from the Indian remote sensing satellite (IRS) comprising high-resolution sensors can be utilized effectively at the various stages of the monitoring of the sustainability aspects of groundwater development.

Acknowledgments

The authors are thankful to the Department of Drinking Water Supply (DDWS), Ministry of Rural Development (MoRD) Government of India, for the financial support, and to the Rajiv Gandhi National Drinking Water Mission project in which the study was undertaken. The authors are also thankful to the Director, NRSC and the Deputy Director, RS and GIS-AA, NRSC, for their encouragement in carrying out this work. The help provided by colleagues from the Hydrogeology Division of the NRSC in preparing this chapter is gratefully acknowledged.

References

Dages, C., Voltz, M., Bsaibes, A. et al., 2009. Estimating the role of a ditch network in groundwater recharge in a Mediterranean catchment using a water balance approach, *Journal of Hydrology*, 375(3–4): 498–512.

Frot, E. and van Wesemael, B., 2009. Predicting runoff from semi-arid hillslopes as source areas for water harvesting in the Sierra de Gador, southeast Spain, *CATENA*, 79(1): 83–92.

Handbook of Hydrology, 1972. Soil Conservation Department, Ministry of Agriculture, New Delhi.

Horton, R.E., 1945. Erosional development of streams and their drainage basins: Hydrophysical approach to quantitative morphology, *Bulletin of the Geological Society of America*, 56: 275–370.

Jenson, S.K. and Domingue J.O., 1988. Extracting topographic structure from digital elevation data for geographic information system analysis, *Photogrammetric Engineering and Remote Sensing*, 54(11): 1593–1600.

Karanth, K.R., 1987. *Ground Water Assessment, Development and Management*. Tata McGraw Hill Publishing Company Ltd., New Delhi, 720pp.

Kulbhushan, B., Kalro, A.H., and Kamalamma, A.G., 2008. Community initiatives in building and managing temporary check-dams across seasonal streams for water harvesting in South India, *Agricultural Water Management*, 95(12): 1314–1322.

Rao, M.S., Adhikari, R.N., Chittaranjan, S., and Chandrappa, M., 1996. Influence of conservation measures on ground water regime in a semi-arid tract of south India, *Agricultural Water Management*, 30(3): 301–312.

RGNDWM, 2008. *Ground Water Prospects Mapping Manual for Rajiv Gandhi National Drinking Water Mission* (RGNDWM), National Remote Sensing Centre, ISRO, Department of Space, Hyderabad, 256pp.

Sang-Ki, M., Woo, N.C., and Lee, K.S., 2004. Statistical analysis of hydrographs and water-table fluctuation to estimate groundwater recharge, *Journal of Hydrology*, 292(1–4): 198–209.

Saraf, A.K. and Choudhury, P.R., 1998. Integrated remote sensing and GIS for groundwater exploration and identification of artificial recharge sites, *International Journal of Remote Sensing*, 19(10): 1825–1841.

Sekar, I. and Randhir, T.O., 2007. Spatial assessment of conjunctive water harvesting potential in watershed systems, *Journal of Hydrology*, 334(1–2): 39–52.

Shah T., 2008, India's master plan for groundwater recharge: An assessment and some 41 suggestions for revision, *EPW*, 43(51): 41–49.

14

Fuzzy Arithmetic Approach to Characterize Aquifer Vulnerability Considering Geologic Variability and Decision Makers' Imprecision

Venkatesh Uddameri and Vivekanand Honnungar

CONTENTS

14.1 Introduction

Anthropogenic land surface activities, such as agriculture and industrialization, can have a deleterious impact on water quality in shallow aquifers and therefore limit the amount of water available at any location. This recognition of the interlinkages between land-use activities and ground-water quality has led to the concept of aquifer vulnerability. Very broadly, aquifer vulnerability is defined as the susceptibility of aquifer to pollution from human activities (NRC 1993; Gogu and Dassargues 2000; Connell and van den Daele 2003). Therefore, delineation of aquifer vulnerability is vital to protect groundwater resources and facilitate environmentally sustainable development.

Multicriteria decision-making (MCDM) approaches are commonly used to define aquifer vulnerability over large regional scales. The DRASTIC approach (Aller et al. 1985) put forth by the U.S. Environmental Protection Agency (USEPA) is a popular MCDM approach for aquifer vulnerability delineation. This approach continues to be used extensively in the United States (Fritch et al. 2000; Panagopoulos et al. 2006; Uddameri and Honnungar 2007) and many other countries including Israel (Melloul and Collin 1998), Egypt (Ahmed 2009), Nicaragua (Johansson et al. 1999), European Union (Lobo-Ferreira 1997; Lobo-Ferreira and Olivera 1997; Ducci 1999), South Africa (Lynch et al. 1997), and South Korea (Kim and Hamm 1999). The DRASTIC approach uses a weighted addition of seven hydrologic and hydrogeologic parameters to establish a composite index (DI) that defines the aquifer's inherent susceptibility to pollution (Equation 14.1):

$$DI = D_R * D_W + R_R * R_W + A_R * A_W + S_R * S_W + T_R * T_W + I_R * I_W + C_R * C_W$$

$$(14.1)$$

where
  D is the depth to water table
  R is net recharge
  A is aquifer media
  S is soil media
  T is topography
  I is impact of vadose zone
  C is hydraulic conductivity
  the subscripts R and W are the corresponding ratings for the area being evaluated and importance weights for the parameter, respectively

The ratings are nondimensional scores assigned to a parcel of land based on its geographic characteristics and therefore reflect the observed conditions in the field. Weights, on the other hand, symbolize the relative importance that a decision maker assigns to each variable. The availability of digital data greatly facilitates the regional-scale mapping of aquifer vulnerability using geographic information systems (GIS).

While DRASTIC provides a pragmatic approach to delineate aquifer vulnerability, its application is hampered by subjectivity. For example, different approaches to establish ratings from known hydrogeologic properties have been provided by various authors (e.g., Aller et al. 1985; Navalur and Engel 1998). In a similar fashion, there are disagreements with regard to the relative importance of different hydrogeological parameters that are used to develop the composite vulnerability score (Massam 1988; Nyerges et al. 1995; Armstrong et al. 1996; Malczewski 1996; Jankowski et al. 1997; Jankowski and Nyerges 2001; Al-Adamat et al. 2003; Babiker et al. 2005).

The subjective preferences of the decision makers with regard to weighting and rate assignments introduce imprecision in the estimated vulnerability characterization. It is likely that different decision makers will emphasize different attributes based on their understanding and value preferences with regard to the system under consideration. In regional-scale assessments, an individual decision maker may also face difficulties in coming up with a single weighting scheme that is appropriate over the entire domain. As aquifer vulnerability is an abstract concept defined using a linear weighted addition scheme, the elimination of subjectivity is not possible. However, it is imperative that it should be properly quantified and considered during the decision-making process.

The concept of fuzzy set theory, first proposed by Zadeh (1965), provides a convenient approach to quantify the subjectivity associated with decision makers' preference. As such, the integration of fuzzy set theoretic concepts with aquifer vulnerability characterization has been undertaken in recent times. In particular, several authors (e.g., Uricchio et al. 2004; Dixon 2005; Martino et al. 2005) have used fuzzy inferencing systems to delineate aquifer vulnerability. In this approach, the extent of the aquifer vulnerability is mapped to measurable hydrogeologic properties using a set of fuzzy IF-THEN rules. In most instances, these schemes cannot be directly integrated into a GIS environment. Also, Uddameri and Honnungar (2007) used a variant of fuzzy set theory called rough sets (Pawlak 2005) to capture the indiscernability in aquifer vulnerability characterization and fully integrated it into a GIS environment. In addition to the above-mentioned fuzzy set-theoretic techniques, the concept of fuzzy numbers and fuzzy arithmetic (Kaufman and Gupta 1985) can be used to characterize the uncertainties in the estimated aquifer vulnerability due to imprecision in ratings and weights. The main goal of this study is to develop methodologies to characterize the uncertainties in aquifer vulnerability arising due to subjective ratings and weights using concepts from fuzzy arithmetic theory and demonstrate its integration within a GIS framework.

14.2 Methodology

14.2.1 Fuzzy Sets and Fuzzy Numbers

In fuzzy set theory, the subjectivity of the decision maker's preference is captured using the mathematical concept called membership function. Figure 14.1 depicts a membership function wherein the domain of a set (variable) is represented on the X-axis and its degree of membership (or the extent to which a particular value belongs to the fuzzy set) is mapped on the Y-axis. A fuzzy number is a fuzzy set that is both normal and convex (Kaufmann and Gupta 1985). The normality assumption requires that at least one value of the domain has a degree of membership of unity. Convexity implies that the membership function has an increasing part and a decreasing part (Uddameri 2003). An α-cut is a subset of a fuzzy set in which all elements belonging to the subset have a membership of at least α and can be described using its upper and lower limits. From Figure 14.1, it is evident that α-cut can assume a value between 0 and 1.

While fuzzy sets and fuzzy numbers can assume any arbitrary shape, the use of triangular membership function is very common due to its simplicity and intuitive appeal (Cox 1995). As can be seen from Figure 14.1, they can be described using the values at the three vertices.

14.2.2 Fuzzy Mathematics

Mathematical operations can be carried out on fuzzy sets and numbers using the extension principle (Zadeh 1965). Dubois and Prade (1979) provided a convenient approach to apply the extension principle using the

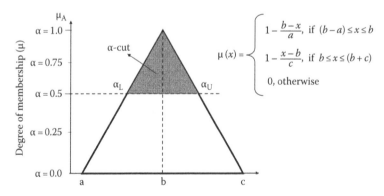

$$\mu(x) = \begin{cases} 1 - \dfrac{b-x}{a}, & \text{if } (b-a) \leq x \leq b \\[2mm] 1 - \dfrac{x-b}{c}, & \text{if } b \leq x \leq (b+c) \\[2mm] 0, & \text{otherwise} \end{cases}$$

FIGURE 14.1
An example of a fuzzy set and its mathematical representation.

concept of α-cuts. Consider the following generic mathematical operation (∘) on two fuzzy sets (*W* and *R*):

$$Y = W \circ R \tag{14.2}$$

Let $Y_{L,\alpha}$ and $Y_{U,\alpha}$ be the lower and upper bounds of *Y* at α-cut equal to α. These values can be obtained from the corresponding upper and lower bounds of variables *W* and *R* at α using the following expressions:

$$Y_{L,\alpha} = \text{Min}\left[W_{L,\alpha} \circ R_{L,\alpha}; W_{L,\alpha} \circ R_{U,\alpha}; W_{U,\alpha} \circ R_{L,\alpha}; W_{U,\alpha} \circ R_{U,\alpha}\right] \tag{14.3}$$

$$Y_{U,\alpha} = \text{Max}\left[W_{L,\alpha} \circ R_{L,\alpha}; W_{L,\alpha} \circ R_{U,\alpha}; W_{U,\alpha} \circ R_{L,\alpha}; W_{U,\alpha} \circ R_{U,\alpha}\right] \tag{14.4}$$

The membership function of the output fuzzy set (*Y*) can be constructed by performing the above calculations at a variety of α-cuts in the range of 0–1. For nonmonotonic membership functions and unwieldy mathematical operations, the minimization and maximization have to be obtained using optimization routines. However, elementary arithmetic operations on monotonic fuzzy numbers (e.g., triangular fuzzy numbers) can be directly obtained from the knowledge of interval arithmetic. The required operations are presented in Table 14.1.

It is important to note that the addition and subtraction of two triangular fuzzy sets also results in a triangular fuzzy set. On the other hand, the multiplication and division of two triangular fuzzy sets does not result in a triangular fuzzy set. However, Kaufmann and Gupta (1988) suggest that the output can be approximately represented as a triangular set using the triplet values corresponding to the extreme points of the membership function (i.e., α-cut of 0 and 1) with little loss of accuracy. Therefore, for triangular fuzzy sets, the output fuzzy set can be developed by performing necessary calculations at the end members and by using linear interpolation.

TABLE 14.1

Arithmetic Operations on Two Fuzzy Sets at a Given α-Cut

Fuzzy Arithmetic Operations	$\tilde{Y}_{L,\alpha}$	$\tilde{Y}_{U,\alpha}$	$\tilde{W}(W_{L,\alpha},R_{U,\alpha}) \circ \tilde{R}(R_{L,\alpha},R_{U,\alpha})$
$\tilde{Y} = \tilde{W} + \tilde{R}$	$W_{L,\alpha} + R_{L,\alpha}$	$W_{U,\alpha} + R_{U,\alpha}$	$[W_{L,\alpha} + R_{L,\alpha}, W_{U,\alpha} + R_{U,\alpha}]$
$\tilde{Y} = \tilde{W} - \tilde{R}$	$W_{L,\alpha} - R_{L,\alpha}$	$W_{U,\alpha} - R_{U,\alpha}$	$[W_{L,\alpha} - R_{L,\alpha}, W_{U,\alpha} - R_{U,\alpha}]$
$\tilde{Y} = \tilde{W} * \tilde{R}$	$W_{L,\alpha} * R_{L,\alpha}$	$W_{U,\alpha} * R_{U,\alpha}$	$[W_{L,\alpha} * R_{L,\alpha}, W_{U,\alpha} * R_{U,\alpha}]$
$\tilde{Y} = \tilde{W}/\tilde{R}$	$W_{L,\alpha} * 1/R_{U,\alpha}$	$W_{U,\alpha} * 1/R_{L,\alpha}$	$[W_{L,\alpha} * 1/R_{U,\alpha}, W_{U,\alpha} * 1/R_{L,\alpha}]$

14.2.3 Fuzzy Aquifer Vulnerability Characterization

If the ratings and weights associated with hydrogeologic parameters are considered to be fuzzy numbers, then the composite aquifer vulnerability score is also fuzzy. Therefore, when weights and ratings are expressed using fuzzy numbers, the DRASTIC approach can be described as follows:

$$DI = \left(\tilde{D}_W * \tilde{D}_R\right) + \left(\tilde{R}_W * \tilde{R}_R\right) + \left(\tilde{A}_W * \tilde{A}_R\right) + \left(\tilde{S}_W * \tilde{S}_R\right)$$

$$+\left(\tilde{T}_W * \tilde{T}_R\right) + \left(\tilde{I}_W * \tilde{I}_R\right) + \left(\tilde{C}_W * \tilde{C}_R\right) \tag{14.5}$$

where the superscript (◦) denotes a fuzzy number, the subscript W and R refer to weightings and ratings, respectively, and the DRASTIC parameters are as defined as in Equation 14.1.

The extension principle and the associated α-cut optimization approach presented in Equations 14.3 and 14.4 provide a general framework to extend the crisp MCDM process to fuzzy inputs. If the weightings and ratings are expressed as triangular fuzzy numbers, then the interval arithmetic operations listed in Table 14.1 can be used. It is important to recall that each multiplication of ratings and weights will result in a nontriangular fuzzy number. However, if the product of each weighting and rating can be approximated as a triangular fuzzy number, then the membership function for the composite vulnerability index (DI) can be constructed by making the necessary calculations at the vertices (i.e., $\alpha=0$ and $\alpha=1$) alone.

14.2.4 Defuzzification of the Composite Fuzzy Vulnerability Index

In decision-making environments, a fuzzy definition of vulnerability is of limited value as each parcel of land has to be characterized as either vulnerable or not. The process of obtaining a representative value from a fuzzy set is termed defuzzification. While there are several defuzzification approaches, the use of centroid of the resultant fuzzy set as a representative crisp value is most common (Wang and Mendel 1990). The centroid of a resultant fuzzy set (Y) can be obtained by performing the following integration:

$$\text{Centroid} = \frac{\int \mu(y) y \, dy}{\int \mu(y) dy} \tag{14.6}$$

where $\mu(y)$ is the membership function of the resultant fuzzy set Y. The integrations are carried out over the entire domain of the resultant Y. If the output fuzzy set is triangular, then the centroid can be calculated analytically from the boundary values as follows:

$$\text{Centroid} = \frac{Y_{L,0} + Y_{M,1} + Y_{U,0}}{3} \tag{14.7}$$

where

$Y_{L,0}$ and $Y_{U,0}$ are the lower and upper bounds at α-cut equal to zero
$Y_{M,1}$ is the value of Y corresponding to α-cut of unity

The fuzzy arithmetic and MCDM approaches discussed here can be tightly coupled with GIS. The implementation of the developed approach will be illustrated using an aquifer vulnerability delineation case study, which is discussed next.

14.3 Illustrative Case Study

14.3.1 Study Area and Data Compilation

An 18-county region of South Texas, previously studied by Uddameri and Honnungar (2007), was used to illustrate the concepts developed here.
Data were compiled from a variety of sources and are listed in Table 14.2. All datasets were reprojected into a common projection system (UTM Zone 14N, NAD 83) and converted to rasters (30 m × 30 m cells) prior to the development of raw DRASTIC maps. The reader is referred to Uddameri and Honnungar (2007) for additional details pertaining to the data compilation and preprocessing. Briefly, the depth to the water table (D) map was developed from interpolation of measured average water table depths in shallow wells (screened within 500 ft of land surface). The recharge (R) map was constructed using the equation provided by Williams and Kissel (1991). The aquifer media (A) was noted to be uniform and as such excluded from the analysis. The soil type (S) was represented using soil texture obtained from STATSGO database. The topographic (T) influences were quantified using slopes computed using 1:250,000 DEM. The soil organic matter was used as a surrogate for the impact of vadose zone (I), and drainage class was used as a proxy for conductivity (C).

14.3.2 Development of Fuzzy Ratings and Weights

To facilitate proper implementation of MCDM scheme, the raw values of DRASTIC parameters need to be converted into a consistent set of

TABLE 14.2

Sources and Types of Input Datasets

Data	Scale	Source	Remarks
Texas County maps	1:100,000	U.S. Census Bureau	Used as base maps
Groundwater database	—	TWDB	Data from 1966 to 2000 were used
Topography	1:250,000	USGS	Used to calculate slope (%) that was used to represent topography
STATSGO	1:250,000	NRCS, USDA	Drainage class was used for conductivity (C); soil organic matter was used for impact of vadose zone (I); surface texture was used for soil type (S)
Precipitation	—	NWS, NOAA	Used to calculate recharge in Kissel–Williams' equation

TWDB, Texas Water Development Board; USGS, United States Geological Survey; STATSGO, State Soil Geographic Database; NRCS, Natural Resources Conservation Service; USDA, United States Department of Agriculture; NWS, National Weather Service; NOAA, National Oceanic and Atmospheric Administration.

dimensionless numbers called ratings. The ratings are also scaled such that the higher the rating of a parameter, the higher the vulnerability (cēterēs paribus). The conversion of raw values to dimensionless ratings is a subjective process and depends upon the risk-preferences of the decision maker, which can be both real and perceived (Daughton 2004). In the context of this study, the risk-preference characterizes the decision maker's perception of the impacts of aquifer vulnerability on the value of the groundwater to humans and other users (Lobo-Ferreira 1997; Ducci 1999; Perles Rosello et al. 2009). In the presence of a contaminant source, a risk-averse decision maker perceives that small levels of vulnerability can greatly diminish the value of the groundwater. On the other extreme, a risk-taking decision maker would assume lower levels of risk even at higher levels of vulnerability for the same pollution source. As a result, a risk-averse decision maker will assign a high rating (implying higher vulnerability) even when the water table is relatively deep. On the other hand, a risk-taking decision maker will specify a low rating even at a very shallow aquifer.

The risk-preferences of the decision maker were captured using exponential utility functions, summarized in Table 14.3 (Kirkwood 1996). The monotonically increasing exponential utility functions (EUF) were used for parameters that have a direct relationship with vulnerability (i.e., recharge, soils ranked as per increasing pore size [clay to gravels], conductivity [increasing drainage classes]). The monotonically decreasing

TABLE 14.3

Exponential Utility Functions for Different Risk Preferences

Decision Makers' Preference	Monotonically Increasing	Monotonically Decreasing
Risk-averse	$X_r = \dfrac{1-\exp\left(\dfrac{-(X_i - X_L)}{\rho}\right)}{1-\exp\left(-\dfrac{(X_H - X_L)}{\rho}\right)} \quad \forall \rho \geq 0$	$X_r = \dfrac{1-\exp\left(\dfrac{-(X_H - X_i)}{\rho}\right)}{1-\exp\left(-\dfrac{(X_H - X_L)}{\rho}\right)} \quad \forall \rho \geq 0$
Risk-neutral	$X_r = \left(\dfrac{X_i - X_L}{X_H - X_L}\right)$	$X_r = \left(\dfrac{X_H - X_i}{X_H - X_L}\right)$
Risk-taking	$X_r = \dfrac{1-\exp\left(\dfrac{-(X_i - X_L)}{\rho}\right)}{1-\exp\left(-\dfrac{(X_H - X_L)}{\rho}\right)} \quad \forall \rho \leq 0$	$X_r = \dfrac{1-\exp\left(\dfrac{-(X_H - X_i)}{\rho}\right)}{1-\exp\left(-\dfrac{(X_H - X_L)}{\rho}\right)} \quad \forall \rho \leq 0$

X_i: actual parameter value at the land parcel, i; ρ: risk-tolerance factor based on decision makers' preference; subscripts H and L: observed high and low values at the ith land parcel.

utility functions were used for parameters that have inverse relationship with vulnerability (i.e., depth, topography, and impact of vadose zone [soil organic matter]). A primary advantage of these functions lies in the fact that the risk characteristics can be captured using a single parameter (ρ).

Three different rating functions, one each for risk-averse, risk-neutral, and risk-taking situations, are developed for each parameter. As depicted in Figure 14.2, at any given parcel of land, the ratings corresponding to these situations are used to designate the bounds of the triangular membership function. The complete rating membership function was then constructed via interpolation.

Fuzzy membership functions can be established to characterize the preference of the decision maker with regard to each parameter in the model. For illustrative purposes, the weights were assumed to be nonsymmetric triangular fuzzy sets here as they are easy to construct with limited data and have an intuitive appeal. The bounds (vertices) of the triangular fuzzy set for each parameter are summarized in Table 14.4. In collaborative decision making, the composite weight membership function can be aggregated from individual membership functions using ordered weighted averaging schemes (Yager 1988), and the output can be approximated as a triangular function.

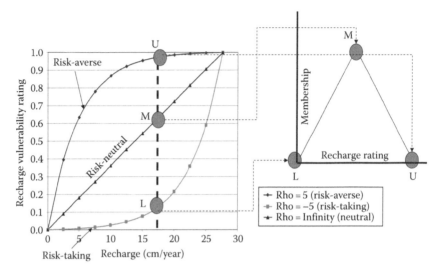

FIGURE 14.2
Developments of rating membership functions from utility functions.

TABLE 14.4

Triangular Membership Function Bounds
for Parameter Weights

Parameter	Lower ($\alpha=0$)	Middle ($\alpha=1$)	Upper ($\alpha=0$)
Depth (D_W)	1	5	5
Recharge (R_W)	1	4	5
Soil media (S_W)	1	3	5
Topography (T_W)	1	1	5
Impact of vadose zone (I_W)	1	2	5
Hydraulic conductivity (C_W)	1	3	5

14.3.3 Model Implementation

Once the rating and weight fuzzy sets are developed for a given parcel of land, they can be used with fuzzy arithmetic operations described in the previous section to develop a fuzzy vulnerability index, which in turn can be defuzzified to obtain a crisp (representative) value.

The required computations were carried out within a GIS environment using Map Algebra routines available in ArcGIS Version 9.3® (ESRI Inc., Redlands, CA). All the calculations were automated using the MODELBUILDER process workflow feature. A flowchart describing

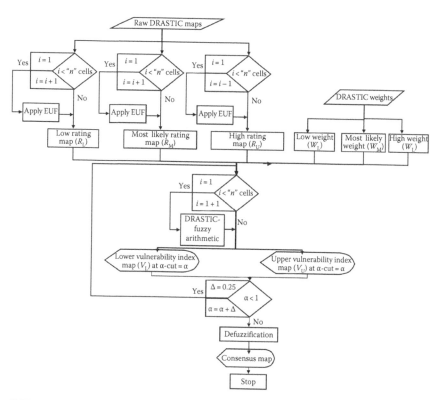

FIGURE 14.3

Flowcharts for implementation of fuzzy DRASTIC model.

the implementation is presented in Figure 14.3. It is important to recognize that fuzzy calculations have to be made at different α-cuts at each parcel of land. Also, the generation of the rating fuzzy membership function at each parcel of land entails additional computations involving the EUF.

In the illustrative case study, the study domain (~16,150 mi²) was discretized into 30 m × 30 m land parcels, resulting in ~46 million cells. The output vulnerability was computed at five different α-cuts, namely 0.00, 0.25, 0.50, 0.75, and 1.00. Separate rating maps corresponding to upper, lower, and most likely values were generated first using the exponential utility functions, each map requiring a little over 6 billion calculations. Once these maps were generated, they were used with fuzzy weights to compute the vulnerability at five different α-cuts, which required more than 30 billion calculations. Clearly, the integration of fuzziness with MCDM DRASTIC approach in a GIS framework is computationally intensive and requires high-end hardware for implementation.

14.4 Results and Discussion

14.4.1 Exact and Approximate Fuzzy Arithmetic Schemes for Vulnerability Calculations

Due to the positive nature of fuzzy multiplication operations, the composite vulnerability index calculated using fuzzy DRASTIC is not a triangular number even when all the ratings and weights are modeled using triangular fuzzy sets. However, following Kaufmann and Gupta (1985), the output from the multiplication operation can be approximated as a triangular set using the values computed at the vertices. The application of the proposed fuzzy DRASTIC approach over large domains entails significant number of calculations. Therefore, the approximation is critical for the successful implementation of the procedure. For example, the application of the proposed fuzzy DRASTIC approach to the study area of interest involved over 36 billion arithmetic operations even when the triangular approximation was invoked. However, any modeling approximation must not result in excessive errors so as to limit the validity of the approach. Therefore, to test the validity of the approximation, the fuzzy DRASTIC index was calculated using both exact and approximate methods at several locations within the domain. Figure 14.4 depicts an example of the fuzzy DRASTIC index computed using exact and approximate approaches.

A statistical comparison was carried out between the two approaches by comparing the centroid and the area of the corresponding membership functions. For the exact fuzzy set, the area was computed numerically using the Newton–Coates formula (Chapra and Canale 2005). The

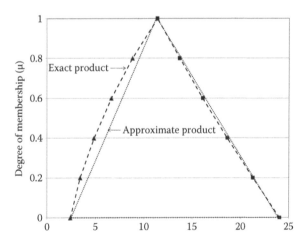

FIGURE 14.4
Illustration of the exact and approximate fuzzy DRASTIC index.

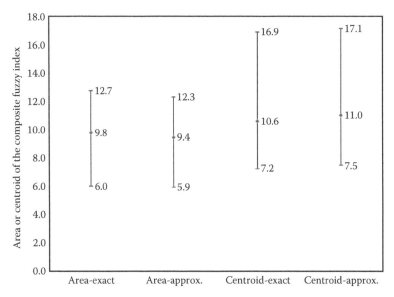

FIGURE 14.5
Statistical comparison of exact and approximate fuzzy arithmetic on aquifer vulnerability calculations.

centroid was calculated using Equation 14.6, where the required integral was again evaluated numerically.

As can be seen from Figure 14.5, the areas and centroids computed using exact and triangular approximation are very small with an average error of <3% and a maximum error of <8%. Figure 14.5 also indicates that the differences between the exact and approximate defuzzification approaches increase with increasing values of vulnerability. However, the centroids of the triangular approximation are noted to be higher than those obtained using the exact arithmetic. Therefore, the use of triangular approximation leads to a slight overestimation of the defuzzified aquifer vulnerability and builds in a factor of safety, while making the calculations computationally tractable. Therefore, the approximate fuzzy defuzzification method using the centroid scheme is recommended here.

14.4.2 Comparison of Fuzzy and Crisp DRASTIC Maps

Fuzzy DRASTIC maps corresponding to five different α-cuts were developed and defuzzified to obtain a crisp DRASTIC map via centroid defuzzification. Figure 14.6 depicts the various maps obtained through implementation of the procedure. As can be seen, there are two maps at each α-cut corresponding to the boundary values. As the ratings are all normalized between 0 and 1 and the weights between 1 and 5, the fuzzy

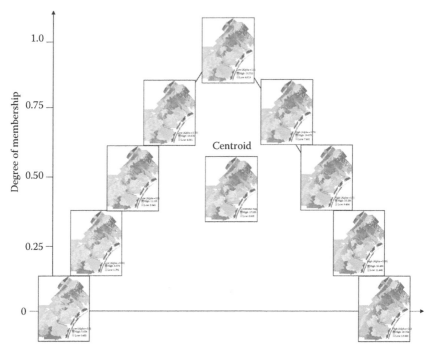

FIGURE 14.6
Calculation of fuzzy aquifer vulnerability at different α-cuts and centroid defuzzification scheme.

vulnerability index theoretically varies between 0 and 30. It was noted that the defuzzified values were in the range of 6.5–17.29. These values were divided into five categories (very low [VL], low [L], medium [M], high [H], and very high [VH]) using the equal interval method to facilitate comparison with traditional DRASTIC schemes. The granularized, centroid defuzzified map of the composite vulnerability index is presented in Figure 14.7, while Figure 14.8 depicts the composite vulnerability index obtained using a crisp (nonfuzzy) DRASTIC model that uses the ratings and weighting scheme of Navalur and Engel (1998). It can be seen that the centroid defuzzification approach provides a more conservative depiction of vulnerability, especially in Kleberg and Duval counties in the South, Goliad and Bee counties in the central, and Lavaca, Gonzalez, and De Witt in the northern portions of the study area. A comparison of relative proportions of the study area being classified under different categories using traditional and fuzzy DRASTIC approaches is presented in Figure 14.9.

The results indicate that the extent of conservatism increases with increasing uncertainty. Most notably, only 3% of the area is classified as

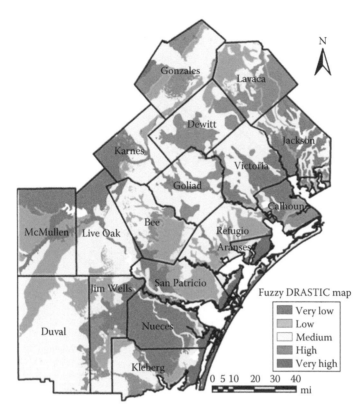

FIGURE 14.7
Composite aquifer vulnerability index using fuzzy DRASTIC approach.

being either "highly" or "very highly" vulnerable in the crisp DRASTIC model; the proportion for these categories increases to 15% when both ratings and weights are considered fuzzy. The fuzzy modeling approach is therefore congruent with the precautionary principle and assumes a parcel to be more susceptible to pollution when the information content is low (i.e., ratings and weights are fuzzy) (Figure 14.9).

14.5 Conclusions

Linear MCDM schemes such as DRASTIC are advantageous to map aquifer vulnerability and can be readily integrated within GIS. However, it is oftentimes difficult to define the required weights and ratings in a

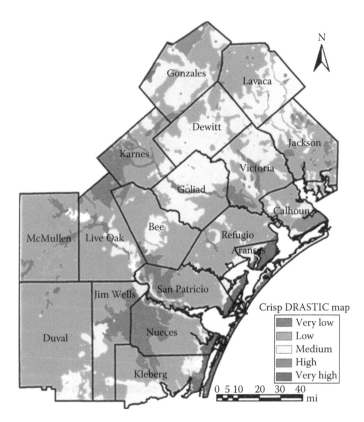

FIGURE 14.8
Composite aquifer vulnerability index using crisp DRASTIC approach.

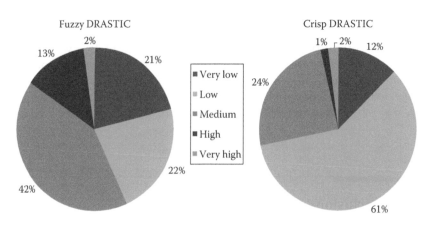

FIGURE 14.9
Comparisons of fuzzy DRASTIC and crisp DRASTIC vulnerability classifications.

crisp (precise) manner. Uncertainties in weight specifications arise due to imprecise preferences on the part of the decision maker, which is further exacerbated in collaborative decision making environments. Ratings are also affected by the risk-taking attitudes of the decision maker as well as due to the variability in the measured data. The fuzzy set theory provides a convenient approach to quantify these uncertainties and incorporate them into the decision-making process. A fuzzy MCDM approach for aquifer vulnerability characterization is developed in this study using approximate fuzzy arithmetic approaches (Kaufmann and Gupta 1985). While this approach is computationally intensive, requiring several billion calculations for the study area considered here, it is mathematically tractable.

A detailed error analysis was carried out to evaluate the accuracy of the approximate fuzzy arithmetic approach by comparing it with the exact approach at several locations in the study area. The results of this comparison indicated that the approximation errors are fairly small (typically <3%, with a maximum error <8%) and as such the use of approximate fuzzy arithmetic is deemed reasonable. The proposed fuzzy aquifer vulnerability approach resulted in a more conservative delineation of aquifer vulnerability than the traditional DRASTIC approach. Therefore, the fuzzy approach is consistent with the precautionary principle paradigm that has been put forth as a basis for environmental decision making (Cameron and Peloso 2005). The proposed fuzzy arithmetic approach provides an innovative decision support tool to assess the aquifer vulnerability issues under data uncertainty and stakeholder preference diversity.

Acknowledgments

This material is based on the work supported by Combined Course and Curriculum Development (CRCD) grant from the National Science Foundation (NSF) and the Center of Research Excellence in Science and Technology—Research on Environmental Sustainability of Semi-Arid Coastal Areas (CREST-RESSACA) at Texas A&M University—Kingsville (TAMUK) through a Cooperative Agreement (No. HRD-0734850) from NSF. Any opinions, findings, and conclusions or recommendations expressed in this material are those of the authors and do not necessarily reflect the views of NSF. The editors and the anonymous reviewers are also thanked for their input, which greatly improved the manuscript.

References

Ahmed, A.A., 2009. Using generic and pesticide DRASTIC GIS-based models for vulnerability assessment of the Quaternary aquifer at Sohag, Egypt. *Hydrogeology Journal* 7(5): 1203–1217.

Al-Adamat, R.A.N., Foster, I.D.L., Baban, S.M.J., 2003. Groundwater vulnerability and risk mapping for the Basaltic aquifer of the Azraq basin of Jordan using GIS, remote sensing and DRASTIC. *Applied Geography* 23(4): 303–324.

Aller, L., Bennett, T., Lehr, J.H., Petty, R.J., 1985. DRASTIC—A standardized system for evaluating groundwater pollution potential using hydrogeologic settings. EPA/600/2-85/018, U. S. Environmental Protection Agency Report.

Armstrong, M.P., Densham, P.J., Kemp, K., 1996. Report from the specialist meeting on collaborative spatial decision making, Initiative 17. National Center for Geographic Information Analysis, University of California-Santa Barbara, Santa Barbara, CA.

Babiker, I.S., Mohamed, M.A.A., Hiyama, T., Kato, K., 2005. A GIS-based DRASTIC model for assessing aquifer vulnerability in Kakamigahara Heights, Gifu Prefecture, Central Japan. *Science of the Total Environment* 345: 127–140.

Cameron, E., Peloso, G.F., 2005. Risk management and the precautionary principle: A fuzzy logic model. *Risk Analysis* 25(4): 901–911.

Chapra, S.C., Canale, R.P., 2005. *Numerical Methods for Engineers*. McGraw-Hill Education, New York.

Connell, L.D., van den Daele, G., 2003. A quantitative approach to aquifer vulnerability mapping. *Journal of Hydrology* 276: 71–88.

Cox, E.D., 1995. *Fuzzy Logic for Business and Industry*. Charles River Media, Inc., Rockland, MA.

Daughton, C.G. 2004. Groundwater recharge and chemical contaminants: Challenges in communicating the connections and collisions of two disparate worlds. *Ground Water Monitoring and Remediation* 24(2): 127–138.

Dixon, B., 2005. Applicability of neuro-fuzzy techniques in predicting groundwater vulnerability: A GIS-based sensitivity analysis. *Journal of Hydrology* 309: 17–38.

Dubois, D., Prade, H., 1979. Operations in a fuzzy-valued logic. *Information and Control* 43: 224–240.

Ducci, D., 1999. GIS techniques for mapping groundwater contamination risk. *Natural Hazards* 20: 279–294

Fritch, T.G., McKnight, C.L., Yelderman, J.C., 2000. An aquifer vulnerability assessment of the Paluxy aquifer, central Texas, USA, using GIS and a modified DRASTIC approach. *Environmental Management* 25(3): 337–345.

Gogu, R.C., Dassargues, A., 2000. Current trends and future challenges in groundwater vulnerability assessment using overlay and index methods. *Environmental Geology* 39(6): 549–559.

Jankowski, P., Nyerges, T.L., 2001. GIS-supported collaborative decision making: Results of an experiment. *Annals of the Association of American Geographers* 91(1): 48.

Jankowski, P., Nyerges, T.L., Smith, A., Moore, T.J., Hovarth, E., 1997. Spatial group choices a SDSS tool for collaborative spatial decision-making. *International Journal of Geographical Information Systems* 11(6): 566–602.

Johansson, P.O., Scharp, C., Alveteg, T., Choza, A., 1999. Framework for ground water protection—The Managua ground water system as an example. *Ground Water* 37(2): 204–213.

Kauffman, A., Gupta, M.M., 1985. *Introduction to Fuzzy Arithmetic: Theory and Applications*. Van Nostrand Reinhold Company, New York.

Kauffman, A., Gupta, M.M., 1988. *Fuzzy Mathematical Models in Engineering and Management Science*. North Holland, Amsterdam, the Netherlands.

Kim, Y.J., Hamm, S., 1999. Assessment of the potential for ground water contamination using DRASTIC/EGIS technique, Cheongju area, South Korea. *Hydrogeology Journal* 7(2): 227–235.

Kirkwood, C.W., 1996. *Strategic Decision Making: Multiobjective Decision Analysis with Spreadsheets*. Local, Duxbury Press, Belmont, CA.

Lobo-Ferreira, J.P., 1997. GIS and mathematical modeling assessment for the vulnerability and geographical zoning for groundwater management and protection. *Proceedings of NATO Advanced Research Workshop on Environment Contamination and Remediation Practices at Former and Present Military Bases*, Vilnius, Lithuania.

Lobo-Ferreira, J.P., Oliveira, M.M., 1997. DRASTIC groundwater vulnerability mapping of Portugal. *Proceedings from the 27th Congress of the International Association for Hydraulic Research*, San Francisco, CA, pp. 132–137.

Lynch, S.D., Reynders, A.G., Schulze, R.E., 1997. A DRASTIC approach to ground water vulnerability in South Africa. *South African Journal of Science* 93(2): 59–60.

Malczewski, J., 1996. A GIS-based approach to multiple criteria group decision-making. *International Journal of Geographical Information Systems* 10(8): 955–971.

Martino, D.F., Sessa, S., Loia, V., 2005. A fuzzy-based tool for modelization and analysis of the vulnerability of aquifers: A case study. *International Journal of Approximate Reasoning* 38: 99–111.

Massam, B.H., 1988. Multi-criteria decision making (MCDM) techniques in planning. *Progress in Planning* 30(1): 1–84.

Melloul, M., Collin, M., 1998. A proposed index for aquifer water-quality assessment: The case of Israel's Sharon region. *Journal of Environmental Management* 54(2): 131–142.

Navalur, K.C.S., Engel, B.A., 1998. Groundwater vulnerability assessment to nonpoint source nitrate pollution on a regional scale using GIS. *Transactions of the American Society of Agricultural Engineers* 41: 1671–1678.

NRC, 1993. *Groundwater Vulnerability Assessment, Contamination Potential under Conditions of Uncertainty*. National Academy Press, Washington, DC.

Nyerges, T.D., Laurini, M.R., Egenhofer, M., 1995. *Cognitive Aspects of Human-Computer Interaction for Geographic Information Systems*. Kluwer Academic, Dordrecht, the Netherlands.

Panagopoulos, G., Antonakos, A., Lambrakis, N., 2006. Optimization of DRASTIC model for groundwater vulnerability assessment, by the use of simple statistical methods and GIS. *Hydrogeology Journal* 14(6): 1431–2174.

Pawlak, Z., 2005. Some remarks on conflict analysis. *European Journal of Operation Research* 166: 649–654.

Perles Rosello, M.J., Vias Martinez, J.M. et al., 2009. Vulnerability of human environment to risk: Case of groundwater contamination risk. *Environment International* 35: 325–335.

Uddameri, V., 2003. Estimating natural attenuation rate constants using a fuzzy framework. *Ground Water Monitoring and Remediation* 23(3): 105–111.

Uddameri, V., Honnungar, V., 2007. Combining rough sets and GIS techniques to assess aquifer vulnerability characteristics in the semi-arid South Texas. *Environmental Geology* 51: 931–939.

Uricchio, V.F., Giordano, R., Lopez, N., 2004. A fuzzy knowledge-based decision support system for groundwater pollution risk evaluation. *Journal of Environmental Management* 73: 189–197.

Wang, L.-X., Mendel, J.M., 1990. Back-propagation fuzzy system as nonlinear dynamic system identifiers. *Fuzzy Systems* 3: 349–358.

Williams, J.R., Kissel, D.E., 1991. *Water Percolation: An Indicator of Nitrogen-Leaching Potential in Managing Nitrogen for Groundwater Quality and Farm Profitability.* Soil Science Society of America, Madison, WI.

Yager, R., 1988. On ordered weighted averaging aggregation operators in multi-criteria decision making. *IEEE Transactions on Systems, Man, and Cybernetics* 18: 183–190.

Zadeh, L., 1965. Fuzzy sets. *Information and Control* 8: 338–353.

15

Remote Sensing and GIS in Petroleum Exploration

D.S. Mitra

CONTENTS

15.1 Basic Concept of Petroleum Occurrences and Entrapment

Petroleum usually occurs in sedimentary basins, which are depressions in the earth's surface that are filled by thick sediment cover. The occurrence of petroleum in the basin requires a source rock, usually a mudrock or shale, rich in organic matter. The source rock must be buried deep enough so that temperature and time can cause the organic matter to mature into petroleum. The petroleum, thus generated, migrates from the source bed to accumulate within a porous and permeable reservoir called traps.

Most or, in some cases, all of the petroleum will be dispersed for lack of a good arrangement of strata to trap it, or it will leak out to the surface for lack of a good impermeable seal or cap rock. Thus, there are five factors that comprise the critical risks to petroleum accumulation: (1) a mature source rock, (2) a migration path connecting source rock to reservoir rock, (3) a reservoir rock that is both porous and permeable, (4) a trap, and (5) an impermeable seal. If any one of these factors is missing or inadequate, the prospect will be dry and the exploration effort will be unrewarded. Hence, tools employed in the petroleum exploration gather information related to these critical factors in the basin.

15.2 Petroleum Exploration

Petroleum exploration involves series of critical information phases. With each step, there is a progressively increasing database to evaluate the petroleum prospects of a region. The surface mapping is the initial stage, where the geologist's role is to obtain a more detailed knowledge of surface structures and evaluate other aspects critical to the exploration task, such as lithology, surface indicators, and possible tectonism and metamorphism. Until recently, aerial photographs assisted field geologists to extend their observations in the regional context, particularly in inaccessible terrains. However, in recent years, satellite-based remote sensing due to its higher ground and radiometric resolution provides this information more rapidly at lower cost. The information can be used for basin modeling to prioritize locales of probable hydrocarbon entrapment for further detailing. For this, the exploration geologist works closely with the geophysicist to relate the surface stratigraphy and structures to the subsurface and recommends areas for seismic survey. The geomorphic input, derived from remote sensing data, renders terrain information for planning the seismic program (Mitra et al., 1992). The seismic survey provides information on depth configuration of potential traps and knowledge of the character and volume of the sedimentary fill and source potential. This in conjunction with other geoscientific information identifies location for drilling. Drilling establishes a detailed sampling of the sediment character (reservoir, source, and cap rock potential), maturation, geothermal regime, reservoir, and the type of hydrocarbon trap. As the production begins, the exploration geologist provides information on the reserve estimate and hydrocarbon potential of the basin, which may help in the development of the oil field. At this stage, data can also be reexamined for the refinement of the developed geological model.

15.3 Remote Sensing and GIS in Petroleum Exploration

A typical petroleum exploration team consists of geologists, geophysicists, chemists, reservoir and production engineers, and many more. Each discipline approaches the problem-solving process in different ways. In addition to the diversity of discipline and problem-solving approaches, the team relied on a wide variety of computer software, hardware, and database, which has added to the need to seek common ground and synergy. The input data, in most of the cases, are spatial in nature, having precise locations and physical descriptions. The common base for referencing these varied data in the spatial domain is geographic referencing. GIS provides valuable data management, graphical display functions, and presentation in many forms. A common GIS technique is to filter data to restrict an analysis or presentation to only the information that satisfies some specific criterion. Map overlaying and data management aspects of GIS facilitate understanding of interrelationship of various data to solve specific geologic problems and decision makings (Mitra, 2002).

The geologic input is the backbone of sedimentary basin analysis. From remote sensing perspective, the sedimentary basins can be grouped as exposed and covered (including partly covered and offshore basins). Due to the varied surface cover the analytical techniques differ, however, the underlying philosophy from hydrocarbon exploration point of view remains the same. Quite a good number of published literatures are available, highlighting utility of air photo and satellite images in onshore/onland hydrocarbon exploration. In recent years, the availability of satellite SAR and altimetry has opened up a new vista in offshore hydrocarbon exploration using remote sensing data. The presentation of material in this chapter is greatly influenced by the work of Berger (1994), wherever possible examples/case studies from Indian sedimentary basins have been cited.

15.4 Remote Sensing in Exposed Basin

Before conventional exploration tools are applied to an exposed basin, it is necessary to firm up an idea about the geological setting of the area. The goal of geologic image analysis is to detect variations in the reflected solar radiation represented on the image and attach geologic significance to those variations, which is meaningful in the context of the problem being addressed.

15.4.1 Recognition of Rock Types

The lithology can be inferred by mapping variations in spectral reflectance, which indicate differences in the composition of surface materials. Spectral characteristics of rocks have been used to identify and map lithologies. However, there are no rigid guidelines for the same, as different rocks may generate the same photorecognition characters, and also the same rock type may display different characters in varying climatic conditions. Sedimentary rocks are usually stratified and display a banded appearance on photographs. Since a number of volcanic and metamorphic rocks also show layering, some previous knowledge of the area is required. Once lithology is identified, the next step is to study the arrangement and succession of strata. Mapping has always been a primary source of basic stratigraphic information. In general, the stratigraphic utility of remote sensing lies in providing maps that match (or improved upon) those produced by field geologists. The field geological maps have their own problems (like mapping scale, concept and data gap, etc.) while compiling for regional analysis. The problems can be overcome by using satellite images, and when the maps are prepared on GIS platform, it provides the needed flexibility during data integration. The new generation of satellite sensors has improved stratigraphic resolution, both in terms of spatial and spectral, thus providing a rapid and cost-effective tool for high-resolution geologic mapping.

Recent researches in hyperspectral remote sensing, also known as imaging spectroscopy, intend to identify gross mineralogical assemblages rather than lithology. The hyperspectral sensor data, substantiated with limited laboratory backup, may provide clue to facies change, reservoir characteristics, and environment of deposition in the basin. They have also been used to identify mineralogical changes associated with hydrocarbon microseepages (Ellis et al., 2001). Thermal inertia images using NOAA-AVHRR data generated in Brahmaputra basin (Mitra and Majumdar, 2004) and Kutch–Cambay basins, India (Nasipuri et al., 2005), have shown the efficacy of these images in gross lithological mapping. Analysis of rock types, substantiated with laboratory backup, provides clue to hydrocarbon generation potential of the basin and its ability to hold the generated hydrocarbons.

15.4.2 Recognition of Exposed Structures

Analysis of dip and strike of inclined bedrock strata is fundamental to the analysis of exposed structures from remote sensing data. Dip and strike of inclined bedrock strata can be recognized by relative slope (back slope vs. dip slope) and interrupted slope in alternating resistant and nonresistant litho units (flatirons pointing away from dip direction). These

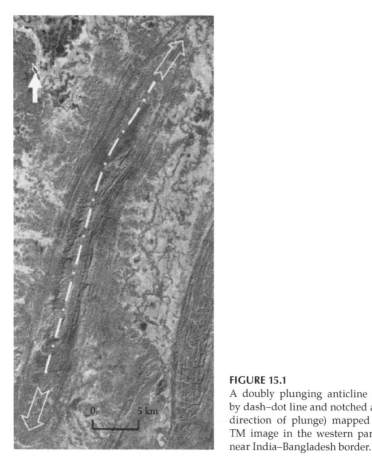

FIGURE 15.1
A doubly plunging anticline (axis shown by dash–dot line and notched arrows show direction of plunge) mapped on Landsat TM image in the western part of Tripura near India–Bangladesh border.

interpretations are generally qualitative in nature, and for more accuracy, stereoscopic investigations are required. The attitude of the beds helps in identifying the nature of the exposed structure. Satellite images have helped in mapping structures in simple (Figure 15.1) as well as complex tectonic settings (Figure 15.2). Exposed folds, domes, and basins are usually partially breached by erosion and manifest detectable topographic expressions, which can be used to recognize orientation, that is, dip and strike, of exposed limbs on remote sensing data. Berger (1994) has shown examples of various structures affected by erosion. The unaffected young folds and domes exhibit a direct relationship between topography and structure (Figure 15.3). With the advancing erosion, the upper layers of the folds get eroded and their geomorphic expression changes.

In the initial stages of erosion, the crest of the fold remains preserved. Figure 15.4a shows a low-amplitude plunging anticline as seen on Landsat ETM FCC image. The top of the anticline is partly eroded, showing the

FIGURE 15.2
Landsat TM image showing structural mapping in a complex tectonic setup in North Cachar, India. From the attitude of the beds (flatirons) tight syncline (axis shown by notched arrow) associated with a tight anticline (axis shown by arrow) whose eastern limb is faulted (fault runs along the river shown by filled arrow) can be mapped.

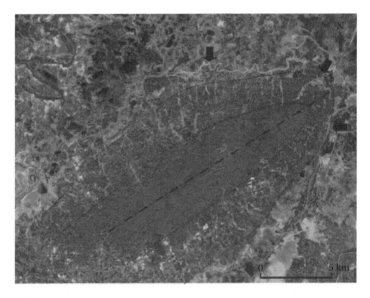

FIGURE 15.3
An elongated dome with a gentle topographic expression, as seen on Landsat TM image, occurring in north of Sylhet, Bangladesh. A river (see arrows) circumnavigates the structure. It is interpreted as a young fold whose axis is shown as dash–dot line and is plunging in the NE direction.

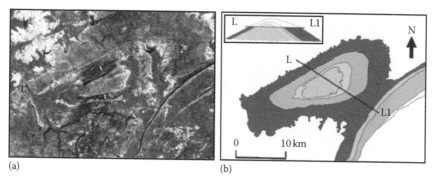

(a) (b)

FIGURE 15.4

(a) A low amplitude plunging anticline as seen on Landsat ETM image, in Vindhyan basin, India. (b) The corresponding interpreted lithological map. The different lithologies are shown in different grey shades as concentric elongated units. The top of the anticline is partly eroded, showing the oldest rock (innermost unit) surrounded by progressive younger rocks (dark grey being the youngest). From the attitude of resistant beds (shown by arrows in a), fold can be reconstructed having plunge toward NE. A schematic cross section of the fold along the line LL1 is shown as inset in (b). The eroded part of the fold is shown as unfilled line on top of the line LL1.

oldest rock (the innermost unit in Figure 15.4b) surrounded by younger rocks. From the attitude of resistant beds (shown by arrows in Figure 15.4a), fold can be reconstructed—a schematic cross section of the fold along the line LL1 is shown as inset in Figure 15.4b. The eroded part of the fold is shown as unfilled lines above the line LL1. In the late breaching stage, folds will undergo a complete removal of their crest but have preserved limbs. In this case, there is an inverse relationship of structure and topography. In highly deformed strata involving thrusting and wrench faulting and regions that were exposed to a long period of erosion and denudation, most of the structural relief will be eliminated leaving only subtle remnants. Mapping of such structures requires support of stereoscopic data (air photo/high-resolution stereoscopic satellite images) for interpretation, as they lack clear surface manifestation and have inverse relationship between fold and topography.

Analysis of faults in relation to the folded strata is an important part of interpretation of remote sensing data. Faults can be identified by the direct displacement of rock strata or by subtle geomorphic expressions. Faults commonly produce strong linear topographic scarps that are easy to recognize on remote sensing data. If the scarp coincides with the fault plane itself, it is referred as fault scarp. This is, however, very rare and requires extremely recent movements along the fault. More commonly, the fault line is eroded and removed from the location of the fault plane by differential erosion, producing a series of scarps referred to as fault-line scarps or fault-line trace. Topographic features that directly reflect the fault-line

FIGURE 15.5
Expression of fault with triangular facets indicating high dip (black arrow) and lithology variation on either side of the fault trace as seen on Landsat TM image of southern Meghalaya, India. Based on banding versus massive nature of beds and discordant nature of bedding traces, faulted contacts can be inferred, marked with white and gray arrows, respectively.

trace as linear scarps and triangular facets, and geomorphic features that develop along the faults such as springs, lakes, sag ponds, linear valleys, offset drainage, alignment of alluvial fans, and linear drainage divides help in identifying faults. Figure 15.5 shows an example of faults identified based on triangular facets, lithology, and nature of bedding traces.

Aerial photographs assisted field geologists to extend their observations in the regional context, particularly in inaccessible terrains. However, satellite images provide a rapid and cost-effective tool in better understanding the correlation of local structures/features in the regional framework, making improvements while developing the geologic model and its continuous updation during the entire project.

15.4.3 Prospect Identification

The recognition and analysis of exposed lithology and structures from image data are initially done independent of surface or subsurface control. Subsequently, with selective field checks, surface geological maps can be prepared. Geological mapping has always been a primary source of spatial information in hydrocarbon exploration. The petroleum exploration industry became an important early market to exploit the map making and management skill of the GIS. The field geological maps have their own problems while compiling for regional analysis. The problems can be overcome by using satellite images, and when the maps are prepared on GIS platform, it provides the needed flexibility during data integration and analysis (Sengupta and Mitra, 2005). These intelligent maps with database attached provide opportunities for query-based search and user-defined output generation.

The surface geological map when integrated, on a GIS platform, with interpretation from key seismic lines and other geophysical data helps in visualizing the subsurface structures. This is done by drawing geological cross sections. A geological cross section is a profile showing geological features in a vertical plane through the earth. The superimposition/disposition of structures as seen on satellite images and geological cross sections helps in identifying the tectonic process causing the deformation and relative timing of the resultant structures, which are useful in the search for potential hydrocarbon traps and prospect identification for further detailing.

When considering the generation, migration, and accumulation of petroleum, the timing of structural growth is very important. The types of folds and faults that develop within a basin are partly due to deformation mechanisms and partly to its sediments. Deformation by compression commonly produces folds and thrust faults, extension leads to normal and block faulting. Preparation of Rose Diagram, representing statistical distribution of major structural trends, helps in understanding relative ages of the trends and major stress direction in the area (Ganju and Khar, 1985). Structural deformation, which occurs at a later stage of basin sedimentation, can also assist the generation of oil, since it may be accompanied by higher-than-average heat flow. However, if structures are formed after petroleum generation and migration have ceased, they may well be barren. The timing of petroleum generation and migration can be estimated by numerical modeling assisted by laboratory backup. Extreme tectonics following basin development can produce adverse effects. They may elevate the reservoir rocks to the near surface causing degradation or loss of oil and gas or overmaturation of hydrocarbons due to metamorphism and hence may be considered as nonprospective.

Petroleum accumulation is often favored by active structural deformation during sedimentation, which leads to rapid changes in sediment facies and thickness. Organic-rich shale can be deposited in deep, structurally low areas, while coarser-grained reservoir facies and combination traps can develop over structural highs. Unconformities and faults are often present to assist migration processes. Such areas are considered to be prospective and can be identified on satellite images. Integrated analysis of remote sensing (including airborne SAR, thermal, and gamma ray spectrometric) and other collateral data have been attempted in many exposed frontier basins of India to provide input for basin modeling and in identifying locales of possible hydrocarbon traps (Agarwal et al., 1991; Dotiwala et al., 1995; Dotiwala and Pangtey, 1997; Bhoj et al., 1999). The technique has also been used in proven petroliferous basin to identify additional prospects based on new geologic model (Saha et al., 2004).

15.5 Remote Sensing in Onshore Covered Basin

Large percentage of world's onshore hydrocarbon resource is either obscured by thick cover of vegetation and soil and areas of low topographic relief or is completely buried under younger and relatively undeformed rock strata. Buried structures in such settings are recognized on imagery data by their subtle influence on the regional topographic setting and related drainage, vegetation, and soil moisture pattern. However, the analysis of such a structure requires the integration of available subsurface data. Hydrocarbon generation and entrapment potential of such a basin is inferred from the study of exposed rocks, occurring mostly near the a basin margin or as isolated outcrops within the basin. Because of the synoptic view, satellite images help in defining the basinal limits and possible extension of exposed structures within the basin (Figure 15.6).

15.5.1 Recognition of Obscured/Buried Structure

The covered basins generally possess low-relief terrain and in such cases the streams tend to flow down the regional slope. However, because of

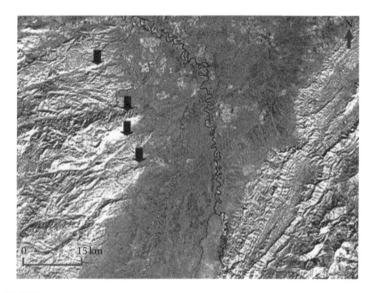

FIGURE 15.6
IRS LISS 2 image showing sedimentary basin covered by alluviums of Dhansiri River, northeastern India, bounded on west by exposed rocks of Mikir Hills (massive in nature) and on east by exposed layered rocks of Naga foothills. Prominent easterly trending linears seen in exposed western margins (black arrows) can be extended into the basin.

their low gradient, such streams are sensitive to changes in surface conditions that often occur over and in the vicinity of obscured and buried structures. In adjusting their channels to such changes, these streams develop unique drainage patterns and features that are anomalous with respect to the regional drainage systems. These local drainage phenomena are often called "drainage anomalies" or "structurally controlled streams." Berger (1994) has identified major structurally controlled stream patterns, which are large enough to be detected on satellite imagery. Burrato et al. (2003) in the Po Plain, Northern Italy, have demonstrated a methodology to identify blind thrusts using anomalous drainage patterns. Mazumder et al. (2009) have used drainage characteristics to decipher deep-seated tectonic elements in Western Ganga basin, India.

Locally perturbed drainage patterns may be formed over areas that are affected by obscured and buried structures. Increased meander frequency, possibly related to cross-trending buried structure, was observed on air photo mosaic in a segment of Vellar River, Cauvery basin, India (Figure 15.7), and the area was recommended for detailing by seismic survey. The survey confirmed the presence of subsurface structure and when probed by drilling, yielded hydrocarbon. Local deflection of stream orientation may occur in streams upon crossing or circumnavigation of obscured and buried structures. One such situation is shown in Figure 15.8, where local deflection in stream orientation of Barak River, North Cachar, India, was observed on Landsat TM image. Subsequent seismic survey confirmed the presence of subsurface structure, later found to be hydrocarbon bearing when drilled. Annular drainage pattern with subsequent marginal and other associated drainages as seen on Landsat image (Figure 15.9) of

FIGURE 15.7
Air photo mosaic showing locally perturbed drainage pattern of River Vellar, Cauvery basin, India. The anomalous area, shown by dashed circle possibly represents a buried structure. Seismic survey confirmed the presence of subsurface structure and the schematic diagram of the structure is shown as inset. Position of two hydrocarbon bearing wells (see arrow) subsequently drilled also shown.

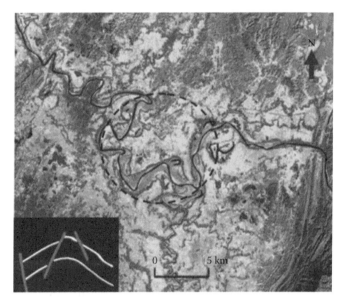

FIGURE 15.8
Local deflection in stream orientation of Barak River, North Cachar, India, as seen on Landsat TM image indicating a buried structure. Seismic survey confirms the presence of a subsurface structure, surface outline shown as dotted circle, and schematic representation of the structure is shown as inset.

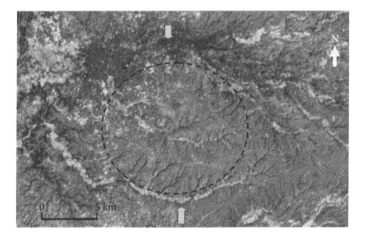

FIGURE 15.9
Annular drainage pattern with subsequent marginal stream (shown by arrows), outbound consequent drainage developed around the periphery (see within the dashed circle), and other associated drainages as seen on Landsat TM image of Agartala area in Tripura, India—typical representation of a domal structure in the subsurface. Seismic survey confirms the presence of subsurface dome.

Agartala area in Tripura, India, found to be surface manifestation of a gas bearing domal structure in the subsurface (Mitra and Agarwal, 1994).

In low-relief areas, subtle topographic ridges formed over structural highs are usually well drained and are characterized by reduced surface moisture content. These ridges appear brighter on satellite images than their surroundings. Similarly, subtle valleys and depressions, which may reflect structural lows, are usually poorly drained and become areas of excessive ground moisture concentrations. These lows appear darker on satellite data than their surroundings. In Krishna–Godavari basin, India, Babu (1973) identified a number of tonal anomalies on air photos and opined that a number of them are possible manifestation of the subsurface structures on the ground surface as drainage anomalies or anomalous humps/depression in an otherwise flat terrain. Most of these anomalies were subsequently picked up on Landsat images. Airborne gamma ray spectrometric surveys (Khar et al., 1994) have shown that many of the gamma anomalies correspond well with earlier defined anomalies. Seismic survey confirmed the presence of subsurface structures associated with most of these anomalies, and oil/gas was encountered in few of them when penetrated by drilling (Mitra and Agarwal, 1994). Similar observation has also been noted in Cauvery basin, India, where anomalies mapped on air photos could be seen on Landsat, IRS, and SAR images. Seismic survey confirmed the presence of subsurface structures associated with some of these anomalies and drilling indicated the presence of oil/gas (Mitra and Agarwal, 1991, 1994). Alignment of small-scale linear surface features such as stream segments or changes in vegetation and tonal patterns, commonly referred as lineament, can be distinctly seen on air photos and satellite images. Such a feature is believed to reflect the surface expression of major obscured and buried faults or zone of weakness in the subsurface strata. An interpreted satellite image may show numerous linear features. Criteria must, therefore, be established to allow the interpreter to identify those surface elements that truly represent major fault systems or zone of weakness in the subsurface (Mitra et al., 1993). Integration with available geophysical data, on GIS platform, helps in the analysis of these linear features (Dotiwala et al., 2002; Sengupta et al., 2004). The superimposition of various lineaments through time and space results in cross-trends and can influence the hydrocarbon migration and entrapment. If any hydrocarbon is generated in the basin at one stage, its migration and entrapment will depend upon the subsequent tectonic disturbances. Hence, structures developed along the younger trends (due to neotectonic movements) attain importance for hydrocarbon entrapment (Mitra and Agarwal, 1991; Mitra et al., 1993). Babu (1994) identified major neotectonic events in some of the sedimentary basins of India and highlighted their significance in hydrocarbon exploration.

15.5.2 Prospect Identification

The hydrocarbon generation potential and reservoir characteristics of the rocks exposed in the basin margins categorize the basin to be prospective for hydrocarbon exploration. The integrated analysis of satellite and other collateral data enables the identification of specific potential structures. To identify prospective structures, it is necessary to make an exploration model describing the relationship of subsurface structures, its surface manifestation, and the known hydrocarbon occurrences of the area (Mitra et al., 1993; Mullick and Mitra, 2002). The leads can be further ranked based on comparison with the developed model and recommended for further detailing by seismic surveys for subsurface exploration.

The process of overlaying maps for integration is the first stage that any data analyst performs, which need not be done by physically overlaying them on transparent media. Maps converted from hard copy (paper or film) to soft copy (in computer) are geo-referenced to user-desired projection system for graphical overlaying to understand and analyze the surface distribution of the data and to discover relationship among mapped variables. The spatial representation of various data sets, through overlaying, using GIS, can be very revealing and trends, patterns, and anomalies can become apparent. The ultimate purpose of GIS is to provide support for making decisions based on spatial data (Dave and Mitra, 2007). Figure 15.10 shows an overlay of various thematic map/information created on GIS platform for a sedimentary basin to identify prospective areas for further detailing. It can be seen that some of the major E–W trending structural trends separate areas with different drainage pattern or density, indicating areas with different morphotectonic setup. The presence of earthquake epicenters indicates the active nature of some of these faults. Identified drainage anomalies can be further prioritized based on their nature, nearness to major structural trends, cross-trends, and lineaments to delineate prospective areas for further detailing.

15.6 Remote Sensing in Offshore Basins

The presence of water column over offshore basin restricts the usage of conventional remote sensing techniques. Satellite altimetry, side-scan sonar, and SAR surveys are some of the tools that are widely used as reconnaissance survey in offshore oil exploration.

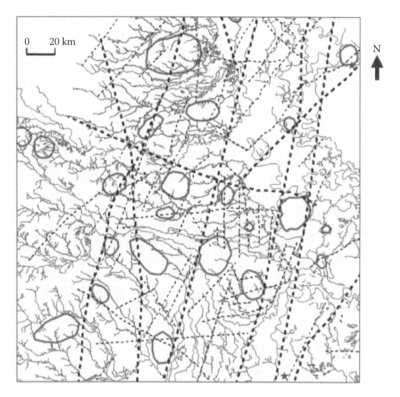

FIGURE 15.10
Overlaying of drainage, anomaly (circular/quasi circular lines) major structural trend (thick dotted lines), lineaments (thin dotted lines) and earthquake epicenters (star) on GIS platform to prioritize the areas for further detailing.

15.6.1 Satellite Altimetry

In recent years, satellite altimetry has emerged as a powerful reconnaissance tool for exploration of sedimentary basins on both the Indian continental margins and in deepwater regions. The altimeter on board satellite transmits a microwave signal to measure the sea surface height, which provides information about the geoid, the equipotential surface, after doing necessary corrections. This can be modeled to derive free air gravity (Rapp, 1983).

Depending upon the resolution of the data it can be used to decipher mega- or mesoscale structures in the offshore, for detailing by marine gravity surveys to identify prospective areas. Majumdar et al. (1998) have shown the efficacy of the satellite altimeter data (ERS-1, having a cross track resolution of 16 km) in Indian Offshore for hydrocarbon exploration. However, with the availability of very high resolution Geosat GM (Geodetic Mission) data along with Seasat, ERS-1/2, and TOPEX/POSEIDON altimeter data, it is now possible to generate very high resolution data with a grid of around 4 km (Majumdar et al., 2006; Chatterjee et al., 2007). This is found to be very useful in delineating geological details over the Indian offshore.

15.6.2 Side-Scan Sonar Surveys

The side-scan sonar system enables the acquisition of images of the seabed through the use of acoustic waves. It can obtain images similar to those of aerial photography, showing changes in the seabed composition as well as physiographic details over a wide area. The sea beds, if associated with oil seepage, may cause small-scale changes in acoustic properties of the sediments, resulting in change in backscatter level against the surrounding area. These images can be used as a reconnaissance tool to delineate oil and gas seep sites on the seafloor and, thereby, reduce the dry-hole risk of petroleum exploration in deepwater frontier areas (Wen and Larsen, 1996; De Beukelaer et al., 2003; Sager et al., 2004).

15.6.3 Offshore Oil Seepage Mapping

Typically, oil and gas escape into the water, from discrete vents within large seep sites on the sea floor. When it reaches the sea surface, it spreads into thin, very elongated layers that coalesce and are recognized as slicks. They can be readily detected by any common remote sensing sensors. Many satellite missions have been acquiring SAR data over nearly all areas of the world. The operating frequency and viewing angle of these sensors are well suited to the detection of surface features related to the presence of films such as pollution, seepage slicks, and natural surfactants (Espedal and Wahl, 1999). These slicks show up as dark areas on the SAR image. This is because of the way an SAR images the sea surface and the effect oil slick has on the sea surface. The multiple coverage can be used to help with the identification of features with respect to ship and other pollution sources. Once detected, the degree of persistence of the slicks can give indicators as to the nature of the hydrocarbon source. When combined with additional geological and geophysical data, on a GIS platform, this becomes an extremely powerful technique for the location of promising sites for more costly high-resolution surveying. Figure 15.11 shows two natural slicks identified on ENVISAT SAR image in the eastern offshore basin, India. When regional subsurface fault trends were superimposed on it, using GIS platform, it was observed that they occur near to a prominent NE–SW trending fault found to be continued from basement to the sea bottom, as seen on seismic data. This observation, coupled with

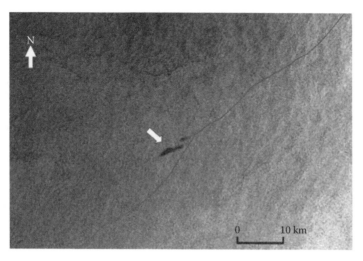

FIGURE 15.11
ENVISAT SAR image showing oil seepages (shown by arrow) occurring near a prominent NE–SW trending subsurface fault (diagonal curvilinear line) as seen in parts of eastern offshore of India.

other geological parameters, helped in understanding the hydrocarbon migration and entrapment condition in the basin and locating areas for further detailing. Sengupta and Saha (2008) and Dave et al. (2009) have used satellite SAR data in conjunction with other collateral data in Indian offshore basin to identify areas for further detailing.

15.7 Conclusions

Petroleum exploration involves series of critical information phases. The surface mapping is the initial stage, where the geologist's role is to obtain a more detailed knowledge of surface structures and evaluate other aspects critical to the exploration task.

Field geological maps have their own problems while compiling for regional analysis. The problems can be overcome by using satellite images and when the maps are prepared on GIS platform, it provides the needed flexibility during data integration. Digital methods make it easier to reuse preexisting data, during renewed phases of fieldwork carried out to test a new concept or hypothesis. Exposed folds, domes, and basins are usually partially breached by erosion and manifest detectable topographic expressions, which can be used to recognize orientation, that is, dip and strike, of exposed limbs on remote sensing data.

The new generation of satellite sensors provides improved stratigraphic resolution, both in terms of spatial and spectral, thus providing a rapid and cost-effective tool for high-resolution geologic mapping. Geological structures whose outcropping units are either obscured by vegetation and soil or completely buried under consolidated sediments are recognized on imagery data by their subtle influence on the regional topographic setting and related drainage, vegetation, and soil moisture pattern. The analysis of such a structure requires the integration of all other subsurface data preferably on GIS platform.

In offshore basins, satellite altimetry and detection of oil seepages by satellite SAR surveys offer promising reconnaissance tool. However, to identify prospective structures, it is necessary to integrate data from other sources. GIS is a particularly effective technology that enables petroleum exploration team to share information, analyze data in new ways, and integrate the evaluation process. Such an approach enhances the overall problem-solving process.

Once a GIS project is constructed, it can yield extremely attractive, colorful maps of all sorts of geoscientific properties and related exploration information. The data used to prepare the map are easily accessible, updatable, and transportable, which are very important criteria in hydrocarbon exploration.

Acknowledgments

The author is thankful to the two unknown referees and his colleagues in RS&G, who have helped in improving the chapter. The views expressed here are those of the author and not necessarily of the organization for which he is working.

References

Agarwal, R. P., Joshi, S. V., Mitra, D. S. et al. 1991. Integrated geological research in Vindhyan basin. In *Proceedings of Integrated Exploration Research: Achievements and Perspective*, KDMIPE, Dehradun, India.

Babu, P. V. L. P. 1973. Morphostructures of Krishna and Godavari delta region. *Photonirvachak* 1: 65–67.

Babu, P. V. L. P. 1994. Neotectonic events and hydrocarbon plays in the sedimentary basins of India. In *Proceedings of Second Seminar on Petroliferous Basins of India*, Indian Petroleum Publishers, Dehradun, India, vol. 3, pp. 237–240.

Berger, Z. 1994. *Satellite Hydrocarbon Exploration*, Springer Verlag, Berlin, Germany.
Bhoj, R., Pangtey, K. K. S., and Mitra, D. S. 1999. Lithological and structural mapping of a part of Spiti sub basin using high resolution satellite data and limited field checks on a GIS platform. In *Proceedings of 14th Himalaya Karakorum Tibet Workshop*, Germany, March 24–26.
Burrato, P., Ciucci, F., and Valensise, G. 2003. An inventory of river anomalies in the Po Plain, northern Italy: Evidence for active blind thrust faulting. *Annals of Geophysics* 46: 865–881.
Chatterjee, S., Bhattacharyya, R., Michael, L., Krishna, K. S., and Majumdar, T. J. 2007. Validation of ERS-1 and high-resolution satellite gravity with in-situ ship borne gravity over the Indian offshore regions: Accuracies and implications to subsurface modeling. *Marine Geodesy* 30: 197–216.
Dave, H. D., Mazumder, S., Samal, J. K., and Mitra, D. S. 2009. Oil seepage detection using SAR technology and seep-seismic studies in KG offshore basin to delineate possible areas of hydrocarbon exploration. In *Proceedings of Explotech*, ONGC, Dehradun, India.
Dave, H. D. and Mitra, D. S. 2007. Integrating onland lineaments with offshore data using GIS approach—A case study from Kutch Basin, India. *International Journal of Remote Sensing* 28: 925–930.
De Beukelaer, S. M., MacDonald, I. R., Guinnasso, Jr. N. L., and Murray, J. A. 2003. Distinct side-scan sonar, RADARSAT SAR and acoustic profiler signatures of gas and oil seeps on the Gulf of Mexico slope. *Geo-Marine Letters* 23: 177–186.
Dotiwala, F., Mullick, A. K., Dotiwala, S., and Mitra, D. S. 2002. A study of lineament fabric of the Mikir hills using satellite-remote sensing data as a GIS for inputs to exploration in Dhansiri valley, Assam, India. In *1st APG Conference*, Mussoorie, India, September 28–29, 2002.
Dotiwala, S. and Pangtey, K. K. S. 1997. Structural setting and hydrocarbon prospects of the Son-Mahanadi basin, India—Emphasis on Son Graben. *Bulletin ONGC* 34(1): 81–97.
Dotiwala, S., Pangtey, K. K. S., and Dey, B. K. 1995. Tectonics of South Rewa Gondwana Basin and its bearing on petroleum prospects-based on remote sensing data. In *Proceedings of Petrotech-95*, New Delhi, India.
Ellis, J. M., Davis, H. H., and Zamudio, J. A. 2001. Exploring for onshore oil seeps with hyperspectral imaging. *Oil and Gas Journal* 99: 49–58.
Espedal, H. A. and Wahl, T. 1999. Satellite SAR oil spill detection using wind history information. *International Journal of Remote Sensing* 20(1): 49–65.
Ganju, J. L. and Khar, B. M. 1985. Structure, tectonics and hydrocarbon prospects of Naga hills based on integrated remotely sensed data. *Petroleum Asia Journal* III: 142–151.
Khar, B. M., Kak, S. N., and Katty, V. J. 1994. Airborne gamma ray spectrometry in petroleum exploration—An emerging exploration tool. In *Proceedings of Second Seminar on Petroliferous Basins of India*, Indian Petroleum Publishers, Dehradun, India, vol. 3, pp. 247–262.
Mazumder, S., Dave, H. D., Samal, J. K., and Mitra, D. S. 2009. Morphotectonic features and its correlation with deep seated tectonic elements in western Ganga basin. *ONGC Bulletin* 44: 11–18.

Majumdar, T. J., Krishna, K. S., Chatterjee, S., Bhattacharyya, R., and Michael, L. 2006. Study of high resolution satellite geoid and gravity anomaly data over the Bay of Bengal. *Current Science* 90: 211–219.

Majumdar, T. J., Mohanty, K. K., and Srivastava, A. K. 1998. On the utilization of ERS-1 altimeter data for offshore oil exploration. *International Journal of Remote Sensing* 9: 1953–1968.

Mitra, D. S. 2002. Geographic information system (GIS) for hydrocarbon exploration. In *Proceedings of Geomatics-2002*, Tiruchirappalli, India, September 18–20, 2002, pp. 34–39.

Mitra, D. S. and Agarwal, R. P. 1991. Geomorphology and petroleum prospects of Cauvery basin, Tamilnadu based on interpretation of Indian Remote Sensing satellite (IRS) data. *Journal of Indian Society of Remote Sensing* 19: 263–267.

Mitra, D. S. and Agarwal, R. P. 1994. Remote Sensing—The primary input towards developing integrated aseismic exploration approach for hydrocarbons—Achievements and perspectives. In *Proceedings of Second Seminar on Petroliferous Basins of India*, Indian Petroleum Publishers, Dehradun, India, vol. 3, pp. 319–327.

Mitra, D. S., Bhoj, R., and Agarwal, R. P. 1993. Hydrocarbon exploration in Shahgarh and Myajlar sub basin, Rajasthan, India using Remote Sensing technique. *Indian Journal of Petroleum Geology* 2(1): 31–42.

Mitra, D. S., Bhoj, R., and Joshi, S. V. 1992. Evaluation of Rajasthan desertic terrain, India, for logistic support in oil exploration using Landast Thematic Mapper data. *International Journal of Remote Sensing* 13: 2773–2782.

Mitra, D. S. and Majumdar, T. J. 2004. Thermal inertia mapping over the Brahmaputra basin, India using NOAA-AVHRR data and its possible geological applications. *International Journal of Remote Sensing* 25: 3245–3260.

Mullick, A. K. and Mitra, D. S. 2002. Morphotectonic analysis and hydrocarbon prospectivity of north bank of Brahmaputra, India. In *Proceedings of Fourth SPG Conference*, Mumbai, India, January 7–9, 2002.

Nasipuri, P., Mitra, D. S., and Majumdar, T. J. 2005. Generation of thermal inertia image over a part of Gujarat: A new tool for geological mapping. *International Journal of Applied Earth Observation and Geoinformation* 7: 129–139.

Rapp, R. H. 1983. The determination of geoid undulations and gravity anomalies from Seasat altimeter data. *Journal of Geophysical Research* 88: 1552–1562.

Sager, W. W., MacDonald, I. R., and Hou, R. 2004. Side-scan sonar imaging of hydrocarbon seeps on the Louisiana continental slope. *American Association of Petroleum Geology Bulletin* 88: 725–746.

Saha, K. K., Mullick, A. K., and Mitra, D. S. 2004. Utility of remote sensing in proven petroliferous basins. In *Proceedings of Second Association of Petroleum Geologists Conference*, Khajuraho, India, September 24–26, 2004.

Sengupta, S., Dave, H. D., and Mitra, D. S. 2004. Lineaments and subsurface data integration for petroleum exploration in a GIS based approach from Cambay basin. In *Proceedings of Second Association of Petroleum Geologists Conference*, Khajuraho, India, September 24–26, 2004.

Sengupta, S. and Mitra, D. S. 2005. GIS as a tool for theme integration and geologic analysis: A case study from a section of Eastern Gondwana basin India. In *Proceedings of Petrotech-2005*, New Delhi, India, January 16–19, 2005.

Sengupta, S. and Saha, K. 2008. Use of satellite imageries to identify potential hydrocarbon microseepage in off-shore areas. In *Proceedings of Seventh Society of Petroleum Geologists Conference*, Hyderabad, India.

Wen, R. and Larsen, S. R. 1996. Mapping oil seeps on the sea floor by Gloria side-scan sonar images—A case study from Northern Gulf of Mexico. *Nonrenewable Resources* 5: 3.

16

Geoinformatics in Terrain Analysis and Landslide Susceptibility Mapping in Part of Western Ghats, India

Siddan Anbazhagan and K.S. Sajinkumar

CONTENTS

16.1 Introduction

Landslides are recorded in different climates, terrains, and involve various assemblage of earth material (Zaruba and Mencl, 1969). The frequency, magnitude, and duration of landslides vary in different parts of the world.

Landslides have caused large number of casualties and huge economic losses in mountainous areas of the world (Dai et al., 2002). The western flank or the windward slope of Western Ghats is one of the major landslide-prone areas of the country (Valdiya, 1987). The Western Ghats witnesses landslides every monsoon season (Sajin kumar and Anbazhagan, 2004). This region is characterized by rugged hills with steep slope associated with loose unconsolidated soil and earth materials. The intensive weathering and saturation of fractured slope, compounded with anthropogenic activities lead to frequent slope failure in this rugged mountain system (Thampi et al., 1998). A map depicting the spatial distribution of landslide susceptibility zone is useful, as this hilly terrain has high natural resources apart from the human population. Moreover, landslide susceptibility map is useful in mitigating landslide hazard for a region (Anbalagan et al., 1993). The susceptibility zonation map is also useful for the planning and implementation of various developmental schemes in hilly areas (Sarkar and Gupta, 2005). For the preparation of landslide susceptibility map, one requires a detailed knowledge of the processes that are or have been active in an area and the factors leading to the occurrence of the potentially damaging phenomenon (Mantovani et al., 1996). The historical records of landslide occurrences, local geology, structure, geomorphology, hydrologic conditions, vegetation, and climatic conditions are required for such landslide hazard zonation study (Chau et al., 2004).

In the present context, the Indian remote sensing (IRS)-1D satellite merged linear imaging spectral sensor (LISS III) and panchromatic (PAN) data are used for interpreting various terrain parameters. Geology, geomorphology, structure, slope, relief, drainage, soil, land use/land cover, and rainfall were analyzed in detail. These parameters were given weightage according to the depth of influence on landslide occurrences in this region. The terrain parameters were spatially integrated using geographic information systems (GIS), and different landslide susceptibility zones were demarcated in the region. The purpose of this chapter is to highlight the zones of landslide-prone areas and derive possible management practice to prevent landslide occurrences.

16.2 Study Area

Western Ghats run almost parallel to the west coast of India from Tapti river in the north to Kanniyakumari in the south, and it is one of the prominent physiographic features of the Indian peninsula. The present landscape carved out by a combination of fluvial, denudational, and tectonic processes can bring out salient geomorphic features in the area and landslide events often witnessed during the monsoon.

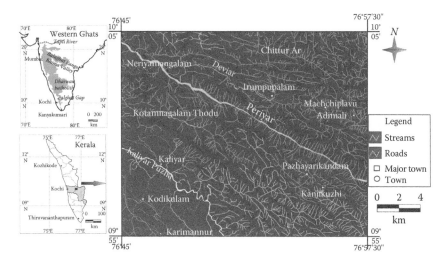

FIGURE 16.1
The study area Neriyamangalam—Adimali region located in part of Ernakulam—Idukki district in Kerala state. The drainage network impregnated on the satellite imagery (IRS-1D PAN+LISS III merged).

The study area, Neriyamangalam region, is one of the nature-rich areas of the Kerala State and located in the windward slope of Western Ghats in Ernakulam and Idukki districts. It lies geographically within north latitudes 09°55′00″ and 10°05′00″ and east longitudes 76°45′00″ and 76°57′30″ (Figure 16.1). Remote sensing based landslide susceptibility mapping was carried out for the aerial extent of 420 km². The area is effectively interconnected by roads and provides good accessibility and connectivity to the rest of the country.

The area experiences southwest and northeast monsoons with an annual precipitation of 450 cm. Neriyamangalam located within the study area receives highest rainfall (>500 cm) in the Kerala State. The temperature in this region varies between 13° and 29°C. Periyar, Kaliyar, and Kothamangalam thodu are (Ar and thodu are synonymous with streams) the three important rivers that traverse through the study area. Besides these rivers and their tributaries, there are numerous rivulets that traverse and crisscross in the study area.

16.3 Materials and Method

The mountainous terrain of the study area needs detailed study of different terrain parameters for delineating the landslide-prone areas and

TABLE 16.1

Data Utilized for Landslide Susceptibility Mapping in Parts of Ernakulam
and Idukki District, Kerala, India

Data	Source
Topographic map 58 B/16 and 58 C/13 on 1:50,000 scale	Survey of India (SOI), Bangalore
IRS-1D LISS III and PAN satellite data in digital format	National Remote Sensing Centre (NRSC), Hyderabad
Geological map of Kerala 1:50,000 scale published 1955	Geological Survey of India
Rainfall data 1992–2002	Indian Meteorological Department (IMD), Trivandrum
Soil map	Soil Survey Organization, Trivandrum
Soil erosion map	Kerala State Land Use Board, Trivandrum

proposing site-specific management practices. In this chapter, the factors
that influence the occurrences of landslides were studied systematically.
The terrain parameters were studied and interpreted with the help of
IRS-1D PAN + LISS III merged satellite data. The IRS-1D LISS III satellite
data have 23.5 m resolution, whereas PAN data have 5.8 m resolution. The
purpose of using merged satellite data for this chapter is to interpret various terrain parameters through high spatial resolution as well as multispectral mode. The merged data provide quality interpretation for terrain
mapping both the spectral and spatial resolution. The various data utilized in this chapter are shown in Table 16.1. A subjective rating analysis is
done under supervised weighting method. Ranks were distributed for different classes within a thematic map, whereas weightages were assigned
for different thematic maps. Multidimensional spatial analysis is carried
out in GIS environment using Geomedia 5.2 Professional software. On
the basis of landslide susceptibility zones and controlling terrain factors,
management practices were proposed for settlement areas and transport
corridors located within the high landslide susceptibility zone.

16.4 Rainfall

Rainfall is an important triggering mechanism in landslide occurrences
in this region. Rainfall-induced shallow landslides, mostly soil slip and
debris flow, are initiated by a transient loss of shear strength either by
reducing the apparent soil cohesion or pressure due to the increase
of pore water, caused by intense rainfall (Terzaghi, 1980; Wilson and
Wieczorek, 1995). Temporal occurrence of landslides and movement

activities is mainly controlled by rainfall pattern with different types of resolution (Van Asch et al., 1999). Western Ghats experiences two monsoons, namely southwest (June to September) and northeast (October to December) monsoon, of which SW monsoon yields the maximum precipitation. Thampi et al. (1998), in their studies on landslides in Western Ghats, concluded that an intense rainfall exceeding 20 cm in 24 h can trigger debris flow in vulnerable slopes. Rainfall data collected from Indian Meteorological Department (IMD) have shown that for an average of 30 days per year, the Neriyamangalam area experiences more than 20 cm rainfall. Neriyamangalam, the place that receives the highest rainfall in Kerala, is situated in the study area. On the whole, the study area experiences heavy precipitation than the state average annual rainfall (300 cm). Hence, the monsoon is considered as the single largest factor on the operation of the geomorphic process in the study area. The geographical location of the study area in the Western Ghats is the reason behind these convulsions. The high mountains, especially Anamudi and Munnar massif, are effective barriers that give rise to pockets of unusual heavy precipitation. Most of the places in the study area receive more than 380 cm annual rainfall. The amount of precipitation increases from southeast to northwest part of the study area.

16.5 Conditioning and Triggering Factors

The relationship between landslides and their causative factors is inherently linear and varies over area and time. Considering the spatial and temporal distribution of the landslides, the subdivision by Crozier (1986) between "preparatory" and "conditioning" factors, which include terrain characteristics and "triggering" factors that are the direct cause of the movements, seems to be a more useful approach. The study of the conditioning factors should be based on a systematic inventory conducive to the relationship between landslides and the geological and geomorphological characteristics of the terrain (Carrara et al., 1982). The temporal distribution of slope movements is determined by the occurrence of triggering factors such as rainfall, earthquakes, human activities, etc.

16.5.1 Geological Setup

The Neriyamangalam–Adimali region forms a part of Madurai granulite block (MGB), which is perhaps the largest single crustal block in the southern high-grade metamorphic terrain of South India (Mohan, 1996). It is sandwiched between Palghat–Cauvery shear zone in the north and Achankovil shear zone in the south (Koshimoto et al., 2004). This Neriyamangalam region is characterized by the outcrops of hornblende gneiss and pink granite gneiss (Munnar granite), traversed by bands of pyroxene granulite, calc granulite, quartzite, and magnetite quartzites of different dimensions (Figure 16.2). Hornblende gneiss is formed by the retrogressive metamorphism of charnockites due to the emplacement of Munnar granite (Rajan et al., 1984). Munnar granite, emplaced within the Precambrian gneisses, is spatially related to the intersection zone of the NE–SW trending Attur lineament and NW–SE trending Idamalayar lineament (Santosh et al., 1987).

16.5.2 Structure and Tectonic Landforms

False color composite (FCC) image was generated through Erdas Imagine 8.6 image processing software using band 2,3,4 of IRS-1D LISS III + PAN satellite data (Figure 16.3). The FCC provides contrast information delineating faults, folds, and lineaments in the study area. Structural features such as faults, lineaments, and folds are the prominent features in the study area and the products of tectonic activity that the Western Ghats witnessed during the Tertiary period. Kerala and Periyar are the two major faults present in the study area, respectively, referred as Kerala lineament and Periyar lineaments (Figure 16.4). These faults are epicenters of earthquakes with magnitude of 4.5, witnessed in the past two decades. Rastogi et al. (1989) interpreted that the movement had occurred due to strike slip faulting on a NW–SE plan along the NW–SE flowing Kallar river, which is the continuation of the Periyar river in the area. Gravity studies by Mishra et al. (1989) have also shown that a shallow fault at a depth of about 4 km coincides with the Periyar fault. The cross-cutting relationship and surface expression have shown that Kerala fault is younger than Periyar fault. The course of Periyar river slightly changes at the location where Kerala fault meets Periyar river.

Lineaments, which are the surface manifestation of underlying linear features, are quite ubiquitous in the study area. Most of the mega and intermediate type of lineaments coincide with emplacement of faults and fractures of the basement rocks (Prabhakara Rao et al., 1985). Two major sets of lineaments were interpreted from satellite data in the area, trending NNE–SSW and NW–SE directions. The study of lineaments has paramount importance in landslide occurrences as these are active sites, where earth materials can topple down. Folds in the study area are significant and a major fold system with NW–SE orientation was interpreted as "Deviar Fold System." This is an important structural feature in the study area, since it is impregnated with number of previous landslides. The morphostructural setup of the study area throws light upon the role played by tectonism in landslide occurrences throughout the length of

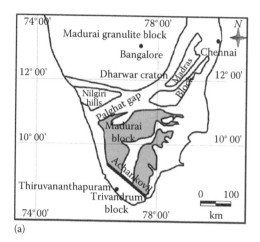

(a)

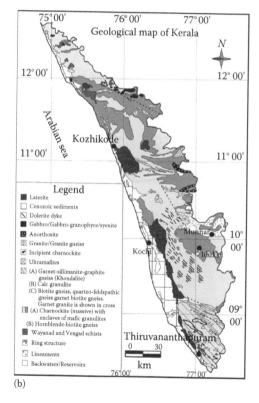

(b)

FIGURE 16.2
Regional and local geology: (a) Geology of Madurai granulitic block, (b) geology of Kerala, and

(continued)

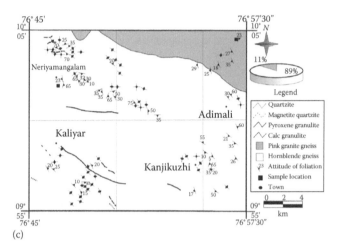

(c)

FIGURE 16.2 (continued)
(c) geology of the study area.

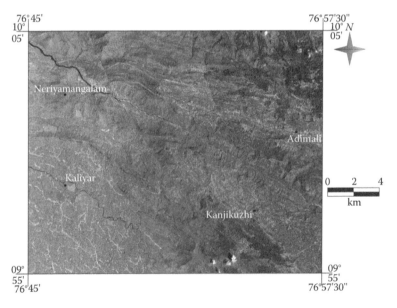

FIGURE 16.3
False color composite (FCC) imagery of IRS-1D PAN+LISS III merged satellite data for Neriyamangalam—Adimali region. Structural, geomorphological, and land use/land cover pattern were interpreted from the FCC output.

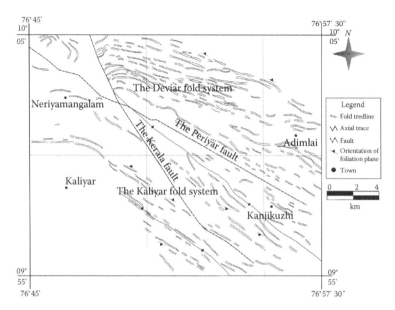

FIGURE 16.4
The major faults and fold systems in the Neriyamangalam—Adimali region interpreted from satellite data.

Western Ghats. The occurrence of earthquakes in the vicinity of "Kerala Fault" could also trigger landslides in this region. The M5 earthquake of 2000 at Erattupetta (20 km south of study area) with intensity VII and the 4.5 magnitude earthquake of 1988 at Nedumkandam (15 km south of study area) with intensity VI warn strong vulnerability of the area for earthquake-induced landslides (Devarajan, 1990).

Escarpments or cliffs are the vertical slopes facing the hills and are ubiquitous throughout the study area. Escarpments have common regional trend in NW–SE direction and are folded at many places. The alignment of escarpments coincides with the trend line of fold and lineament system, and it indicates the role of tectonism in the evolution of these landforms. Escarpments, which coincide with the structural trend lines, are prone to landslides because of its steep slope with barren surface.

Western Ghats is characterized by steeper and precipitous slopes and the presence of major interior plateaus. Structural hill system is the dominant feature of the study area. The structural hill system is developed on particular structure of mappable dimensions and is characterized by a recurring pattern of dip slopes, escarpment slopes, debris slopes, and a crest (Devarajan, 1990). Most of the landslide scars are associated with structural hill system, which needs to be mitigated with strong measures.

16.5.3 Geomorphology

The present day landscapes of the study area are carved out by a combination of fluvial, denudational, and tectonic processes (Figure 16.5). These landforms are fragile and sensitive and often become vigorous once triggering mechanisms, like rainfall or earthquake operating upon these, open geomorphic systems. The digitally processed IRS-1D PAN + LISS III merged satellite data provide clear picture for interpretation of geomorphology in the study area.

Plains are evolved by erosional and depositional activity of fluvial processes and are relatively flat areas composed of alluvium. Plains are mostly confined to midlands. These plains may be evolved by coalescence of palaeochannels of earlier streams that have shifted their courses simultaneously with any one of the tectonic activities. Kaliyar river, called "Muvattupuzha" in its latter traverse, flows in a zigzag pattern through this plain. Plains do not have any contribution for mass movement.

Intermontane valleys are the landforms situated between or surrounded by mountains and mountain ranges. In the study area, these valleys are observed along two stretches, where Periyar and Kothamangalam rivers traverse. The intermontane valley usually receive high precipitation due to its enclosed nature. Neriyamangalam town, which receives the

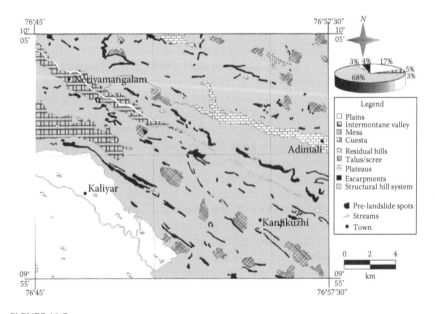

FIGURE 16.5
Geomorphology of Neriyamangalam—Adimali region in Ernakulam—Idukki districts, Kerala. The geomorphic features and the previous landslide spots were interpreted from satellite data.

highest rainfall in Kerala, is located in the intermontane valley region. The highly weathered materials and boulders along intermontane valley slope and high-intensity rainfall could trigger landslide occurrences in such landforms.

Mesas, the flat-topped hills with steep-sided escarpments on all sides, are the result of the presence of hard capping strata that have resisted erosion. Laterite acts as a capping in this geomorphic feature. Grits, the products of tropical weathering, are randomly distributed in mesa-type flat-topped hills. Cuestas are the landforms with gentle slope on one side and steep slope on other side. Cuestas are rarely distributed and associated with escarpments. Colluvial materials are usually deposited at the base of the steep side slope of a cuesta. The steep slope of cuesta is always prone to landslide occurrences.

Talus scree are the landforms developed by the result of deposition of the materials due to mass movements at the base of the mountains, hills, hillocks, and escarpments. Such landforms are widespread in the study area. During high intensity rainfall, the plane of contact between rock and overlying soil acts as a slippery path for the materials to come down in such landform.

Plateaus represent old planation surfaces modified by erosional process, but elevated to various altitudes due to tectonic movements during different geological ages (Soman, 2002). Similar high relief plateaus are characteristic of Western Ghats. The major plateau interpreted in the study area extends from Machichiplavu and traverse to ENE direction toward Munnar. Except the elevation and highly weathering condition, plateau region is less significant for landslide occurrences. Residual hills are the end products of peneplantation that reduces original mountain masses into a series of scattered knobs standing on the peneplain (Thornbury, 1990). Residual hills are randomly distributed in the plains situated in the southwestern part of the study area, which have no role in the present context of landslide occurrences.

16.5.4 Slope

Slope is, perhaps, the single most important attribute to be considered in any landslide analysis. The presence of slope is the prerequisite for landslide occurrences, since a component of the gravitational force tangential to the surface is required to generate the shearing stress (Panickar, 1995). A landslide occurs when downslope component of the force exceeds the shearing strength of the material. It is a critical factor controlling the distribution of landslides, as failure will occur on slopes exceeding the critical angle for the material to be moved (Nilsen, 1986). According to Thomas (1974), in tropical areas, mass movements are generally confined to slopes between 30° and 60°. As the slope angle increases, shear stress in soil or other unconsolidated material generally increases, as well. Gentle slopes are expected to have a low frequency of landslides because of lower shear stress associated with low gradients. In the present context, contour configurations were studied from topographic map and the homogeneous unit of the slope angle and aspect were delineated. From this, a slope vector map was generated and classified into seven groups adopting Young and Young (1974) scheme of classification (Figure 16.6).

The slopes were further regrouped into valley-head slope, valley-side slope, and spur-end slope (Panickar, 1995). The valley-head slopes include precipitous and very steep slope regions, whereas the valley-side slopes covered the entire range of moderate, moderately steep, and steep slope (5°–30°) regions; spur-end slopes are associated with gentle slope regions (<5°). Valley-head slopes were developed in the originating place of streams in the high altitudes and spur-end slopes are confined where the streams enter into the midland or midland–lowland interface. Valley-side slopes are located in between these two categories. The valley-head slopes, and to a less extent the valley-side slopes, are seen mainly in the east–north eastern part of the study area.

Based on slope morphology, the terrain is further categorized as convex, concave, and planar slopes and plain surfaces. This classification is done on the basis of slope angle and shape of contours. Concave slopes are

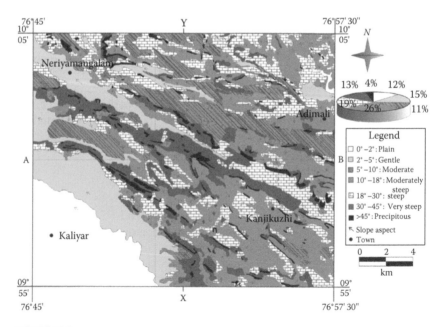

FIGURE 16.6
Slope map depicts various slope category, more than two-thirds of the terrain is covered by moderate to precipitous slope.

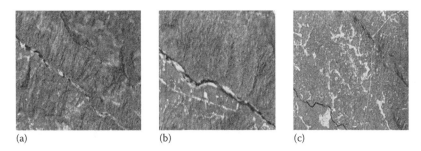

FIGURE 16.7
Slope stability unit (a) active, (b) passive, and (c) dead slope category interpreted from satellite data.

slopes that have high elevation in the periphery and low elevation in the center. Progressing down the steep side from the periphery, the contour lines in the topomap will first be spaced close together and then further apart on the nearly flat bottom. Convex slopes have just the reverse. Planar slopes are equally spaced and parallel to each other. Plain surfaces are gentle and occupy a vast area in the SSW part of the study area.

On the basis of the stability of slope, the region is classified into active and passive slope. The slope stability unit is classified based on steepness of slope and land cover pattern. The factors like drainage, texture, and lithology can be incorporated for such classification if slope steepness and land cover parameter are inappropriate. Active slope is a combination of steep to precipitous slope with barren rock, whereas passive slope has plain to moderate slope category with vegetation, and barren slope has less influence on landslide occurrences (Figure 16.7). The active slope shows preferred alignment with regional structural trend in the area of study.

16.5.5 Relative Relief

Relative relief is the difference between the maximum and minimum elevation in an area—a measure of the ruggedness of the terrain. The relief of a terrain reflects, in general sense, the geomorphological history of that area. The relative relief does not take into consideration the dynamic potential of the terrain, but it is closely linked with slope and is useful in understanding morphogenesis (Thampi et al., 1998). The more the intensity of dissection (particularly vertical erosion), the greater is the relative relief. The occurrences of landslide are directly related to the relief. The region is divided into seven classes with $100\,m/km^2$ intervals (Figure 16.8). The present relief pattern has shown the extremity of terrain ruggedness of the area. The appraisal of the relative relief of the study area brings out four distinct regions: the northeasterly plateau region, the southwesterly

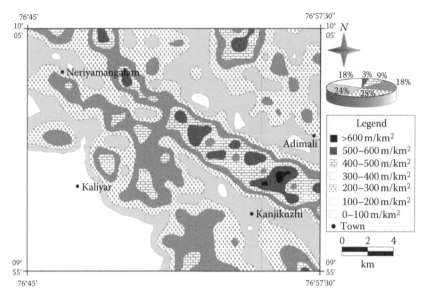

FIGURE 16.8
Relative relief map of the study area.

midlands, and the middle region consisting of plateau margin and dissected marginal hills and, the plains. The northeasterly plateau region witnessed the maximum number of landslide occurrences. The southwesterly plains are free of landslides. The middle zones have wide contrast in the relief, which need attention to mitigate landslide occurrences.

16.5.6 Drainage Analysis

Periyar, Kaliyar, and Kothamangalam thodu are the major rivers that flow in the study area (Figure 16.1). Periyar is the longest river in the state of Kerala and traverses diagonally through the region of study. It is bordered by escarpments and has a straight traverse through "Periyar Fault." Deviar, a tributary of Periyar, originates at high altitude and traverses through a number of falls and cascades, indicating structural breaks. Kaliyar river flows through the plains in a very sinuous way and attains its base level. However, Kannadi Ar, tributary of Kaliyar, originates from high altitude (1100 m) with steep profile during its short traverse. Kothamangalam thodu has steep profile in its initial stage and flows through low altitude in the later stage, showing the maturity of development. Chittur Ar has its origin in highlands with steep profile. The entire drainage system in this part of the region is characterized by trellis pattern with NNW orientation. It shows the influence of structure on drainage pattern. The pattern, density, and frequency of drainages provide clue to the underlying structure, lithology, besides slope and vegetation cover. For GIS spatial analysis, drainage density map was prepared and classified into five categories as very high, high, moderate, low, and very low drainage density domain. The role of drainage texture in landslide occurrences is contradictory. Few workers have positively correlated the high drainage texture with landslide occurrences (Franks, 1999) and others (Sarkar and Kanungo, 2002) are negatively correlated. In the present context, drainage texture has shown positive correlation. It means, areas of high drainage texture are associated with high incidence of mass movements, especially rock falls.

16.5.7 Soils

Soil, the loose unconsolidated and weathered material, is an important conditioning factor in the occurrence of landslides. The knowledge of the spatial distribution and variation of soil thickness with its properties, coupled with detailed measurements of rainfall, would allow for the temporal and spatial prediction of slope instability in a region. The region of study is covered by four different types of soil, viz., lateritic soil, forest loam, riverine alluvium, and brown hydromorphic soil. Lateritic soil in reddish brown to yellowish red color is the most widespread soil type. The morphological character of this soil varies to some extent, depending on the underlying rock type. The genesis of the soil is closely linked with laterite formation. Forest loams are developed over the crystalline rocks of the Western Ghats. These are shallow immature soils under canopy of vegetation, mantling the gneissic/granulitic parent rocks at various stages of chemical weathering (Soman, 2002). Forest loams are dark reddish brown to black in color with loam to silty loam in texture and have wide variation in depth. Riverine alluvium is confined to the banks of Periyar and shows wide variations in the physicochemical properties, and the arrangement of layers depends on province features. Predominance of sand and silt fraction, presence of mica flakes, and deposition of organic matter are the characteristics of this soil. Brown hydromorphic soil is occupied at valley bottom. These soils are generally very deep brown in color and surface texture varies from sandy loam to clay. It also exhibits wide variation in physicochemical properties and morphological features, possibly resulting from transportation and deposition under varying conditions and provenances. The soil cover over the Precambrian rocks and the contact between these two acts as a pathway to bring down the material during landslides. Moreover, the resistance to erosive forces and depth of potential failure surface is primarily determined by soil properties. Soil erosion in the region of study is examined using the data collected from Kerala State Land Use Board. The region is delineated under four soil erosion

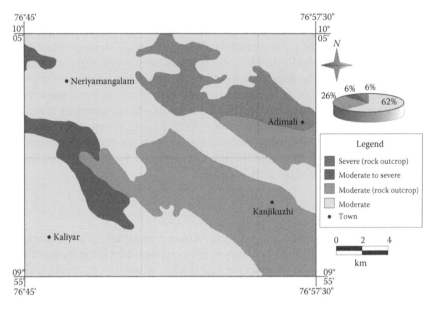

FIGURE 16.9
Soil erodability map of Neriyamangalam—Adimali region. (From Soil Survey Organisation, Trivandrum, India.)

categories as severe with rock outcrop, moderate to severe, moderate with rock outcrop, and moderate soil erosion (Figure 16.9). This classification is derived on the basis of the extent of soil erosion and volume of materials brought down. The soil erosion is significant in severe with rock outcrop category and equally important in occurrence of landslides.

16.5.8 Land Use/Land Cover Mapping

The land use and land cover in an area is more relevant to landslide studies. Improper land use practices such as agricultural development on steep slopes, quarry activities, and constructions of roads, buildings along unfavorable slope contribute to landslide occurrences. Deforestation is another important factor to be considered in landslide studies. In Idukki district, about 2049 km^2 area was covered by natural vegetation in 1998, when compared to 4352 km^2 area in 1966–1967 (Jha et al., 2000). There are eight land use and land cover categories in the study area interpreted with the help of IRS-1D PAN + LISS III satellite data and topomap. The categories are settlement with mixed cultivation, cardamom, rubber, teak and eucalyptus plantations, open scrub, barren land, and dense forest (Figure 16.10). Settlement with mixed cultivation is confined to plains and

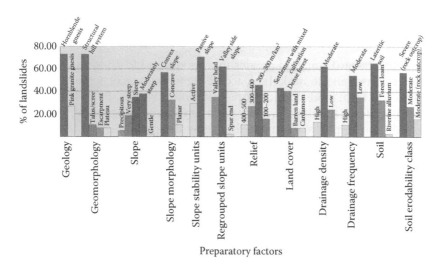

Preparatory factors

FIGURE 16.10
Correlation of previous landslide locations with terrain parameters.

partly in the mountainous regions. This category is a product of reclamation of paddy fields, which is mostly restricted with flood plains. The township of Kaliyar in the plains and Neriyamangalam and Adimali in the hilly tract essentially constitute the inhabited area. Rubber and cardamom plantations are seen in the midlands and highlands, respectively. These are conversion products of natural vegetation. The teak and eucalyptus plantations are afforestation practices implemented for preserving natural forest habitat planted under Social Forestry Plantation Scheme by the Government of Kerala. Open scrub are the product of deforestation observed at few locations. Barren rocks are the exposed rock surfaces, including the escarpments. Dense mixed forest occupies the major part of the study area and is degrading at a fast rate due to lateral spread of population.

Vegetation provides both hydrological and mechanical effects that are generally beneficial to the stability of slopes. The vegetation clearance for new road construction, development of residential areas, or industrial purposes often allows rapid soil erosion and weathering to occur, which increases the potential for landslides (Smyth and Royle, 2000). The trend of decrease of dense forest cover for plantations is less in the study area; however, the encroachments continue for new settlements. Sometimes, landslide occurrence is significant along the transitive zones of different vegetation types (Zhou et al., 2002). Such types of vegetation transition zones are common in the mountainous terrain in the study area and those areas need more attention for the prevention of landslides.

16.6 Spatial Analysis for Landslide Susceptibility Mapping

Landslide hazard zonation applies in a good sense to divide an area into discrete zones and to rank them according to degrees of actual or potential hazard from landslides or slope instability (Varnes, 1984). For the last three decades, a number of inputs were considered in assessing landslide hazards and in preparing maps that show its spatial distribution. There is no general agreement, which method is suitable for producing landslide hazard zonation mapping (Carrara et al., 1982). In this chapter, all the terrain parameters were interpreted in detail and subjected for further spatial analysis through GIS. The Geomedia professional 5.2 GIS software was used for spatial analysis.

16.6.1 Reclassification of Thematic Maps

A combination of different terrain parameters influences the landslide occurrences and each parameter has its own contribution. Herein lies the importance of reclassification of thematic maps. Reclassification involved a type of sorting of different classes within a parameter on the basis of their contribution toward landslide occurrences. On the basis of field evidences and petrography, it is found that geology has no direct influence on the occurrence of landslides in the study area. However, the structures associated with various lithology show relation with landslides. So, the geology in the region is broadly classified into geology with structural component and without major structures.

Slope vector, slope morphology, and slope stability units have different degree of influence on landslide occurrences in the terrain. However, in view to include the entire slope theme as single parameter for spatial integration, a reclassified slope map was prepared on the basis of overlay analyses of all three themes. The resultant slope map is broadly classified into two broad classes—hazard and nonhazard slope category. On the basis of characteristics relative relief and contributions to landslide occurrences, the region is divided into three zones as northeasterly plateau region, middle zone, and southwesterly midlands and plains. The land use/land cover map was also reclassified into four units. The broad group plantations include rubber, cardamom, teak, and eucalyptus, whereas barren rock and open scrub were clubbed together. Mixed cultivation with settlements and dense mixed forests are retained as a single unit.

Geomorphology was retained as such for spatial integration because the different landforms have different modes of influence on landslide occurrences. Drainage density and drainage frequency have similar influence on landslide occurrences. In this chapter, drainage density is incorporated for further analysis. Soil erosion is an important parameter to be considered in landslide analyses and all four categories are retained as such for spatial integration.

16.6.2 Ranking and Weightage Assignment

Landslide susceptibility mapping usually involves predicting and expressing the probability of a landslide occurrence. The approaches vary from qualitative to quantitative. Normally, those factors that are responsible for landslide occurrence are given weightages according to certain algorithms. In the present context, the influence of different terrain parameters is site specific, a "heuristic approach" or "subjective rating analysis" is done for the distribution of ranks for different classes within a thematic map and weightage for different thematic maps. Defining "subjective rating" requires a high degree of specific local knowledge and experience. The rating, however, is not rigid and it requires continual use of skilled judgment. This method would reflect the relative importance of each map and the relative contribution of each class within a single map. Moreover, there would be a bias in determining the higher probable contributing elements, and it purely depends upon the ground truth. The product of rank and weight gives the score. The distribution of ranks, weights, and scores for different classes and thematic maps are given in Table 16.2. The scale of rank varies from 0 to 10. A rank "10" means the probability of landslide occurrence with respect to particular class is high. Assigning weightage to each thematic map is according to the expected order of importance, i.e., interparameter contribution. For example, slope is an important terrain parameter on landslide occurrence than drainage density. So, slope will have high weightage when compared to drainage density. The total weightage for all themes is 100.

16.7 Results and Discussion

In order to understand the role of different terrain parameters, the locations of previous landslides were correlated with various conditional parameters. There are about 37 landslide locations that were identified through satellite data and verified during field visits. The number of landslide occurrences was plotted against the individual terrain parameters and presented in the form of histogram (Figure 16.10). Though the geology has not directly contributed to landslide occurrences in the area of study, about 75% of landslide locations fall in hornblende gnessic terrain. This is because a larger area is covered by gnessic rock. Similarly, high landslide incidence (75%) is associated with structural hill system. The various

TABLE 16.2

Assignments of Rank and Weights for Different Thematic Maps on the Basis of Subjective Rating Approach

Thematic Maps and Its Classes	Rank (for Classes)	Weight (for Thematic Maps)	Score (Rank × Weight)
Geology			
Geology with structure	8	15	120
Geology without structure	0		0
Geomorphology			
Structural hill system	9		135
Escarpments	8		120
Intermontane valley	5		75
Talus/scree	5		75
Plateau	4	15	60
Cuesta	2		30
Mesa	1		15
Residual hill	1		15
Plain	1		15
Slope			
Hazard slope	10	30	300
Nonhazard slope	0		0
Relative relief			
North easterly plateau region	6	15	90
Middle zone	8		120
South westerly midlands and plains	1		15
Drainage density			
Very high	3		15
High	4	5	20
Moderate	3		15
Low	2		10
Very low	1		5
Soil erodability			
Severe (rock outcrop)	9		90
Severe to moderate	8	10	80
Moderate (rock outcrop)	8		80
Moderate	6		60
Land use/land cover			
Mixed cultivation with settlement	5		50
Plantation	7	10	70
Barren rock and open scrub	8		80
Dense mixed forest	3		30

Note: Rank in the scale of 0–10, weight in percentage.

studies across the world show that the landslides are mainly concentrated with steep to moderately steep slope categories. In the present case, the maximum number of landslides occurred along the moderately steep slope followed by the steep slope region. The high incidence of landslides also correlated with valley-side slope. In case of relief, the middle zone with 200–300 m/km^2 relative relief shows high correlation with landslide occurrences. As far as the land use and land cover is concerned, the settlement zone correlated with landslide incidences. Similarly, high landslide occurrences coincide with moderate to low drainage density zone, where failure happened due to high infiltration and pore water pressure. As far as soil is concerned, maximum landslide occurrences are associated with lateritic soil and severe with exposed rock under soil erosion category. The correlation has provided valuable inputs for understanding the influence of various terrain parameters on landslide occurrences.

Landslide susceptibility mapping for the study area was done on the basis of landslide inventory and detailed analysis of different terrain parameters, followed by reclassification and assignment of ranks and weights to various themes. The final landslide susceptibility map was obtained through multidimensional spatial analysis of seven parameters in GIS environment (Figure 16.11). The entire area is broadly divided into three zones, viz., high landslide susceptibility (Zone I), moderate

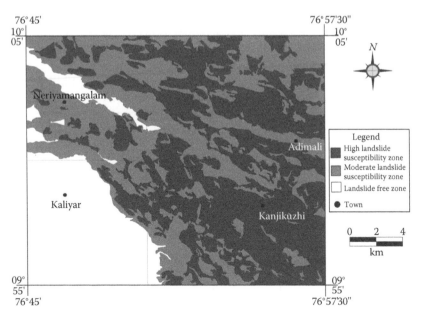

FIGURE 16.11
Landslide susceptibility zones for region of study generated through spatial integration of terrain parameters.

landslide susceptibility (Zone II), and landslide free zone (Zone III). As the name implies, the high landslide susceptibility zone is highly prone to landslide occurrences and moderate zone prone to moderate to least chances of landslide occurrences. These findings were verified and validated with the help of previous landslide incidences in this region. As discussed earlier, there are about 37 previous landslide locations interpreted and identified from satellite data and field investigations. Out of the 37 landslides, 35 landslide locations fall within Zone I and II in the region of study.

Zone I includes inhabited regions of Adimali, Kanjikuzhi, and Pathayanikandam. The Periyar and Kerala lineaments experiences with previous seismic activity traverse through this zone. This zone is also covered by the structural hill system, escarpments, talus scree, and steep to precipitous slope. Moreover, high incidence of rainfall is a common phenomenon. Zone II differs both in terms of aerial extent and terrain characteristics. Though geomorphic features are almost similar in both zones, this region is less inhabited when compared to Zone I. The major part of plateau region is associated with moderately steep to steep slope, dense mixed forest cover, and moderate to very high drainage density.

The detailed analysis of different terrain parameters and spatial intersection has shown that the region of study is prone for frequent occurrence of landslides during monsoon. It suggests that the area needs much attention in terms of management practices and preventive measures. In this chapter, preventive and management practices were proposed for settlement area and communication corridors located within the high landslide susceptibility zone. For this purpose, a 200 m buffer zone was created. The buffer zones fall within high landslide susceptibility zone, and are spatially integrated with all terrain parameters. It gives 15 different combinations of terrain factors, which influence the occurrences of landslides (Table 16.3). Site-specific management practices were proposed for this combined output.

16.8 Conclusion

The region of study belonging to parts of Western Ghats comprises of various tectonic and geomorphic landforms evolved through external and internal agents that acted upon this system. Landslides, the common phenomena in this part of the region, are witnessed particularly during southwest monsoon from June to September, where rainfall is the major triggering mechanism for occurrence of landslides. In order to understand the role of each and every tectonic and geomorphic features

TABLE 16.3

Landslide Controlling Terrain Parameter in High Hazard Zone, Type of Landslides, and Proposed Management Practices

Sl No	Controlling Terrain Parameters[a]	Location	Possible Type of Landslide	Proposed Management Practices
1	VS+L	Near Orakkuravu Mala	Debris flow	Surface and subsurface drainage
2	SHS+VS+L	North and south of Neriyamangalam	Debris flow	Surface and subsurface drainage and rock buttress
3	SHS+VS+R	North of Neriyamangalam and Kanjikuzhi	Debris flow	Surface and subsurface drainage and rock buttress
4	MCS+VS+L	SE of Neriyamangalam	Debris flow and debris slide	Surface and subsurface drainage
5	SHS+VS+F	Near Irumpupalam	Debris flow	Surface and subsurface drainage
6	MCS+(VS+VH)+L	Near Orakkuravu Mala	Debris flow and rockfall	Netting and buttress fill
7	SHS+(VS+VH)+L	Near Kuzhimattam	Debris flow and rockfall	Netting and buttress fill
8	SHS+PT+VH+L	Near Kaitappara	Rock fall	Netting
9	SHS+PT+VS+F	Near Orakkuravu Mala	Debris flow and debris slide	Surface and subsurface drainage and rock buttress
10	SHS+MCS+VS+L	South of Machichiplavu	Debris flow	Surface and subsurface drainage
11	SHS+MCS+(VS+VH)+L	North of Plamala Estate	Debris flow and rockfall	Surface and subsurface drainage, netting, and rock buttress
12	SHS+PT+(VS+VH)+L	Near Kaitappara	Debris flow and rockfall	Scientific method of cultivation, netting, and rock buttress
13	SHS+PT+VH+(R+L)	North of Kanjikuzhi	Debris flow and rockfall	Scientific method of cultivation, netting, and pile works
14	SHS+MCS+(VS+VH)+(L+F)	North and south of Adimali plateau	Debris flow and rockfall	Surface drainage, netting, pile works, and anchor systems
15	(SHS+ES)+MCS+(VS+VH)+L	Kanjikuzhi panchayat	Debris flow and rockfall	Surface and subsurface drainage, netting, and anchor systems

[a] SHS, structural hill system; ES, escarpments; MCS, mixed cultivation with settlements; PT, plantations; VS, valley-side slope; VH, valley-head slope; L, lateritic soil; R, riverine alluvium; F, forest loam.

and occurrence of landslides, the IRS-1D PAN + LISS III merged satellite data are used. The satellite data provide valid input for the preparation of thematic maps on structural, tectonic, geomorphological, and land use/land cover theme and spatial intersections for the generation of landslide susceptibility mapping. The structural hill system, moderately steep slope, escarpments, moderate relief, settlement with mixed cultivation practices, moderate drainage density, lateritic soil cover, severe rock outcrop (soil erodability unit) are the important terrain features that control major landslide occurrences in this region. It indicated that the high landslide susceptibility zone is inhabited with settlement at important locations and transport corridors. These regions need priority based preventive measures to mitigate catastrophic events. The probable locations of landslide occurrences and type of landslides are given in Table 16.3. Based on the information, the Public Works Departments and highways department should take necessary preventive measures every year before the onset of monsoon. The possible types of landslides in this region are debris flow and rock fall. So, the preventive measures should be based on these aspects. The rainfall data in the study area have shown that an average of 30 days in a year has more than 20 cm rainfall, which indicate that the frequency of landslide occurrence is more.

Acknowledgments

The necessary data were collected from the Indian Meteorological Department (IMD), National Remote Sensing Centre (NRSC), Geological Survey of India, Soil Survey Organization, and Kerala State Land Use Board is acknowledged. Sajin Kumar acknowledges the Council for Scientific and Industrial Research (CSIR) for providing fellowship for his research work.

References

Anbalagan, R., Sharma, L., and Tyagi, S., 1993. Landslide hazard zonation mapping of a part of Doon Valley, Garhwal Himalayas, India. In *Environmental Management Geo-Water and Engineering Aspects*, Eds. R.N. Chowdhury and M. Sivakumar, pp. 253–260. Balkema, Rotterdam.
Carrara, A., Valvo, S.M., and Reali, C., 1982. Analysis of landslide form and incidence by statistical techniques, Southern Italy. *Catena* 9: 35–62.

Chau, K.T., Sze, Y.L., Fung, M.K., Wong, W.Y., Fong, E.L., and Chan, L.C.P., 2004. Landslide hazard analysis for Hong Kong using landslide inventory and GIS. *Computers and Geosciences* 30: 429–443.

Crozier M., 1986. *Landslides: Causes, Consequences and Environment*. Croom Helm, London

Dai, F.C., Lee, C.F., and Ngai, Y.Y., 2002. Landslide risk assessment and management: An overview. *Engineering Geology* 64: 65–87.

Devarajan, K.M., 1990. Geomorphology and morphotectonics of Idukki catchment, Kerala, using Remote Sensing techniques. PhD dissertation, IIT, Bombay, India.

Franks, C.A.M., 1999. Characteristics of some rainfall-induced landslides on natural slopes, Lantau Island, Hong Kong. *Quarterly Journal of Engineering Geology* 32: 247–259.

Jha, C.S., Dutt, C.B.S., and Bawa, K.S., 2000. Deforestation and land use changes in Western Ghats, India. *Current Science* 79(2): 231–238.

Koshimoto, S., Tsunogae, T., and Santosh, M., 2004. Sapphirine and corundum bearing ultra high temperature rock from the Palghat-Cauvery shear system, Southern India. *Journal of Mineralogical and Petrological Sciences* 99: 298–310.

Mantovani, F., Soeters, R., and VanWesten, C.J., 1996. Remote sensing techniques for landslide studies and hazard zonation in Europe. *Geomorphology* 15: 213–225.

Mishra, D.C., Singh, A.P., and Rao, M.B.S.V., 1989. Idukki earthquake and its tectonic implications. *Journal of Geological Society of India* 34: 147–151.

Mohan, A., 1996. The Madurai granulite block. In *The Archean and Proterozoic terrains in Southern India within East Gondwana*, Eds. M. Santhosh and M. Yoshida, Gondwana Research Group Memoir 3, pp. 223–242. Field Science Publishers, Osaka, Japan.

Nilsen, T.H., 1986. Relative slope stability mapping and land use planning in the San Fransico Bay region, California. In *Hillslope Processes*, Ed. A.D. Abrahams, pp. 389–413. Allen and Unwin, Boston, MA.

Panickar, S.V., 1995. Landslides around Dehradun and Mussoorie: A geomorphic appraisal. PhD dissertation, IIT, Bombay, India.

Prabhakara Rao, P., Nair, M.M., and Raju, D.V., 1985. Assessment of the role of remote sensing techniques in monitoring shoreline changes: A case study of the Kerala coast. *International Journal of Remote Sensing* 6(3–4): 549–558.

Rajan, P.K., Santosh, M., and Ramachandran, K.K., 1984. Geochemistry and petrogenetic evolution of the diatexites of Central Kerala, India. *Proceedings of Indian Academic Science, Earth and Planetary Science* 93(1): 57–69.

Rastogi, B.K., Chadha, R.K., Kumar, N., Satyamurthy, C., Sarma, C.S.P., and Raju, I.P., 1989. Report on the Idukki earthquake of magnitude 4.5 on June 7, 1988 as an example of reactivation of NW-SE wrench fault in Peninsular India. NGRI Technical Report No. ENVIRON 61, 78pp.

Sajin kumar, K.S. and Anbazhagan, S., 2004. Delineation of landslide hazard zones in parts of Western Ghats using IRS ID Satellite Data and GIS. In *Remote Sensing and GIS Applications*, Eds. H.T. Basavarajappa, K.N. Prakash Narasimha, P. Madesha, B. Suresh, and A. Balasubramanian, pp. 54–63. Bellur Prakashana, Mysore.

Santosh, M., Iyer, S.S., and Vasconcellos, M.B.A., 1987. Rare earth element geochemistry of the Munnar carbonatite, Central Kerala. *Journal of Geological Society of India* 29: 335–343.

Sarkar, S. and Gupta, P.K., 2005. Techniques for landslide hazard zonation—Application to Srinager-Rudraprayag area of Garhwal Himalaya. *Journal Geological Society of India* 65: 217–230.

Sarkar, S. and Kanungo, D.P., 2002. Landslides in relation to terrain parameters—A remote sensing and GIS approach. *Map India 2002: GIS Development.* http://www.gisdevelopment.net/application/natural_hazards/landslides/nhls0010.htm

Smyth, C.G. and Royle, S.A., 2000. Urban landslide hazards: Incidence and causative factors in Niterói, Rio de Janerio State, Brazil. *Applied Geography* 20: 95–117.

Soman, K., 2002. *Geology of Kerala.* Geological Society of India, Bangalore.

Terzaghi, K., 1980. Mechanism of landslides. In *Application of Geology to Engineering Practice*, Ed. S. Paige, pp. 83–123. The Geological Society of America, Berkeley, CA.

Thampi, P.K., Mathai, J., Sankar, G., and Sidharthan, S., 1998. Debris flow in Western Ghats—A regional evaluation. In *Final Proceedings of Tenth Kerala Science Congress*, Kozhikode, India, pp. 73–75.

Thomas, M.F., 1974. *Tropical Geomorphology: A Study of Weathering and Landform Development in Warm Climates.* MacMillan, London, U.K.

Thornbury, W.D., 1990. *Principles of Geomorphology.* Wiley, New York.

Valdiya, K.S., 1987. *Environmental Geology—Indian Context.* Tata McGraw Hill, New Delhi.

Van Asch Th, W.J., Buma, J., and Van Beek, L.P.H., 1999. A view of some hydrological triggering systems in landslides. *Geomorphology* 30: 25–32.

Varnes, D.J., 1984. Landslide hazard zonation: A review of principles and practices. Commission on landslides of the IAEG Natural Hazards 3, UNESCO, Paris. 63pp.

Wilson, R.C. and Wieczorek, G.F., 1995. Rainfall thresholds for the initiation of debris flows at La Honda, California. *Environmental Engineering and Geosciences* 1: 11–27.

Young, A. and Young, D.M., 1974. *Slope Development.* MacMillan Education, London, U.K.

Zaruba, Q. and Mencl, V., 1969. *Landslides and Their Control.* Elsevier, Amsterdam, the Netherlands.

Zhou, C.H., Lee, C.F., Li, J., and Xu, Z.W., 2002. On the spatial relationship between landslides and causative factors on Latau Island, Hong Kong. *Geomorphology* 43: 197–207.

17

Impact of Tsunami on Coastal Morphological Changes in Nagapattinam Coast, India

E. Saranathan, V. Rajesh Kumar, and M. Kannan

CONTENTS

17.1 Introduction

The east coast of India is more prone to natural hazards like cyclones, storm surges, and, in recent times, a new hazard in the form of tsunami. A tsunami, on the other hand, has nothing to do with atmospheric disturbances. Tsunamis were recorded during 1881, 1883, 1941, 1945, and 2004 along the coast of India (Chadha et al. 2005). The December 26, 2004, tsunami was the most disastrous and caused heavy loss of life and property. The tsunami triggered due to undersea mega thrust earthquake that occurred with an epicenter at N 3.7° and E 90° with a magnitude of MW 9.3, near the west coast of Sumatra, Indonesia, and is referred to as the Sumatra–Andaman earthquake (Lay et al. 2005). The devastating tsunami killed over 230,000 people in 14 countries. This second largest tsunami

had an impact of sudden changes in the geomorphology and geomorphological processes along the east coast of Tamil Nadu, India. This has
resulted in transformation in the coastal landforms and quality of life as
well. The catastrophic event has eroded the beach, foreshore, and backshore areas and completely generated new outlook in the coastal geomorphology, thereby resulting in the need of new sets of coastal management
practices. Many coastal morphological changes have occurred all over the
world in response to the December 26, 2004, tsunami (Ella et al. 2006).
Nagapattinam district in the east coast of India, located in Tamil Nadu
state, is one of the areas that was severely affected by the 2004 tsunami
waves (Ramasamy et al. 2005). The severity of damage varies depending
on the location of the villages, altitude above mean sea level (MSL), presence of sand dunes, and the distance from the coast.

Geomorphological mapping is an essential prerequisite in many circumstances to problem appraisal and it is often a necessary precursor to
evaluate the nature of changes. Remote sensing and GIS techniques are
useful in providing practical and cost-effective output for geomorphology,
input environmental protection, and management. In the present context,
village-level tsunami inundation mapping for three villages of Sirkazhi
taluk along the Nagapattinam coast was carried out. In order to delineate
and characterize the coastal geomorphological changes due to the tsunami, the IRS-P6 (LISS-III and IV) merged satellite data of November 2004
aerial photographs acquired during 1985, and data collected through field
survey investigation in the year 2005 were used. The results highlighted
the geomorphic changes due to denudation and depositional process that
took place pre- and post-tsunami.

17.2 Region of Study, Nagapattinam District

The coastal stretch of about 12 km length between Sevanar River in the
south and Kilmuvakkarai in the north near Poombukar, Nagapattinam
district was selected to interpret coastal landform changes due to tsunami. The region of study is located between 11°06′38″ and 11°12′47″
northern latitudes and 79°48′53″ and 79°51′39″ eastern longitudes. It
covers three revenue villages in Sirkazhi taluk, namely, Peruntottam,
Vanagiri, and Kilaiyur, with an aerial extent of 29 km². The digitized
boundary of the revenue villages imposed on the IRS-1D LISS-III and IV
merged satellite imagery (Figure 17.1). Subtropical climate prevails in the
study area. The maximum and minimum atmospheric temperatures are,
respectively, 32°C and 25.3°C. The normal rainfall ranges from 1168 to
1500 mm. According to the census of India 2001, about 28% of the present

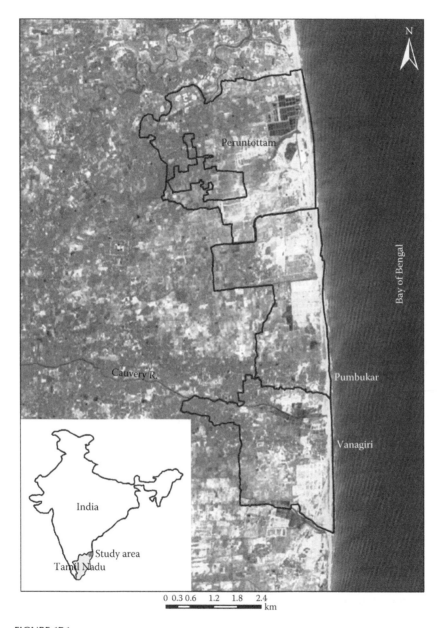

FIGURE 17.1
IRS-P6 LISS-III and IV merged satellite data, acquired in November 2004. The village boundaries are shown in black color.

population is located in the coastal areas within 10 km from the seashore. The total population in the revenue villages as per 2001 census is 21,345. Since the coastal region is witnessing "blooming revolution" in aquaculture, it has well-connected network with the rest of the country.

The coastline is represented by undulating topography with a seaward slope and sandy silt in nature. The coastal plain of study area comprises various morphological units like sand dune complexes, mud flats, and backwaters. There are no prominent relief features observed in the study area. The eastern part of coastal plain is level terrain gently sloping toward the sea and the west has slightly steep slope of about 1 m/4.8 km along the coast line. The land has an elevation of 1–10 m above MSL.

17.3 Geomorphological Mapping

The geomorphological mapping is to record information on the surface form, materials (soil and rock), surface processes, and the age of landforms for selected locations. As such, it provides a base for terrain assessment, which is useful in the context of many environmental problems. The most successful approach for such mapping is to combine the remote sensing data with field survey and investigation. Remote sensing and GIS techniques were effectively utilized for coastal morphological mapping (Ojeda Zújar et al. 2002; Seker et al. 2003).

17.3.1 Pre-Tsunami Geomorphological Mapping

The IRS-P6 LISS-III and IV merged satellite data were used to prepare coastal geomorphology of the region of the study before the tsunami event. The satellite data dated November 2004 were collected from National Remote Sensing Centre (NRSC), Hyderabad. The spatial resolutions of the satellite data are 23.5 and 5.8 m, respectively, for IRS-P6 LISS-III and IV sensors. The merged satellite data provide enhanced information both in spatial and spectral context. In addition to satellite data, the aerial photographs acquired in the year 1985 were also used.

The methodology mentioned in the manual of Rajiv Gandhi National Drinking Water Mission (Anon 2008) is adopted for mapping of various landforms in the study area. As per guidelines, the landforms are categorized under third level of geomorphic classification under coastal plain (Table 17.1).

The Survey of India (SOI) toposheets provide topographic information required to interpret the geomorphology from the satellite imagery.

TABLE 17.1

Coastal Landform Classification

Geomorphology	Geomorphic Unit	Landform
1. Plains	Coastal plain	Older/upper
		Shallow
		Moderate
		Deep
		Younger/lower
		Shallow
		Moderate
		Deep
		Beach
		Beach ridge
		Beach ridge and swale complex
		Swale
		Offshore bar
		Spit
		Mud flat
		Salt flat
		Tidal flat
		Lagoon
		Sand dune
		Channel island
		Palaeochannel
		Buried channel

Source: Anon, Rajiv Gandhi National Drinking Water Mission methodology manual for preparation of ground water prospects map, National Remote Sensing Centre, Hyderabad, 2008.

The synoptic view of satellite imagery facilitated better appreciation of geomorphology and helped in mapping of different landforms and their assemblage. Initially, the entire image was classified into three major zones, as hills and plateaus, piedmont zones, and plains with reference to physiography and relief of the terrain. Then, within each zone, different geomorphic units were mapped based on the landform characteristics, areal extent, depth of weathering, and thickness of the deposition. Subsequently, within the coastal plains, individual landforms were mapped. The geomorphic units and landforms interpreted from the satellite imagery and aerial photographs were checked during field investigation by observing stream cuttings and inventory of bore well litho–log

data. The details were incorporated and the final geomorphologic map
was prepared.

The landform features mapped for the revenue villages Peruntottam,
Vanagiri, and Kilaiyur through interpretation of the merged satellite data
and aerial photographs are discussed in the following.

The major landforms in the Peruntottam village are deltaic plain, palaeochannels, beach ridges, mudflat, and sand sheet (Figure 17.2). The
deltaic plain covers most part of Vanagiri village (Figure 17.3) and the
river Cauvery flows in the northern side of the village. The other geomorphic features are the flood plains, palaeochannel, beach ridges, mudflats, swales, and sand sheet. The Kilaiyur village mostly occupied the
deltaic plain (Figure 17.4). The other landforms are beach ridges, mudflats,
swales, and sand sheet.

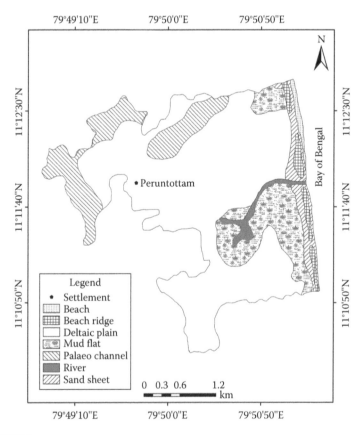

FIGURE 17.2
Coastal geomorphology of Peruntottam village, in Nagapattinam district, India.
Palaeochannel and mudflats are prominent.

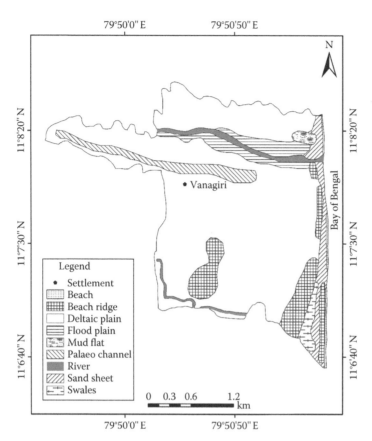

FIGURE 17.3

Coastal geomorphology of Vanagiri village, in Nagapattinam district, India.

17.3.2 Field Survey and Mapping after Tsunami

In order to delineate the detailed morphological changes due to tsunami on the coastal plain area of the revenue villages, an extensive field survey was conducted using global positioning system (GPS) and total station and documented the impact of tsunami waves on the coast. Measurements were made on land elevations, beach slope, tsunami flow distance, and tsunami run-up height. In addition, these phenomena were also documented through systematic interviews with eyewitnesses in the villages.

17.3.2.1 Determination of Land Elevation

The total station instruments Leica 1105 and Sokkia 305R were used for obtaining elevation of scattered points and the distance of inundation

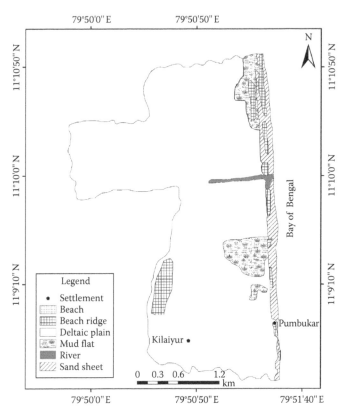

FIGURE 17.4
Coastal geomorphology of Kilaiyur village, in Nagapattinam district, India.

from the coast in the study area. The instruments were supplemented with prism (target). The instrument was kept at observer's position and the prism was kept on a tripod placed at the point up to which the distance and elevation required. The total station works with the principle of electromagnetic (EM) waves. The EM waves from the instrument hit the target (prism) and reflected and received by the instrument traveling two times the distance. The distance acquired divided by two will give the exact distance using the following formula:

$$2D = n\lambda + \Delta\lambda$$

$$D = \frac{n\lambda + \Delta\lambda}{2}$$

where

D is the distance between instrument and target
λ is the wavelength
n is the whole number of wavelength traveled by wave
Δλ is the fraction of the wavelength traveled by the wave

This single instrument is used as a substitute for all the conventional survey. It avoids all human errors such as reading the values, calculations, and plotting, since all these works are done by the instrument itself.

The vertical and horizontal angle accuracy of the Leica instrument is 5 seconds, and for Sokkia instrument, 3 seconds. The instruments were set right by making some temporary adjustments such as setting up the instrument with optical plummet or laser plummet and leveling up with the electronic bubble. Then the back sighting approach of orientation was followed by putting the northing and easting values acquired by GPS and by keeping the target (prism) in the tripod to ensure the stability of target. The input of GPS values eliminates the flaws encountered due to the arbitrary values assumed for northing and easting. The stability of target improves the accuracy of data acquired by the total station.

About 1500 points for three villages were targeted by the instruments and the elevations of different features for the study area were obtained with their locations in terms of northing and easting. The data were then transferred through PC card from instrument memory and loaded in PC through omnidrive. The northing and easting values were converted to their respective latitudes and longitudes using the usual conversion procedure.

17.3.2.2 Determination of Tsunami Run-Up Heights

Run-up heights are the maximum vertical distance reached by tsunami waves, measured relative to the MSL. Tsunami inundation and damage were not uniform along the coast. This may be due to the geographical orientation of the coast, geomorphology of the landmass, shallow water bathymetry and orientation of approaching waves, etc. The run-up heights and inundation limit were measured with the help of traces left out by the tsunami waves such as watermarks on the buildings, trees, and debris lines along the coast or vegetation damage. In addition, information was also gathered from eyewitnesses. The information is subject to errors on the order of few tens of centimeters and generally within 10%–15% deviation from the measured data.

17.3.2.3 Contour Mapping

The ArcMap and Geomatica GIS software were used in deriving the contour map for the regions of study. The point data consist of locations

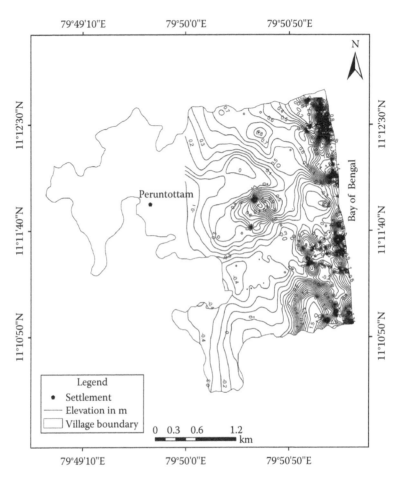

FIGURE 17.5
Relief map of Peruntottam village. The contour depicts 0.1 m interval.

(latitude and longitude) and elevations from ArcMap, which were given as input to the Geomatica GIS software and converted into pix file. Digital elevation models (DEM) were developed from the pix file using Geomatica software, and contour maps were derived from it (Figures 17.5 through 17.7). The pattern of the contours provide vital information to infer the morphological features present in the coastal plain. For example, the contour values increase from bottom to top indicating valley, and if it is reversed, it is referred to as ridge. This basic knowledge is implied to demarcate the erosional and depositional portions in the coastal plains.

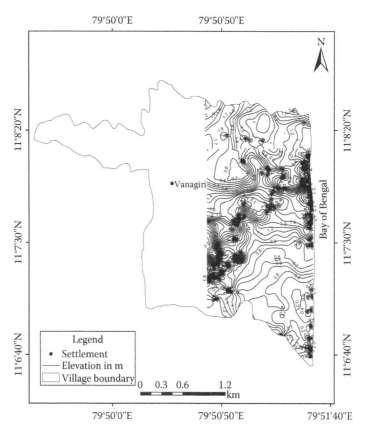

FIGURE 17.6
Relief map of Vanagiri village. Elevations contour with intervals of 0.1 m.

17.4 Results and Discussion

The contour map of **Peruntottam** village is shown in Figure 17.5. The range of elevation of the profile is between −0.7 and 4.5 m above MSL. The maximum elevation of 4.5 m above MSL is measured in the eastern side beach ridge. The tsunami inundated distance from the coast amounts to 1.8 km from the coast. From the contour map, it is inferred that the landforms at **Vanagiri** village are slightly elevated than Peruntottam (Figure 17.6). The elevation ranges between −0.9 and 5.3 m above MSL. The southern part of the beach ridge has a maximum elevation of 5.3 m above MSL. The tsunami waves have inundated for a distance of about 1 km from the coast. The elevation of **Kilaiyur** village is mostly recorded between −0.7 and 5.1 m

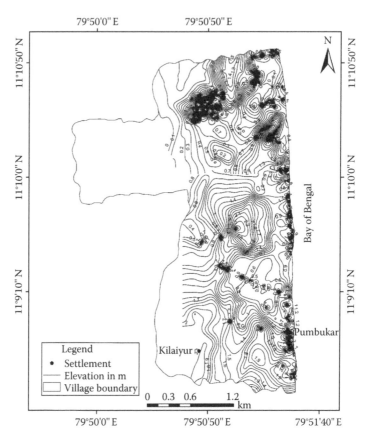

FIGURE 17.7
Relief map of Kilaiyur village, with elevations contouring with intervals of 0.1 m.

above MSL (Figure 17.7). The maximum elevation 5.1 m above MSL was noticed along the palaeo beach ridge. The inundation of tsunami waves in this village is recorded as about 1.6 km from the coast. The contour maps were generated with an interval of 0.1 m to get accurate results on morphological changes. The characteristics of contours and field information supplemented with GPS data were used to identify the erosion and deposition zones after tsunami event and digitized from contour maps. The central ridge of Peruntottam village was subjected to high erosion (Figure 17.8). Moderate erosion is noticed in the northern side of the ridge than the ridge located in the southern side. The total area subjected to erosion is 74,860 m², and about 54,807 m² of the area newly deposited in the central and southern part of the village. The tsunami washed out about 605,022 m² of sand deposits along the coast in the village.

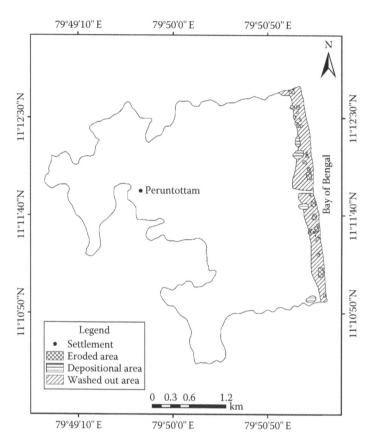

FIGURE 17.8
Erosion and deposition in Peruntottam village, after tsunami event.

In Vanagiri village, significant erosion was found in the central ridges (Figure 17.9) when compared to the ridge located in the southern side. The total area subjected to erosion is about 64,149 m². About 178,654 m² area deposition was taken place in the central part of the village. The tsunami washed out about 419,412 m² of sand deposits along the coast in the village. The northern ridges of the Kilaiyur village were subjected to high erosion due to tsunami inundation (Figure 17.10) and less extent in the ridge located in the southern side. The total area subjected to erosion in the village is 136,487 m², and 146,970 m² of aerial extent in the central and southern part of the village is deposited. The tsunami washed out about 427,594 m² area of sand deposits along the coast in the village (Figure 17.11).

The presence of aquaculture ponds has restricted the entry of tsunami waves into Vanagiri. Hence, the inundated distance is also less in this village compared to the other villages, where the number of aquaculture

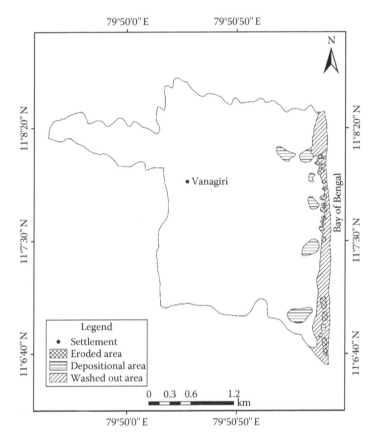

FIGURE 17.9
Erosion and deposition in Vanagiri village, after tsunami event.

ponds is less. Overall, the plantations in all these villages were minimum coverage and hence the inundation distance was significantly high. The variation of levels at different points in landform is indeterminate, so quantification of volume of erosion, deposition, and washed out sand is difficult.

17.5 Conclusion

The run-up levels the prevalence of low elevations for a distance of 2 km, and major settlements very close to the coast need elevation-based setback lines in the settlement areas of these coastal villages. It is recommended to include the adoption of elevation-based setback lines in human settlement planning. The present run-up levels and inundation distance can be used

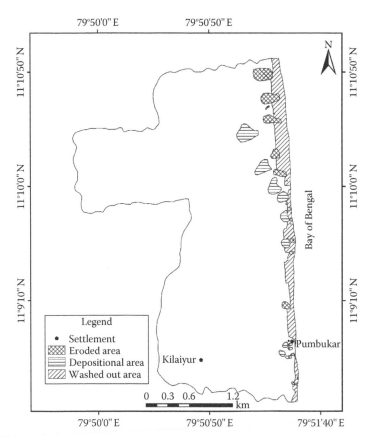

FIGURE 17.10
Erosion and deposition in Kilaiyur village, after tsunami event.

as guidance to determine the safe locations for resettlement of affected population. This micro-level information could be useful for planners and researchers to understand coastal geomorphic problems and predict the future course of changes in landforms.

Acknowledgments

The Earth System Science Division, Department of Science and Technology, Government of India, supported this work. The Department of Survey and Land Record, Tamil Nadu, has provided the village maps. We are thankful to Prof. R. Sethuraman, vice chancellor, SASTRA University, for providing necessary facilities and and for supporting the work.

FIGURE 17.11
Field photographs of various landforms before and after tsunami event. (a) Beach ridge at Peruntottam before tsunami in October 2003 taken during heavy mineral exploration project of CMRI field investigation. (b) Eroded ridge of Peruntottam after tsunami in January 2005. (c) Sand sheet of Vanagiri before tsunami in October 2003 under heavy mineral exploration project of CMRI field investigation. (d) Eroded sand sheet of Vanagiri after tsunami in January 2005. (e) Beach ridge at Kilaiyur village before tsunami in October 2003 under heavy mineral exploration project of CMRI field investigation. (f) Eroded beach ridge of Kilaiyur after tsunami in January 2005.

References

Anon, 2008. Rajiv Gandhi National Drinking Water Mission methodology manual for preparation of ground water prospects map, National Remote Sensing Centre, Hyderabad.

Chadha, R. K., Latha, G., Yeh, H., Peterson, C., and Katada, T., 2005. The tsunami of the great Sumatra earthquake of M 9.0 on 26 December 2004—Impact on the east coast of India, *Current Science* 88(8): 1297–1301.

Ella, M., Marjolein, D. J., and Poh Poh, W., 2006. Indication of coastal morphological changes on Banda Aceh coast as responses to the tsunami on 26 December 2004 [presentation]. In: *NCK-Days 2006*, March 23–24, 2006, Kijkduin, the Netherlands.

Lay, T., Kanamori, H., Ammoric et al., 2005. The great Sumatra—Andaman earthquake of Dec 26, 2004, *Science* 308: 1127–1133.

Ojeda Zújar, J., Borgniet, L., Pérez Romero, A. M., and Loder, J. F., 2002. Monitoring morphological changes along the coast of Huelva (SW Spain) using soft copy photogrammetry and GIS, *Journal of Coastal Conservation* 8(1): 69–76.

Ramasamy, S. M., Kumanan, C. J., Palanivel, K. et al., 2005. Geomatics in tsunami damages Nagapattinam coast, India. In: *Geomatics in Tsunami*, eds. Ramasamy, S. M., Kumanan, C. J., Sivakumar, R., and Singh, B., New India Publishing Agency, New Delhi, pp. 107–134.

Seker, D. Z, Goksel, C., Kabdasli, S., Musaoglu, N., and Kaya, S., 2003. Investigation of coastal morphological changes due to river basin characteristics by means of remote sensing and GIS techniques., *Water Science and Technology* 48(10): 135–142.

18

Remote Sensing for Glacier Morphological and Mass Balance Studies

Pratima Pandey and G. Venkataraman

CONTENTS

18.1 Introduction

A glacier is a moving body of ice, which moves under its own weight. It moves continuously under the force of gravity. Glaciers form when snow that falls in winter does not entirely melt away in summer. Snow that survives in summer is called firn, which has a higher density ($550\,kg/m^3$) and is less bright than snow. Glacier forms over land by compaction and recrystallization of snow over thousands of years. They are highly sensitive to climate change and are good indicators of the energy balance (Oerlemans, 2005). Their capacity to store water for extended periods exerts significant control on the surface water cycle. Their accelerating mass wastage over the last few decades has significant implications for the ongoing rise in global sea level, water resources, and hydropower potential in many regions of the world (Khalsa et al., 2004). There is a strong coupling between energy exchange, melt of snow and ice, and albedo.

The glacier study is important in the sense that it has a direct relation with climate change. Glaciers are mirror reflector of climate change and the information about the climate and changing behavior can be inferred by studying glacier response. Any large-scale changes in the climate can affect the mass balance of glaciers. Mass balance is the difference between the amount of snow and ice gained by the glacier and the amount of snow and ice lost from the glacier. The mass balance of glacier reflects weather conditions and climatic changes and further induces the glacier retreat or advance, which in turn influences snow melt. The climate change affects the glacier shape and geometry by affecting its mass balance, advance/retreat, and snow melting. In that way, the climate change and glacier are strongly coupled.

Glaciers are mostly located in remote and inaccessible places and at higher altitude with extreme environments, which makes it difficult to monitor them by conventional and ground based methods, both in terms of logistics and financial basis. The satellite remote sensing offers the only practical approach to monitor glaciers due to their synoptic view, repetitive and latest coverage. In this chapter, the potential of remote sensing technique for the monitoring of glaciers and glacier parameters with a focus on Himalayan glaciers is discussed.

The glaciers in the Himalayas cover a total of 33,200 km² area (Flint, 1964). In the Indian Himalaya, the glaciers cover approximately 23,000 km² areas and this is one of the largest concentrations of glacier-stored water apart from the polar regions (Kulkarni and Buch, 1991). Glaciers in the Indian Himalaya provide an important environmental and economic service by releasing melt water for the northern and northeastern part of the country during the dry season, i.e., May–September, when little to no rainfall occurs. The glaciers effectively buffer the runoff by storing much of the precipitation falling as snow on the glaciers during the wet season, i.e., October–April, and releasing it throughout the year, including during the dry season when it is most needed. Timely information about glacier properties and their temporal and spatial variability is an important factor in climatology, local weather forecasting, and for the hydropower production in high mountainous terrain. Therefore, mapping the extent of glaciated areas and deciphering the characteristics of glacier help in the estimation of glacier melt runoff. Here, the main focus is the study of glacier morphological features, mapping of glacier, and estimation of mass balance of the glacier using remote sensing techniques.

Advanced Wide Field Sensor (AWiFS) and Indian remote sensing (IRS)-1D satellite LISS-III sensor as well as Advanced Space borne Thermal Emission and Reflection Radiometer (ASTER) are some of the recent sensors that have been found suitable for glacier study. In this chapter, IRS-1D LISS-III satellite data have been used for monitoring various glacier parameters. The IRS-1D LISS-III provides high-resolution (23.5 m) images of the Earth in four different wavelengths of the electromagnetic spectrum, ranging from visible to short wave infrared (SWIR) (0.52–1.75 μm). The preprocessing of images is done before using it for analysis. The preprocessing of data involves conversion of digital numbers (DN) into reflectance, orthorectification of images, and topographic correction.

18.2 Remote Sensing in Glacier Morphological Study

The application of remote sensing techniques has been established in glacier morphological studies. Remote sensing technique is effectively applied for detecting and mapping of the various glaciers zones and features (Ostream, 1975). Ostream (1975) has interpreted the glacier morphology with optical sensors, passive microwave sensors, and synthetic aperture radar (SAR). The major geomorphological features observed in Indian Himalaya are snow cover area, terminal, lateral and medial moraines, moraine-damned lakes, and the snout of the glacier. Moraines are the glacier deposits, which consist of dirt, rock floor, and boulders. Worcester (1965) has defined three types of moraines as follows:

- *Terminal moraines*—huge ridges or belts of debris deposited at the end of the valley glaciers form terminal moraines. In Figure 18.1, the field photograph shows the terminal moraine and the snout of Patsio glacier located in Lahaul-Spiti district of Himachal Pradesh.
- *Lateral moraines*—they exist on the two sides of glaciers. In many cases well-developed lateral moraines join with terminal moraine.
- *Medial moraines*—when the ice wastes are deposited away from the moraines, but well-mixed with the ground morainal debris or with outwash plain. Medial moraines stand out in the middle of the valley glacier.

The extent of moraine cover over glaciers is its rate of melting; recession and degeneration are very sensitive indicators of the global climatic changes and glacier health in the area (Michalcea et al., 2006; Stokes et al., 2007). To have a complete understanding of glacier mass balance and estimate, the study of moraine cover is very important. Thick moraine cover, which characterizes the glacier snout, often led to the formation of moraine-dammed lakes, which is another important reason behind their extensive studies (Gupta et al., 2008).

FIGURE 18.1
Field photograph showing terminal moraine and snout of Patsio glacier located in Lahaul-Spiti, Himachal Pradesh.

18.2.1 Interpretation of Glacier Features

Glacier features can be identified in an image by using standard band combinations. Standard combination of band 2 (0.52–0.59 μm), band 3 (0.62–0.68 μm), and band 4 (0.77–0.88 μm) is used for preparing false color composite (FCC). Blue, green, and red colors are attributed to band 2, 3, and 4 of LISS-III satellite data. FCC image is useful for identifying various glacier features such as ablation area, accumulation area, glacier boundary, equilibrium line, snow line, moraines, and moraine-dammed lakes (Kulkarni and Bahuguna, 2001). Identification of these features on satellite images depends upon spectral reflectance.

- *Glacial boundary:* During August–September season, grass appears on lateral moraines and terminal moraines. Due to the growth of grass it appears red in FCC image around the glacier snout and makes it easy to delineate lower boundary of glacier. In the upper part of ablation area, glacier edges are characterized by dirty snow, which gets accumulated due to avalanche from adjoining cliffs. This gives a distinctly higher reflectance along the edge.

- *Accumulation area:* Accumulation area involves the area where accumulation of snow takes place. Accumulation normally takes place at or near the glacier surface (Paterson, 1994). Accumulation area has very high reflectance. Spectral reflectance is higher in all three bands. Hence, it appears white in the FCC image and easily delineated.

- *Ablation area:* Ablation includes all processes by which snow and ice are lost from the glacier. Melting followed by runoff, evaporation, removal of snow by wind, and the calving of icebergs are examples (Paterson, 1994). In ablation area, total summer melting is more than winter snow accumulation, therefore glacier ice, along with debris, gets exposed on the surface. Glacier ice has substantially lower reflectance than snow, but higher than rocks and soil of the surrounding area. It has a rough texture and gives green–white tone on FCC and can easily be differentiated from the accumulation area and surrounding rock and soil.

- *Transient snow line/equilibrium line:* Equilibrium line is the snow line at the end of the snow-ablation season, usually September month in Himachal Pradesh (Kulkarni, 1992). This line can be easily marked on FCC, if suitable cloud free images are available.

- *Moraine-dammed glacier lake:* These lakes can be marked due to their dark blue and black color and characteristic shape. Initially, all lakes that are not completely detached from a glacier can be treated as unsafe (Gansser, 1983).

18.2.2 Samudra Tapu Glacier: A Case Study

Samudra Tapu is the second largest glacier after Bara Shigri in Chandra subbasin, Himalaya. The morphological features of Samudra Tapu glacier have been studied by using IRS-1D LISS-III satellite data of September 1999. The image of September month was selected because during this period the glacier is fully exposed and snow cover is at its minimum. September month is considered to be the end of the ablation season. The glacier boundary is mapped on the satellite imagery using standard band combinations. The image enhancement technique is used to enhance the difference between glacial and nonglacial areas. In Figure 18.2, the different morphological features of the Samudra Tapu glacier on satellite imagery are shown. The accumulation area is differentiated from the ablation zone by snow line, as shown in Figure 18.2. The aerial extent of terminal and lateral moraine was detected on the glacier in the imagery. A moraine-damned lake located near the terminus of the glacier and snout of the glacier are identified and demarcated in the image.

18.3 Remote Sensing in Snow Cover Mapping

The snow cover mapping is useful to estimate the total snow cover extent, which is a very essential input for the projection of snow melt and runoff

N
Λ

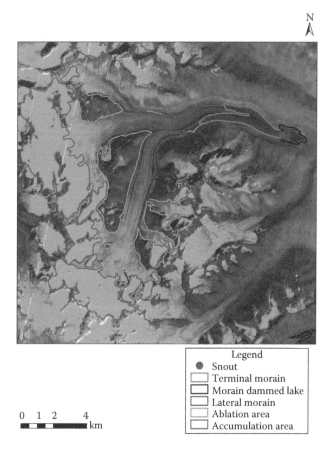

Legend		
●		Snout
☐		Terminal morain
☐		Morain dammed lake
☐		Lateral morain
☐		Ablation area
☐		Accumulation area

0   1   2   4
■■■■■■■ km

FIGURE 18.2
FCC of IRS-1D LISS-III satellite data show different glaciomorphological features of Samudra Tapu glacier (Himachal Pradesh).

estimation. The cloud and snow have similar reflectance characteristics in visible and near-infrared (NIR) region. Therefore, it is difficult to map seasonal snow in the presence of cloud. The discrimination between snow and cloud can be done using various techniques including band ratios. One such method is discussed in the following.

18.3.1 Snow/Cloud Discrimination Using NDSI

The basic principle behind snow and cloud discrimination is that snow reflects less energy in SWIR band and higher in visible and NIR band. There are various ways for discriminating snow and cloud, like texture analysis, association with shadow, and multitemporal analysis (Kulkarni,

2001). The reflectance of snow in SWIR band is very low because the absorptive coefficient of ice is very high. Water clouds and ice clouds, on the other hand, reflect more strongly than snow in the SWIR region because water is less absorptive than ice, and small ice crystals in cirrus clouds are more reflecting than the larger snow grains (Dozier, 1984). The most efficient way of snow/cloud discrimination is by using SWIR band. The technique developed for snow/cloud discrimination is based on normalized difference snow index (NDSI).

$$NDSI = \frac{Green\,Reflectance - SWIR\,Reflectance}{Green\,Reflectance + SWIR\,Reflectance}$$

The utility of NDSI for snow cover mapping is based upon snow reflectance characteristics. Snow reflectance is high in visible and low in SWIR region. The utilization of middle infrared has additional advantage, as cloud reflectance is high in this band. This helps in discriminating between snow and cloud. Additional advantage of NDSI is the delineation and mapping of snow in mountain shadow region. However, field and satellite observations suggest that NDSI values in shadow and nonshadow region are same. This is possible due to reflectance from diffuse radiation in shadow areas. In Figure 18.3, IRS-1D LISS-III satellite data show the snow cover map for Gangotri glacier.

18.4 Remote Sensing in Monitoring of Glacier Retreat

Remote sensing appears to be the only means of monitoring the retreat if it is to be monitored for a large number of glaciers, where field methods are difficult to implement due to tough weather and terrain conditions. Satellite images for the month of September are generally selected for studying the glacier retreat, because during this period snow cover is at its minimum and glaciers are fully exposed. In satellite image, glacial boundary can be mapped using standard combinations of bands as discussed earlier in this chapter. The image enhancement techniques can be used to enhance the difference between glacial and nonglacial areas. For case study, monitoring has been done for the glaciers of Bhaga basin, Himachal Pradesh using IRS-1D LISS-III images from 2001 to 2006. From this study, it was found that the total basin area was 205.89 km^2 in the year 2001 and it has changed to 200.44 km^2 in 2006. The percentage loss of the basin area is about 2.64 during the study period. The rate of loss of the area is 0.529 km^2/year (Figure 18.4).

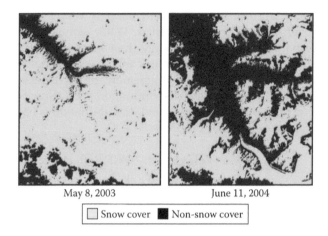

| May 8, 2003 | June 11, 2004 |

☐ Snow cover ■ Non-snow cover

FIGURE 18.3
IRS-1D LISS-III images show snow cover area for Gangotri glacier. (a) Snow cover on May 8, 2003. (b) Snow cover on June 11, 2004.

18.5 Remote Sensing in Glacier Mass Balance Studies

Mass balance is the difference between the amount of snow and ice accumulation of the glacier and the amount of snow and ice ablation (melting and sublimation) lost from the glacier. Traditionally, the glacier mass balance is measured by placing a network of stakes and pits on the glacier surface and determining the change of surface level at the end of ablation season. The density of the snow and ice are assumed to be constant. Due to the rugged terrain of glacier, it is difficult to monitor the mass balance by this field method. Remote sensing provides an alternative for the estimation of glacier mass balance. Two very commonly used methods of estimating mass balance by remote sensing methods are the accumulation area ratio (AAR)-equilibrium line altitude (ELA) method and the digital elevation model (DEM) method.

18.5.1 AAR–ELA Method

The AAR–ELA method of estimating mass balance is well-studied for Himalayan glaciers as well for other alpine glaciers in the world. The AAR is the ratio between the accumulation area and the area of an entire glacier (Meier and Post, 1962). The accumulation area is defined at the end of hydrological year and separated by the equilibrium line from the ablation area.

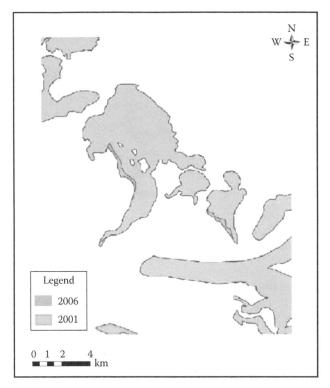

FIGURE 18.4
Glacier retreat between 2001 and 2006 in one of the Bhaga glacier basin.

$$\mathrm{AAR} = \frac{\text{Accumulation area}}{\text{Total area of glacier}}$$

The accumulation area of the glacier is separated from ablation area by the equilibrium line at the end of the ablation season, i.e., in September. The glacier is exposed maximum at the end of the ablation season. The difference in the reflectivity of the ice (ablation area) and snow (accumulation area) is used to delineate the transition between the accumulation and ablation zones. The characteristics of the glacier surface in the accumulation zones are distinct from the ablation zone in imagery, allowing a possibility to determine the ELA by remote sensing. Ice and firn have an albedo of 50% or less and the seasonal snow has an albedo of 90%. So the snow line dividing the two zones can be determined remotely. Figure 18.5 shows the accumulation and ablation zone of Samudra Tapu glacier differentiated by snow line on the satellite imagery.

Accumulation area   Snow line   Ablation area

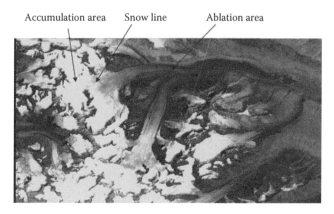

FIGURE 18.5
IRS-1D LISS-III satellite image of September 1999 shows accumulation area, snow line, and ablation area of Samudra Tapu glacier, Himalaya.

If measured at the end of ablation season, the snow line altitude (SLA) is approximately coincident with ELA. The yearly ELA can easily be extracted by overlaying the satellite imagery on a DEM.

The AAR–ELA method proposed by Kulkarni (1992) is based on the generally observed close relationship between annual area–averaged net mass balances, and ELA or AAR for glaciers with a distinct summer ablation season. Such a relationship can be established from long-term field measurements. Once the relationship between field mass balance and AAR is established, the mass balance can be computed from the AAR obtained from the remote sensing data. The yearly AAR for a given glacier is determined from the ELA, and the glacier area is delineated from satellite images by calculating the area above the ELA and below the ELA. The AAR value for a given glacier varies from year to year depending on changes in its mass balance. The flowchart of methodology for the calculation of AAR from satellite imagery is shown in Figure 18.6.

18.5.2 Chhota Shigri Glacier: A Case Study for AAR/ELA Method

The mass balance of Chhota Shigri glacier, Himachal Pradesh, has been estimated by using this method. The Chhota Shigri glacier has been monitored by a team from Jawaharlal Nehru University (JNU) and the data are published in the PhD thesis of Mr. Anurag Linda (2008). The field data of mass balance and AAR/ELA method were used from the PhD thesis of Mr. Anurag Linda to obtain a relationship between the field mass balance and remotely derived AAR and then the relationship is established to estimate mass balance. The relationship between field derived

FIGURE 18.6
Flowchart for mass balance estimation method.

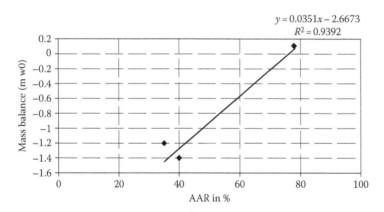

FIGURE 18.7
The relationship between field derived mass balance and AAR derived from remote sensing.

net mass balance for a period of 2003–2006 and remotely derived AAR gives a regression line with $R^2 = 0.9392$ (Figure 18.7). AAR is a good representative of mass balance. The relationship obtained from regression plot can be used to estimate mass balance of Chhota Shigri for the years for which field data are not available and also for the other glaciers of the same region.

18.5.3 Glacier Mass Balance Estimation Using Time Series DEM

The volume change estimation of mountain glaciers can be done using high-resolution multitemporal DEM and with the extent measurement of the glacier. Shuttle Radar Topography Mission (SRTM) C-band data and a DEM generated from topographic maps have been used for measuring the changes in glacier height. ERS-1/2 TanDEM data are very useful for the generation of DEM of the glaciated terrain. Multitemporal DEM derived from ERS-1/2 TanDEM data can be used for estimating the volume change in glacier or ice. Cartosat-1 stereo data based DEM can be generated for estimating change in glacier height. Change in area and height of the glacier generally gives the information of change in total volume of glacier/ice. New emerging radar satellite viz. Cryosat-2 and TanDEM-X can be used to obtain high-precision DEM variation in the glacier elevations and relate these variations to gain or loss in mass balance. The methodology adopted for mass balance change estimation using DEM is shown in Figure 18.8.

18.5.4 Case Study for DEM Method

In this study, Berthier et al. (2007) have developed a methodology to derive mass balance by comparing a 2004 DEM to the 2000 SRTM topography. They derived the 2004 DEM from SPOT 5 satellite images without any ground control points. Before comparison on glaciers, the two DEMs were analyzed on the stable areas surrounding the glaciers, where no elevation change is expected. Two different biases were detected. A long wavelength bias affected the SPOT 5 DEM and was correlated to an anomaly in the roll of the SPOT 5 satellite. A bias was also observed as a function

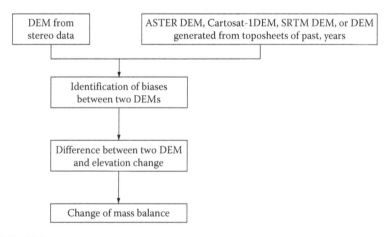

FIGURE 18.8
Flowchart for estimating mass balance change from DEM.

of altitude and had been attributed to the SRTM dataset. Both biases were modeled and removed to permit unbiased comparison of the two DEM on the 915 km² ice-covered areas digitized from an ASTER image. In the case of benchmark glacier Chhota Shigri, they found good agreement between satellite observations and the mass balance measured on the field during hydrological years 2002–2003 and 2003–2004.

18.6 Conclusion

In this chapter, the remote sensing technique has proved its credential for identification and mapping of various glacier morphological features such as accumulation zone, ablation zone, equilibrium line, proglacier lakes, and glacier boundary. The AAR/ELA and DEM methods were successfully adopted for estimation of glacier mass balance in parts of Himalayan region.

Acknowledgments

We wish to acknowledge Dr. A.V. Kulkarni and Dr. Anurag Linda for their support to carry out this work.

References

Berthier, E., Arnaud, Y., Kumar, R., Ahmad, S., Wagnon, P., and Chevallier, P., 2007. Remote sensing estimates of glacier mass balances in the Himachal Pradesh (Western Himalaya, India). *Remote Sensing of Environment*, 108(3): 327–338.

Dozier, J., 1984. Snow reflectance from LANDSAT-4 TM. *IEEE Transactions on Geosciences and Remote Sensing*, GE-22(3): 323–328.

Flint, R.F., 1964. *Glacial and Quaternary Geology*. John Wiley & Sons Inc., New York, pp. 63–85.

Gansser, A., 1983. Geology of the Bhutan Himalaya. *Memoirs de la Societe des Science Naturelle*. Birkhauser, Basel, pp. 149–154.

Gupta, R.P., Shukla, A., Arora, M.K, and Kulkarni, A.V., 2008. Mapping moraine cover in Himalayas using IRS 1C- LISS-III data. In *Proceedings of International Workshop on Snow, Ice, Glacier and Avalanches*, Mumbai, 2008, pp. 155–164. IIT Bombay, Mumbai.

Khalsa, S.J.S., Dyurgerov, M.B., Khromova, T., Raup, B.H., and Barry, R.G., 2004. Space-based mapping of glacier changes using ASTER and GIS tools. *IEEE Transactions on Geosciences and Remote Sensing*, 42(10): 2177–2183.

Kulkarni, A.V., 1992. Mass balance of Himalayan glaciers using AAR and ELA methods. *Journal of Glaciology*, *38:* 128.

Kulkarni, A.V., 2001. Effect of climatic variations on Himalayan glaciers: A case study of upper Chandra river basin of HP. In M.G. Srinivas (ed.), *Remote Sensing Applications*. Narosa Publishing House, New Delhi.

Kulkarni, A.V. and Bahuguna, I.M., 2001. Role of satellite image in snow and glacial investigations. *Geological Survey of India, Special Publication*, *53*: 233–240.

Kulkarni, A.V. and Buch, A.M., 1991. Glacier atlas of Indian Himalaya. SAC/RSA/ RSAG-MWRD/SN/05/91.

Linda, A., 2008. Snow and ice mass budget of Chhota Shigri glacier, Lahaul-Apiti Valley, Himachal Pradesh 2003–2007. PhD dissertation, Jawaharlal Nehru University, New Delhi, India.

Meier, M.F. and Post, A.S., 1962. Recent variations in mass net budgets of glaciers in western North America. In *Symposium at Obergurgl 1962—Variations of Glaciers*, International Association of Scientific Hydrology, Publication 58.

Mihalcea, C., Mayer, C., Diolaiuti, G., Lambrecht, A., Smiraglia, C., and Tartari, G., 2006. Ice ablation and meteorological conditions on the debris covered areas of Baltoro glacier, Karakoram, Pakistan. *Annals of Glaciology, 43*: 292.

Oerlemans, J., 2005. Extracting climate signals from 169 glaciers records. *Science, 308*: 675.

Ostream, G., 1975. ERTS data in glaciology, an effort to monitor glacier mass balance from satellite imagery. *Journal of Glaciology, 15*: 403–415.

Paterson, W.S.B., 1994. *The Physics of Glaciers*, 3rd edn. Pergamon, Oxford.

Stokes, C.R., Popovnin, V., Aleynikov, A., Gurney, S.D., and Shahgedanova, M., 2007. Recent glacier retreat in Caucasus mountains, Russia, and associated increase in supraglacial debris cover and supra-/proglacial lake development. *Annals of Glaciology, 46*: 95–213.

Worcester, P.G., 1965. *A Text Book of Geomorphology*. Van Nostrand East West Press, New York.

19

Geomorphology and Development Mechanism of Sinkholes in Arid Regions with Emphasis on West Texas, Qatar Peninsula, and Dead Sea Area

Fares M. Howari and Abdulali Sadiq

CONTENTS

19.1 Introduction

Sinkholes are a common geologic hazard in many parts of the world. It is considered as an obvious geomorphologic feature of karst terrains and results from the subsidence or collapse of surficial material or subsurface cavities. Its distinctive landscape topography is largely formed through the dissolving of carbonate and/or sulfate bedrocks such as limestone and gypsum by groundwater, and may differ in shape and size from less than a meter to several hundred meters both in diameter and depth. It varies in form from soil-lined bowls to bedrock-edged chasms. The development process in arid regions is profoundly influenced by natural hydrogeological systems. However, human impacts can cause sinkholes to occur in many ways, especially where they might not naturally have happened. Human impacts might cause sinkholes to form catastrophically than under natural conditions. There are many human-related activities that are believed to cause or be linked to sinkholes formation, e.g., decline of groundwater level due to pumping, leaking water especially from sewer pipes, injection of water, and others.

Various methods were used to study sinkholes. Remote sensing and geophysical method are considered as an important geoinformatic tool to detect and map this dangerous phenomenon. This chapter focuses on the mechanisms of sinkhole formation in arid regions and highlights natural and man-made causes, based on examples from three regions: West Texas, Qatar, and Dead Sea—(1) sinkholes in limestone and evaporite (West Texas), (2) limestone, dolomite in combination with gypsum (central and northern parts of Qatar Peninsula), and (3) evaporates (halite of the Dead Sea shore area). The chapter also explains the basic relations between sinkholes and rock type.

19.2 Geoinformatics in Sinkholes Studies

Studies from European remote sensing (ERS) satellite, interferometric synthetic aperture radar (InSAR), and differential interferometric synthetic aperture radar (DInSAR) in relation with results from precise gravity measurements and field investigations were proved to be successful in detecting sinkholes (Bear et al., 2002; Schoor, 2002; Abelson et al., 2003; Kruse et al., 2006). Ground-based geophysical methods have been successfully used to detect sinkholes, too (Ezersky et al., 2008; Legchenko et al., 2009; Ezersky et al., 2010). Traditionally, electrical geophysics, which determines the resistance of the ground to the passage of an electric current, was used in sinkhole investigations. Resistivity is increased or conductivity decreased by the presence of air-filled voids, but opposite characteristics are created where bedrock voids are filled with wet clay soils. Therefore, it is necessary to identify and interpret areas of anomalous apparent resistivity.

Another important geophysical technique to study sinkholes and subsurface cavities is gravity and microgravity. This technique involves the measurements of small variations in the Earth's gravitational field that are produced by localized changes in soil and rock mass as well as density. They are particularly valuable investigation of karst and sinkholes, because negative anomalies represent missing mass. Hence, it can be interpreted either as open- or water-filled ground cavities or as caves. Magnetic measurements were also used by geotechnical and environmental geology specialists to study sinkholes. This method records variations and distortions in the earth's magnetic field produced by the presence of underlying rocks with different magnetic properties. However, they are generally unsuitable for the detection of natural cavities and sinkholes, where magnetic contrasts are low or absent in limestone and soils. Seismic diffraction imaging was also applied successfully for detecting near-surface inhomogeneities (Keydar et al., 2010).

The methods noted earlier were applied on the investigated areas by several researchers, and this chapter reviewed related major findings with some new insights. However, in the present study multiyear aerial photos and satellite images coverage were integrated with field investigations and observations reported in the literature in order to understand the mechanisms of formation of sinkholes in the three different areas noted earlier. The remote sensing data used, in this study, were obtained from USGS Earth Resources Observation and Science (EROS) Center, NASA Earth Observatory, and Google Earth.

19.3 Sinkholes and Rock Types

Limestone: the rate of dissolution of a rock generally depends on the solubility, the degree of saturation of the solvent, the area presented to the solvent, and the movement of the solvent. The solubility of limestone in pure water is extremely low. The key factor that can impact the solubility is aqueous solution of carbon dioxides. Major concentrations of carbon dioxide are released when organic material decays and are taken into solution by rainwater that percolate through the soil. Another important factor is the availability of soil carbon dioxide (Bischoff et al., 1994), which is largely a function of biological activity and, therefore, increases at higher temperatures in terrains at low latitudes or altitudes.

Chalk: although the porosity of chalk may be as high as 50%, the primary pores are very small. Most of the water that flows through chalk is, therefore, concentrated along a system of discontinuities. Dissolution along these produces open fissures, subvertical solution pipes, and small cave passages that are a type of buried sinkhole. Many chalk terrains form a variety of fluviokarst, with dendritic systems of dry valleys largely inherited from periglacial erosion, but sinkholes are also a feature of chalk karst as well as that occurring in areas where chalk lies beneath a thin cover of clastic sediments or soils (Waltham et al., 2004).

Evaporate: deposits are formed by precipitation from saline water that is concentrated beyond the levels of mineral saturation by evaporation in lagoonal or lacustrine environments. Total solubility and dissolution rate are both much greater for gypsum in water than they are for limestone. In massive gypsum, a fissure 0.2 mm wide and 100 mm long would have widened by dissolution within a 100 years, so that a block $1\,m^3$ in size could be accommodated in the tapering entrance to the fissure (James and Lupton, 1978). If the fissure exceeded 0.6 mm, larger caverns would form and a runway situation would develop in a short time. In long fissures, the rate of flow is less, so that the solutions become saturated and little or no material is removed. Under the extremely high hydraulic gradients generated beneath dams, dissolution rates for gypsum may be as high as 10 mm/year (Dreybrodt et al., 2002).

It becomes obvious that sinkholes may be formed in several geological formations and can happen gradually or suddenly and are found world-wide. However, in arid and semiarid regions, it is generally the result from the upward subvertical movement of cavities that developed because of the dissolution of underlying salt layers, as well as due to growth of joint and features, as will be explained in subsequent sections (Arkin and Gilat, 2000; Taqieddin et al., 2000). The results and analysis of data on sinkholes development mechanisms of different geographical locations are described in the following mechanisms.

19.4 Sinkholes of West Texas

In Texas, sinkholes are associated with shallow salt domes of eastern Texas, the Coastal Plain limestone karst in central Texas, and bedded Paleozoic salt beneath the High Plains of northern Texas and within the Permian Basins. Arguably the major and frequent sinkholes that appear in the Permian Basin thus will be emphasized on. The Permian basin is located in west Texas and southeastern New Mexico. It is one of the most productive petroleum provinces of North America. The area holds one of the thickest deposits of rock from the Permian period, which lasted from approximately 290 to 251 million years ago. The basin is a large depression in the bedrock surface along the southern edge of the North American craton, an ancient core of continental crust. The basin is filled with thick layers of sediment of the Paleozoic Era (about 545–251 million years ago) as the region was alternately covered by shallow oceans, or exposed as coastal salt flats. The sediments are of organic-rich carbonate and minerals such as common table salt. Later activity in the earth's crust caused folding of the sedimentary layers, creating ideal conditions for the formation, trapping, and storage of petroleum (http://eol.jsc.nasa.gov/).

The major sinkholes of the Permian Basin exist in a location close to Wink city. Figure 19.1 shows satellite images of Winker County, and also shows Wink and Kermit City. This area is very flat with some outcrops in the north and northeast. This location has three sinkholes, one formed

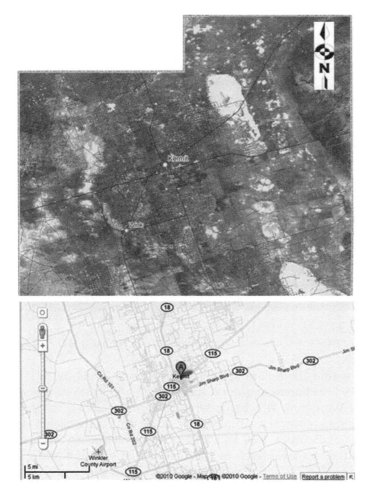

FIGURE 19.1
Satellite relief and shaded relief maps of Winkler.

in 1980, another in late 1990s, and one in 2002. The largest sinkholes have cracks radiating out from the original sinkhole as shown in Figure 19.2 and named Wink Sink #1 and 2. Wink Sink 1 opened in 1982 and is approximately 300 ft in diameter. At about 1.5 acres, Wink Sink 2 is about 1/12 the areal extent of Wink Sink 1. Wink Sink 1 has associated sag cracks that extend well beyond the actual hole. During the time of aerial photo collection (Figure 19.2) and up to 1980s, there were no obvious surface manifestation other than few surface sagging from fissures.

Since 1980, the ground, just east of Wink Sink #1, has subsided about 8.5 m, and a series of concentric fissures and faults now break the ground

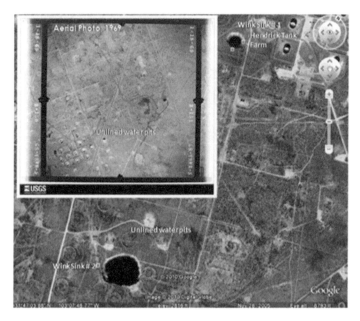

FIGURE 19.2
This is the first of two sinkholes in Winkler County Texas. (Courtesy of Google Earth.)

surface within about 85 m of the sink, as shown in Figure 19.3. In 1999, two more broad areas of ground subsidence were observed about 1500 m south of Wink Sink #1, when a series of earth fissures developed along with sagging power lines, tilted poles, and a dip on a well pad (Johnson et al., 2003). Between 1980 and 1999, ground subsidence reached 7 and 8.5 m in the two areas, each of which is at least several hundred meters across. On May 21, 2002, a second major sinkhole developed 400 m west of the existing subsidence depressions, and 1500 m south of Wink Sink 1 (Johnson, 1989; Johnson et al., 2003). By the end of the 1st day, Wink Sink #2 was about 140 m long, 90 m across, and 30 m deep to the water level. The sink continued to enlarge its vertical banks material into the water-filled hole until it reached about 240 × 185 m across in March 2003, and is still expanding (Waltham et al., 2004).

The geology of the Permian Basin is illustrated in Figure 19.4. As can be anticipated, the most vulnerable geologic formation to dissolution is the Salado salts. Another observation one should bear in mind is that Capitan reef can allow water to pass through and reach the Salado salt. In the subsurface, natural dissolution of salt in the Salado formation has possibly developed in many parts of the Delaware basin from Permian time up through the Cenozoic. Such development is attested by the abrupt thinning of Salado salt units above and just to the west of the buried

FIGURE 19.3
Fissures and faults breaking the ground surface in the vicinity of the Permian Basin Sinkholes.

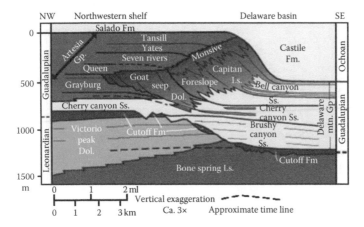

FIGURE 19.4
Cross section through the Permian Basin. Standard stratigraphic nomenclature of the Permian strata exposed in the Guadalupe mountains. (After Kind, 1948; Hayes, 1946; Tyrell, 1969; Parry, 1988.)

Capitan Reef, and also by the presence of a great, sediment-filled dissolution trough directly above the area, where the Salado is anomalously thin (Johnson et al., 2003).

It is believed that water related petroleum activities were instrumental in bringing about the dissolution cavity and the collapse that created the sink. Figures 19.2 and 19.5 show circular earth pits. Those were used to dispose oilfield water from Cenozoic alluvium, but, unfortunately, the earth pits were unlined. Freshwater in the Cenozoic alluvium and the Santa Rosa Sandstone near the Wink Sink might have caused the dissolution of salt in two different ways. The first of this is by the seepage of great quantities of waste oilfield waters from unlined earthen pits. Oilfield brines also have been disposed of in the Hendrick Field through injection wells, and the disposal zones commonly were in permeable sand or gravel units or into the Rustler formation.

FIGURE 19.5
Aerial view of Wink Sink 1 and 2, and the nearby unlined water pit. (Courtesy of CEED, UTPB.)

An abandoned oil well (The Hendrick Well 10-A) was located at the site of the Wink Sink #1, and it appears likely that it was a pathway for water to come in contact with the Salado salt. In all likelihood, the well was drilled using a freshwater drilling fluid that enlarged or washed out the borehole within the salt sequence. Ineffective cement sealing, and possible fractures in the cement lining, may have opened pathways for water movement up or down the bored hole outside the casing. Because of undoubted borehole enlargement during drilling in the Salado salts, the small amount of cement reportedly used to set the casing in the hole was probably enough to cement only the lower part of the hole. This would have left most of the salt exposed behind the uncemented casing.

Wink Sink #2 is centered on the site of a former water-supply well, the Gulf WS-8 (Johnson et al., 2003). This well was completed in September 1960, drilled into the Capitan Reef to a total depth of 1092 m. It met the top and base of the Salado formation at depths of 412 and 686 m, so the Salado salt is 274 m thick. The relation between this well and the sinkhole is similar, the first situation in which the well could have been a conduit for unsaturated water to reach the Salado salt beds, where solution cavity was created and then migrated upward to cause collapse in the land surfaces (Waltham et al., 2004). Figure 19.6 shows a sag area to the east of Wink Sink 2, and 6 in. gas pipeline serving Wink has been exposed for over 2 years. The depth of the fracture is >13 ft. The trend of this crack and the one through the electric substation seems to be independent of Wink Sink #2, thus adequate remote sensing and geoinformatics monitoring are needed.

19.5 Sinkholes of Qatar

The surface of Qatar is low to moderate relief with a slightly undulating surface and several scattered depressions. Most of the depressions are filled with aeolian sand deposits. The central part is formed by a plateau covered with limestone and colluvial soils. Calcareous beach sands are principally along the present coastline. They also occur at the edge of or within the scattered sabkha in the study area, whose outlining areas were previously covered by the Quaternary Sea (Figure 19.7). This location is one of the driest regions on the earth with aridity capable of evaporating 200 times the amount of precipitation received. In summer, it is very hot and the maximum daily average temperature in August reaches 45°C. The minimum is 25°C and the mean is 37°C. The winter in the study area is mild with an average minimum temperature in January of 7°C and average maximum of 26°C and a mean temperature of 15°C. Winds during

FIGURE 19.6
Sag area to the east of Wink Sink 2 and 6 in. gas pipeline serving Wink have been exposed for over 2 years.

the winter, directed mainly toward the southeast, can reach 50–100 km/h (Sadiq and Howari, 2009).

The state of Qatar has about 9700 large and small depressions, as well as several exposed sinkholes and caves, which have not been studied adequately (Sadiq and Nasir, 2002). Our analyses presented in this study found that about 80% of the surface depressions (locally known as Rodhas) or deformation features show NE-SW and NW-SE orientations. These have similar orientation with the joint and fracture systems in Qatar.

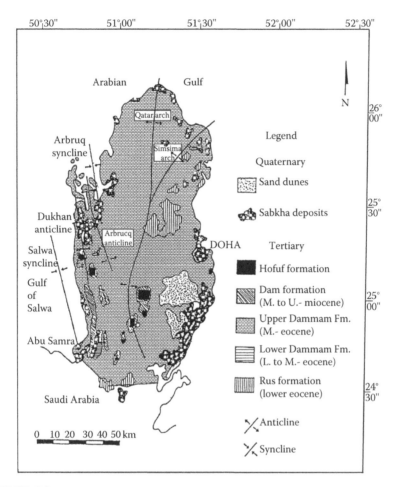

FIGURE 19.7
Geologic map of Qatar. (Modified after Cavelier, C., *Geologic Description of the Qatar Peninsula (Arabian Gulf)*, Publication of the Government of Qatar, Department of Petroleum Affairs, 1970, 39pp.)

In the investigated area, sinkholes are believed to be resulted from a combination of factors that were presented in an earlier work (Sadiq and Nasir, 2002). Those factors include the upward subvertical movement of cavities that developed because of the dissolution of underlying formations, especially those that include salt layers. It is believed that the sediment or salt layers, which were previously surrounded by saturated water in respect to salt, were exposed to aggressive dissolution by unsaturated water mainly due to the middle Pleistocene wet climatic conditions and consequent subsidence. The earlier work also indicated the accelerated dissolution of carbonate and sulfate deposits. The recent joint flow drainage is another additional factor that may account for differential dissolution and enhance the development and migration of the cavities. This has created underground cavities, some of which had their ceilings collapsed producing sinkholes such as in Dahl Al-Hammam area (Figure 19.8).

In Qatar, sinkholes are, as noted earlier, concentrated in the central and northern parts of the country. Analysis of sinkholes and the major axis orientations of depressions show a structural control on the karst

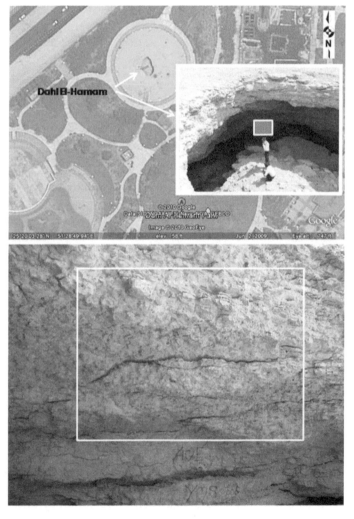

FIGURE 19.8
Sinkhole near Dahl Al-Hammam in Qatar.

development. The long-axis orientations of sinkholes have pronounced NW-SE and NE-SW orientations. This corresponds with the joint and fracture orientations in the study area. The wall rocks of sinkholes are highly jointed and fractured especially in the northern parts, with many stylolites, voids, and selective dissolution cavities. Thin-section studies of wall rocks from different sinkholes in northern Qatar show that the upper 1–2 m of the walls are dolomitic in composition, whereas the lower portion of the walls is biomicritic to biosparitic limestone with a significant proportion of dissolutional pores that are mostly filled with gypsum (Sadiq and Nasir, 2002). The rocks belong to the upper Dammam unit. In central Qatar, marl underlies the upper calcrete, whereas thick gypsum beds underlie the limestone. The rocks composing the sinkholes of central Qatar belong to the upper and lower Dammam units.

Most sinkhole entrances in northern Qatar are vertical and formed by dissolution along planes of structural weakness that extend to the surface. Sinkholes in central Qatar are generally larger in size and depth than those of northern Qatar. They occur as vertical shafts connected to steeply inclined passages. Most sinkholes reach the gypsum layers of the lower Dammam unit and the Rus Formation.

The rate at which limestone and gypsum dissolution proceeds within the surficial meteoric environment is dependent on several factors, including rainfall regime, temperature, distribution of soil cover, and biological activity, and structural weakness as well as the lithology of the carbonate substrate (White, 1984; Trudgill, 1985; Ford and Williams, 1989; Smart and Whitaker, 1991; Hose, 1966; Miller, 1996; Reeder et al., 1996; Frank et al., 1998; Hill, 2000; Sadiq and Nasir, 2002). Of these, dissolution is directly impacted by the amount of rainfall (White, 1984); an increase in precipitation always results in an increase in the rate of limestone dissolution. Temperature primarily influences limestone dissolution through its effect on the level of biological activity. Lithology of the carbonate substrate also has a profound effect on karstification, calcite dissolving more readily than dolomite due to its higher solubility (Martinez and White, 1999). Karstification can take place where dolomite and gypsum are in contact with the same aquifer (Bischoff et al., 1994). Gypsum dissolution drives the precipitation of calcite, thus consuming carbonate ions released by dolomite.

19.6 Sinkholes in Dead Sea–Jordan Valley

The area adjacent to the Jordan River is called the Jordan Valley. It is a part of the geological fault, which extends from Syria down to the Red Sea, as part of the Great African Rift extending from Ethiopia through the Red Sea into Jordan and Syria, creating a new base level for the surrounding surface and groundwater. The geology of the study area was affected by two major factors: (1) the transgression and the regression of the old Tethys Sea during the geological history and (2) the formation of the Jordan–Red Sea transform faults along which a northward displacement of the Arabian Plate has taken place with a total amount of 107 km left-lateral shear. In the late Pleistocene age the Jordan River came into existence cutting into the Lisan marls and gradually worked backward till it captured the Yarmouk River and Tiberia's lake for the Dead Sea drainage (Quennell, 1958; Banat and Howari, 2003).

The Dead Sea lies in a pull-apart basin that is part of the Syro-African rift valley. The left-lateral strike–slip movement between the Arabian plate and the Sinai subplate is approximately 100 km. It is geologically very well known (Quennell, 1958; Freund et al., 1970; Garfunkel et al., 1981). The rift valley is filled with approximately 10 km of overlapped sediments coming from the erosion of the nearby plateaus and from salt deposits of lagoons and ancestral lakes of the Dead Sea, during the past million years (Al-Zoubi and Ten Brink, 2001). As salt deposits are potentially everywhere in this region, the Dead Sea coastal areas can be considered as part of a salt karst system. The Sedom Formation (Zak, 1967) is a particular kilometer-thick salt layer involved in the origin of salt domes affecting the whole area.

Since 1960, the Dead Sea has a size by about a third because of artificial diversions of its feeders. The terminal lake is now mainly fed by the surrounding water tables to compensate for the lowering of the level (Salameh and El-Naser, 1999, 2000a, b). Figure 19.9a shows a view of salt evaporation pans on the Dead Sea, taken in 1989 from the Space Shuttle, Columbia; the southern half is separated from the northern half at what used to be the Lisan Peninsula because of the fall in the level of Dead Sea. Whereas, Figure 19.9b shows a view of the mineral evaporation ponds almost 12 years later. A northern and small southeastern extension was added and the large polygonal ponds subdivided. During the last 40 years, its level decreased by 26 m, and the yearly decline averaged 1 m (Yechieli et al., 2006). The size of the remaining lake is about 55 × 16 km. This is believed to be as one of the main causes for the frequent serious subsidence and sinkhole hazards.

Sinkhole activities have occurred repetitively and were observed in open farms, a cross roads, near dwellings, and near an existing factory, thus causing a serious threat to the locals and farmers of the area and their properties (Closson and Abou Karaki, 2009) (Figures 19.10 and 19.11). The development of subsurface cavities is associated mainly with the variation in the level of the Dead Sea over the past four decades, the presence of regional salt intrusion under the surface of salt beds, the fluctuation of the

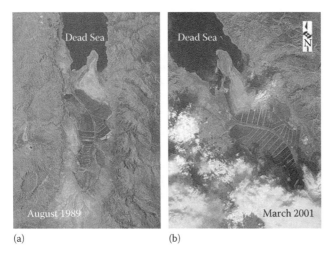

(a)   (b)

FIGURE 19.9
View of salt evaporation pans on the Dead Sea, taken in 1989 and 2001. (Courtesy of NASA Earth Observatory.

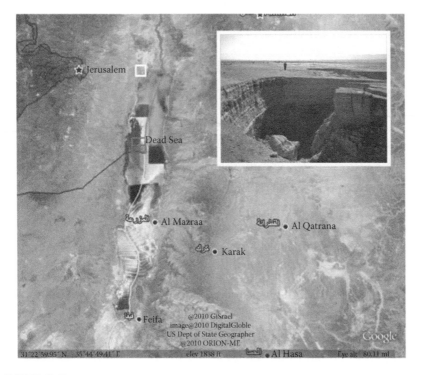

FIGURE 19.10
Overview of the Dead Sea sinkholes and an associate sinkhole.

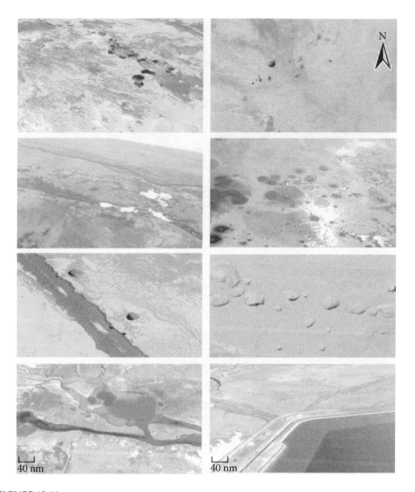

FIGURE 19.11
Different views of Dead Sea sinkholes. (Courtesy of YouTube video.)

water table, and the continuous dissolution and the active tectonism of the area (Taqieddin et al., 2000). Closson et al. (2005) presented data suggesting similar mechanism in which the drop of the water level, in conjunction with the particular tectonic setting of this area, is at least one of the factors that led to the frequent sinkhole disaster in this region, especially in the northern part of the Lisan Peninsula and Ghor Al Haditha, which are the two places undergoing the most intense deformations along the Jordanian Dead Sea coast.

Primarily, the causes for the Dead Sea sinkholes can be considered as a direct consequence of the falling water level of the Dead Sea, the interface between the Dead Sea water and the ground water, which is falling, too. The subsurface Dead Sea salt layers, previously surrounded by saturated water in respect to salt, are now exposed to aggressive dissolution by unsaturated ground water. Subsidence phenomena affect the Dead Sea shore. This subsidence appears to be structurally controlled by faults, seaward landslides, and salt domes (Bear et al., 2002). The drop of water level described earlier could also yield a low hydrostatic pressure, which makes the subsurface formation more vulnerable to collapse.

Abelson et al. (2003) indicated that it appears that the decline of the Dead Sea level affects the formation of sinkholes in three ways: (1) opening the way to eastward migration of the fresh–saline water interface and thus to the undersaturated groundwater, (2) generating differential compaction of fine-grained sediments, and (3) destabilization of underground cavities, which catalyzes their collapse. More related explanation of the Dead Sea sinkhole mechanism based on additional real observations and geoinformatic measurements, as well as additional facts on sinkhole formation, is provided by Ezersky et al. (2010), Legchenko et al. (2009), and Ezersky et al. (2008).

Another sinkhole formation mechanism in the Dead Sea is partially due to rainfall. Freshwater from rainfall most likely have reacted with the surface and subsurface rocks, dissolving salt layers. As a result of this, rain erodes and carves out underground spaces. This with time has created underground cavities, some of which have had their ceilings collapsed. This process gets intensified due to the presence of joints. This facilitates the movement of water and further weakens the subsurface rocks, which makes the subsurface environment more vulnerable to the formation of caverns, which produces a collapse and forms a sinkhole.

19.7 Environmental Impacts of Sinkholes

Sinkholes are costly and damaging natural hazards with frequent occurrences in many parts of the world. This phenomenon causes not only personnel casualties, but it also affects the environmental quality. The main environmental impacts of landslides are loss of human lives, destroying residential and industrial developments, damaging agricultural and forest lands, and negative impacts on the quality of water resources. It could also develop fragmentation in the ecosystem, which could impact both flora and fauna. In the Dead Sea area, sinkholes are considered a serious problem, threatening hotels, resorts, potash and mineral plants. Whereas, in west Texas it threatens the underground gas lines and the petroleum production infrastructure like power lines and county road in the Permian Basin. In Qatar, sinkholes present serious problems to human lives, public parks, highways, and residential areas.

19.8 Conclusion

Recent and historical geoinformatic data from USGS Earth Resources Observation and Science (EROS) Center, NASA Earth Observatory, and Google Earth can be utilized efficiently in various stages of sinkhole mapping and investigations. The analyses from the presented data in this chapter revealed that the development mechanisms of sinkholes are similar. There are several possible causes of a cavity in the Permian Basin of west Texas, which eventually results in a collapse, and the majority were triggered by human activities. The possible causes are (1) fresh waters from shallow aquifers that enter borehole and cause dissolution in Salado Salt, (2) existing Cenozoic collapse in Salado is enlarged by water(s) and collapses, (3) brine water from surface pits enters along fractures in shallow horizons causing dissolution, (4) water moving up from Capitan Reef causes dissolution, and (5) water lost through casing into Salado causes dissolution (Trentham, 2010).

In Qatar, it is believed that the sediment or salt layers, which were previously surrounded by saturated water in respect to salt were exposed to aggressive dissolution by unsaturated water mainly due to the middle Pleistocene wet climatic conditions. This accelerated dissolution of carbonate and sulfate deposits. The recent joint flow drainage is another factor that may account for differential dissolution and enhance the development and migration of the cavities. Primarily, the causes for the Dead Sea sinkholes can be considered as a direct consequence of the falling water level of Dead Sea, the interface between the Dead Sea water and the depleting ground water table. The subsurface Dead Sea salt layers, previously surrounded by saturated water in respect to salt, are now exposed to aggressive dissolution by unsaturated groundwater.

References

Abdulali, S. and Howari, F. 2009. Remote sensing and spectral characteristics of desert sand from Qatar Peninsula, Arabian Gulf. *Remote Sensing* 1(4):915–933; doi:10.3390/rs1040915.

Abelson, M., Baer, G., Shtivelman, V. et al. 2003. Collapse-sinkholes and radar interferometry reveal neotectonics concealed within the Dead Sea basin. *Geophysics Research Letters* 30(10):52.1–52.4.

Al-Zoubi, A. and Ten Brink, U.S. 2001. Salt diapirs in the Dead Sea basin and their relationship to quaternary extensional tectonics. *Marine Petroleum Geology* 18(7):779–797.

Arkin, Y. and Gilat, A. 2000. Dead Sea sinkholes—An ever-developing hazard. *Environmental Geology* 39(7):711–722.

Banat, K.M. and Howari, F.M. 2003. Pollution load of Pb, Zn, and Cd and mineralogy of the recent sediments of Jordan River/Jordan. *Environment International* 28(7):581–586.

Bear, G., Schattner, U., Wachs, D., Sandwell, D., Wdowinski, S., and Frydman, S. 2002. The lowest place on Earth is subsiding—An InSAR (interferometric synthetic aperture radar) perspective. *Geological Society of America, Bulletin* 114(1):12–23.

Bischoff, J.L., Julia, R., Shank, W.C III, and Rosenbauer, R.J. 1994. Karstification without carbonic acid: Bedrock dissolution by gypsum derived dedolomitization *Geology* 22:995–999.

Cavelier, C. 1970. *Geologic Description of the Qatar Peninsula (Arabian Gulf)*, Publication of the Government of Qatar, Department of Petroleum Affairs, 39pp.

Closson, D. and Abou Karaki, N. 2009. Human-induced geological hazards along the Dead Sea coast. *Environmental Geology* 58:371–380.

Closson, D., Abou Karaki, N., Klinger, Y., and Hussein, M.J. 2005. Subsidence hazards assessment in the Southern Dead Sea area, Jordan. *Pure Applied Geophysics* 162(2):221–248.

Dreybrodt, W., Romanov, D., and Gabrovsek, F. 2002. Karstification below dam sites: A model of increasing leakage from reservoirs. *Environmental Geology* 42:518–524.

Ezersky, M., Legchenko, A., Camerlynck, C., and Al-Zoubi, A. 2008. Identification of sinkhole development mechanism based on a combined geophysical study in Nahal Hever South area (Dead Sea coast of Israel). *Environmental Geology* 58(5):1123–1141.

Ezersky, M., Legchenko, A., Camerlynck, C. et al. 2010. The Dead Sea sinkhole hazard–new findings based on a multidisciplinary geophysical study. *Zeitschrift fur Geomorphology* 54(2):69–90.

Ford, D.C. and Williams, P.W. 1989. *Karst Geomorphology and Hydrology*, Unwin Hyman, London, U.K., 320pp.

Freund, R., Garfunkel, Z., Zak, I., Goldberg, M., Weisbrod, T., and Derin, B. 1970. The shear along the Dead Sea rift. *Philosophical Transaction Royal Society of London* 267:107–130.

Garfunkel, Z., Zak, I., and Freund, R. 1981. Active faulting in the Dead Sea rift. *Tectonophysics* 80:1–26.

Hayes, P.T. 1964. Geology of the Guadalupe Mountains, New Mexico: U.S. Geological Survey, Professional Paper 446, 65 pp.

Hill, C. 2000. Overview of the geologic history of cave development in the Guadalupe Mountains, New Mexico. *Journal of Cave and Karst Studies* 62:60–71.

Hose, L. 1966. Geology of a large, high-relief, sub-tropical cave system: Sistema purification, Tamaulipas, Mexico. *Journal of Cave and Karst Studies* 58:6–21.

Earth Sciences Web Team; Yates Oilfield, West Texas (accessed July 2010); http://eol.jsc.nasa.gov/EarthObservatory/Yates_Oilfield,_West_Texas.htm

Shanbell's Channel; Dead Sea Sinkholes from the Air, August 20, 2009 (accessed August 2010); http://www.youtube.com/watch–v=KC8X4GVqaXE

James, A.N. and Lupton, A.R.R. 1978. Gypsum and anhydrite in foundations of hydraulic structures. *Geotechnique* 28:249–272.

Johnson, K.S. 1989. Development of the Wink Sink in West Texas due to salt dissolution and collapse. *Environmental Geology and Water Science* 14:81–92.

Johnson, K.S., Collins, E.W., and Seni, S.J. 2003. Sinkholes and land subsidence owing to salt dissolution near Wink, West Texas, and other sites in western Texas and New Mexico. In: Johnson, K.S. and Neal, J.T. eds., *Evaporite Karst and Engineering/Environmental Problems in the United States*, Oklahoma Geological Survey Circular 109, OK, pp. 183–195.

Keydar, S., Pelman, D., and Ezersky, M. 2010. Application of seismic diffraction imaging for detecting near-surface inhomogeneities in the Dead Sea area. *Journal of Applied Geophysics* 71(2–3):47–52.

King, P.B. 1948. Geology of the southern Guadalupe Mountains, Texas, Geology of the southern Guadalupe Mountains, Texas, p. 183.

Kruse, S., Grasmueck, M., Weiss, M., and Viggiano, D. 2006. Sinkhole structure imaging in covered Karst terrain. *Geophysical Research Letters* 33:L16405, 6pp.

Legchenko, A., Ezersky, M., Kamerlynck, C., Al-Zoubi, A., and Chalikakis, K. 2009. Joint use of TEM and MRS method in complex geological setting. *Comptes Rendus (C.R.) Geosciences* 341:908–917.

Martinez, M.I. and White, W. 1999. A laboratory investigation of the relative dissolution rates of the Lirio limestone and the Isla de Mona dolomite, and implications for cave and karst development on Isla de Mona. *Journal of Cave and Karst Studies* 61:7–12.

Miller, T. 1996. Geologic and hydrologic controls on karst and cave development in Belize. *Journal of Cave and Karst Studies* 58:100–120.

Pray, L.C. 1988. Geologic guide for U.S. Highway 62(180) from El Paso to southern Guadalupe Mountains at the junction of US 26–180 with Texas 54 from Van Horn. In: Sarg, J.F., Rossen, C., Lehmann, P.J., and Pray, L.C. eds. *Geologic guide to the western escarpment, Guadalupe Mountains, Texas: Society for Sedimentary Geology (SEPM), Permian Basin Section*, Publication 88–30, pp. 9–13.

Quennell, A. 1958. The structural and geomorphic evolution of the Dead Sea rift. *Quaternary Journal of Geological Society of London* 114:1–24.

Reeder, P., Brinkmann, R., and Alt, E. 1996. Karstification on the northern Vaca Plateau, Belize. *Journal of Cave and Karst Studies* 58:121–130.

Sadiq, A.M. and Nasir, S.J. 2002. Middle Pleistocene karst evolution in the State of Qatar, Arabian Gulf. *Journal of Cave and Karst Studies* 64(2):132–139.

Sadiq, A. and Howari, F. 2009. Remote sensing and spectral characteristics of desert sand from Qatar Peninsula, Arabian/Persian Gulf. *Remote Sens.* 1, 915–933.

Salameh, E. and El-Naser, H. 1999. Does the actual drop in Dead Sea level reflect the development of water sources within its drainage basin? *Acta Hydrochimica et Hydrobiologica* 27:5–11.

Salameh, E. and El-Naser, H. 2000a. Changes in the Dead Sea level and their impact on the surrounding groundwater bodies. *Acta Hydrochimica et Hydrobiologica* 28:24–33.

Salameh, E. and El-Naser, H. 2000b. The interface configuration of the Fresh-/Dead Sea water—Theory and measurements. *Acta Hydrochimica et Hydrobiologica* 28:323–328.

Schoor, M.V. 2002. Detection of sinkholes using 2D electrical resistivity imaging. *Journal of Applied Geophysics* 50(4):393–399.

Smart, P.L. and Whitaker, F.F. 1991. Karst processes, hydrology and porosity evolution. In: Wright, V.P. ed. *Palaeokarsts and Palaeokarstic Reservoirs*, PRIS Occasional Publication, Series No. 2, University of Reading, Reading, U.K., pp. 1–55.

Taqieddin, S.A., Abderahman, N.S., and Atallah, M. 2000. Sinkhole hazards along the Eastern Dead Sea shoreline area, Jordan: A geological and geotechnical consideration. *Environmental Geology* 39(11):1237–1253.

Trentham, B. 2010. The Wink Sinks Winkler County, Texas CEED, UTPB. http:// ceed.utpb.edu/media/files/Powerpoint_PPT.pp

Trudgill, S.T. 1985. *Limestone Geomorphology*, Longman, London, U.K., 196pp.

Tyrrell, W.W., Jr. 1969. Criteria useful in interpreting environments of unlike but time-equivalent carbonate units (Tansill–Capitan–Lamar), Capitan reef complex, west Texas and New Mexico. In: Friedman, G.M. ed. *Depositional Environments in Carbonate Rocks: SEPM*, Special Publication no. 14, pp. 80–97.

Waltham, A., Bell, F., and Culshaw, M. 2004. *Sinkholes and Subsidence: Karst and Cavernous Rocks in Engineering and Construction*, Springer-Verlag, Berlin Heidelberg, and Praxis Publishing Ltd., Chichester, U.K.

White, W.B. 1984. Rate processes: Chemical kinetics and karst landform development. In: LaFleur, R.G. ed. *Groundwater as a Geomorphic Agent*, Allen and Unwin, London, U.K., pp. 227–248.

Yechieli, Y., Abelson, M., Bein, A., and Crouvi, O. 2006. Sinkholes "swarns" along the Dead Sea coast: Reflection of disturbance of lake and adjacent groundwater systems. *Geological Society of America Bulletin* 118(9/10):1075–1087.

Zak, I. 1967. The geology of Mount Sedom. PhD thesis, The Hebrew University, Jerusalem.

Index